Cognitive Systems Monographs

Volume 15

Editors: Rüdiger Dillmann · Yoshihiko Nakamura · Stefan Schaal · David Vernon

Ralf Der and Georg Martius

The Playful Machine

Theoretical Foundation and Practical Realization of Self-Organizing Robots

Foreword by Rolf Pfeifer

Springer

Rüdiger Dillmann, University of Karlsruhe, Faculty of Informatics, Institute of Anthropomatics, Humanoids and Intelligence Systems Laboratories, Kaiserstr. 12, 76131 Karlsruhe, Germany

Yoshihiko Nakamura, Tokyo University Fac. Engineering, Dept. Mechano-Informatics, 7-3-1 Hongo, Bukyo-ku Tokyo, 113-8656, Japan

Stefan Schaal, University of Southern California, Department Computer Science, Computational Learning & Motor Control Lab., Los Angeles, CA 90089-2905, USA

David Vernon, Khalifa University Department of Computer Engineering, PO Box 573, Sharjah, United Arab Emirates

Authors

Prof. Dr. Ralf Der
Max-Planck-Institute for
Mathematics in the Sciences
Inselstr. 22
04103 Leipzig
Germany
E-mail: ralfder@mis.mpg.de

Dr. Georg Martius
Max-Planck-Institute for
Mathematics in the Sciences
Inselstr. 22
04103 Leipzig
Germany
E-mail: martius@mis.mpg.de

ISBN 978-3-642-20252-0 e-ISBN 978-3-642-20253-7

DOI 10.1007/978-3-642-20253-7

Cognitive Systems Monographs ISSN 1867-4925

Library of Congress Control Number: 2011942882

Typeset by Georg Martius

Printed in acid-free paper

5 4 3 2 1 0

springer.com

To our families

Foreword

This book is about intelligence, embodied intelligence—and it is about a paradigm shift, a shift from a computational view of intelligence, thinking and cognition, to one that takes the complete organism—brain, body and environment—into account. As we know from Thomas Kuhn's famous book, "The structure of scientific revolutions", paradigm shifts do not occur instantaneously, but the process is a long and tedious one that happens at different levels and it takes a lot of time before a new consensus can be reached. There have been a large number of books and publications arguing, essentially from a conceptual, philosophical, or a biological stance, why computation is in many ways inappropriate to explain the behavior of systems in the real world and that completely novel approaches are needed. 25 years after the appearance of Rodney Brooks' seminal article with the innocuous title "A robust layered control system for a mobile robot," in a sense the manifesto for the famous 'subsumption architecture,' which, in the fields of Artificial Intelligence and robotics, marked the starting point of the "embodied turn," there is increasing consensus that embodied intelligence is the new paradigm, not only in AI, but in psychology, philosophy, and the neurosciences.

While most of the arguments in favor of an embodied perspective on intelligence are intuitively plausible and easy to understand, there is a definite lack of scientific theory, rigor, and methodology. Take, for example, the concept of sensory-motor contingencies as introduced by O'Regan and Noë a decade ago, the idea that there are law-like relations between the actions of an agent and associated changes in sensory stimulation (which depend on the environment as well as the morphological and material properties of the organism), which is intuitively very plausible and is by now generally accepted. However, at this point, it is still open how an in-depth mathematical treatment that captures all these relationships could be developed. Moreover, in the literature, in particular in robotics and artificial intelligence, rather superficial notions of embodiment are floating around that might be characterized by the slogan "intelligence requires a body." Many researchers were and still are convinced that if you simply use a robot into which you embed your otherwise traditional control algorithms, you have an embodied system. But embodiment is much more than that: it is not merely about having a body, but it is about the interrelationship between body, interaction with the environment, and information. After all, the body with its motor and sensory systems is the only way in which we can learn something about the environment, in other words, any interaction with the world is mediated by the body. In addition, rather than having to be controlled by a centralized "brain" or computational system, the body can be exploited for movement and locomotion: the elasticity in the muscles, and the compliance in a robotic device can take over part of the functionality of coping with impact in walking or running. Or the morphology and the material properties of the hand-arm-shoulder system, in particular the soft, deformable tissue in the palm and the finger tips, account for easy grasping of hard objects like glasses and cups in front of us with little control. We could go on for a long time.

But now that we are convinced that there is a real need for an "embodied turn," we have to think about how to proceed from here, how to develop a better understanding and a methodology of how to design and build embodied systems. Here, we can draw on knowledge in the areas of biology, in particular evolutionary theory, biomechanics, behavioral ethology, and neuroscience. Powerful concepts such as self-organization, complex systems, emergence, and the idea of sensory-motor contingencies mentioned above, promise to enhance our understanding of biological systems and bear the potential of being transferred to the design of artificial ones. However, the application of these principles to the design of robots and other devices, has proven much more difficult than initially anticipated. One of the big conundrums of research in the cognitive sciences and artificial intelligence has been and still is how creatures are motivated to do things and alternatively how we can design artificial systems that behave in seemingly goal-directed ways, without having to program the goal-directed behavior explicitly into the artifacts (as we would have done in the traditional approach). In the late 1980s, Luc Steels coined the term "design for emergence": Given a particular set of desired functionalities, e.g. a group of agents behaving in a swarm-like fashion, how can we design the system such that its behavior emerges as it interacts with its environment, including other agents? Guided self-organization is a potent principle that enables the exploitation of phenomena of self-organization to achieve desired functionalities in robotic devices, an efficient and robust engineering method—one solution to "designing for emergence."

Ralf Der and Georg Martius with their book "The playful machine" target precisely these issues by mustering the entire gamut of concepts and formal mathematical "machinery" like complex dynamics, self-organization, embodiment, homeostasis—and its later development, homeokinesis which, in addition to stability, provides a notion of intrinsic motivation—bifurcations, and various oscillator regimes, in order to design and build robots that are self-exploratory, self-motivated, situated, and adaptive. These machines are shown to display many surprising behaviors that are truly emergent from control, morphology, and environment, in the sense that they have not been programmed into their behavioral algorithms. The spirit of designing robots in this way, is completely different from the standard way where we have a particular goal, a set of tasks, in mind and we design the devices such that they are most likely to actually perform the tasks in an efficient, cheap and robust way: it is indeed the playful robots. And why playful robots? There are many hypotheses, but one of the main purposes of play seems to be to practice and explore novel skills in a protected environment so that they can be used in the real – serious – world at a later point in time. Because of the missing top-down goal-directedness normally imposed by the engineers, the robots need to be designed such that they are self-motivated to engage in play and that they can learn novel behaviors. In order for the playful robot scenario to work, the robots themselves must be motivated to engage in tasks and behaviors of ever increasing complexity. If done right, the competences acquired during playful behavior will generalize and transfer to other environments precisely because they are not task-specific. Note that this is very different from merely avoiding being damaged or coping with external perturbations; in this latter case, often the best strategy is to simply do nothing, a phenomenon called the "lazy robot" effect—to be prevented by all means.

In a sense, the book reflects its title: it is itself playful and self-motivating. In addition to the theoretical framework and conceptual discussions, the mathematical background and many examples of model robots, of real devices, and simulated creatures are presented. But even more, the whole book comes with lots of software with simulations, examples, and case studies that invite the reader to play around in order to get into the spirit of the "playful machine." With their seminal contribution, Der and Martius have not only substantially advanced the field of embodied intelligence, design for emergence, and robotics in general, but they have given it the rigor which it deserves as a scientific discipline: through their work, they have on the one hand accelerated the paradigm shift towards embodied intelligence, and on the other hand they have made inroads by demonstrating how existing methods in mathematics, dynamical systems, and engineering design, can be applied to this new and exciting perspective. We may in fact be at the beginning of a new revolution: What Bill Gates forcefully requested in his famous article in Scientific American in 2007: "A robot in every home!" (just as he demanded 20 years ago: "A computer in every home"), may in fact materialize sooner than we think. Der and Martius have decisively moved the field in this direction. What we, the community, need to do now, is to popularize it and make it accessible to a broad audience not only of engineering and computer science experts, but of biologists, neuroscientists, psychologists, philosophers, teachers at all levels, and to people simply interested in novel ideas and technological developments. I'm sure that you, the reader, will be thoroughly enjoying the book and that you will draw intellectual benefit and satisfaction from it. But beware—it may change the way you have been thinking about the world and yourself so far in unanticipated ways—and this process will be irreversible!

Someo, Valle Maggia, Switzerland, September 2011 *Rolf Pfeifer*

Preface

Imagine a world of artificial creatures living in a kind of paradise. These creatures would have unlimited access to resources (electrical current), they may live in a richly structured environment, are potentially immortal, and are not subject to any external pressure for development. The cushy situation in this robotic "paradise" gives rise to a nasty question that is the actual seed for this book: without any given task, goal, purpose, or other external pressures, why should such a creature do anything at all? Moreover, if there is no goal, no purpose, no plan, what can we expect the system to do? Will the resulting behaviors (if there ever is one) be arbitrary or will they relate to the specific nature of the physical system?

When trying to find an answer many routes are possible. You may focus on philosophical questions regarding free will or the existence of a machine-self; you may come across important challenges of modern robotics concerning intrinsic motivation, self-learning, or artificial curiosity; and you may end up with thinking about the very roots of autonomy and self-determination. The book faces the problem in a practical way by formulating a general principle—homeokinesis—that is unspecific, unbiased, surprisingly simple, completely internal to the agent, and fully operational on concrete robotic systems of high complexity.

As you will see when reading the book or doing the experiments, this general principle makes machines discover their behavioral variety in a playful individual development. The emerging activities, while being many and varied, are seen to be related to the physical properties of the body and environment so that the robot discovers the most natural modes that are accessible to him by minimal control. While this completely self-determined behavior generation may also help in the future to answer the more philosophical questions, it will in the short term provide very down-to-earth consequences. In particular, as we have learned from the field of embodied AI, behavior generation in complex robotic objects is improved and stabilized by taking brain, body, and environment as a whole. The playful unfolding of behavioral patterns offers a new way of getting the embodiment of the agent involved.

In a wider context, potential benefits are foreseeable when taking a new attitude to robotics, leaving the fixation on a specific goal aside in the beginning of a behavioral design process. Instead, comparable to a liberalist concept of education, we let the machines play freely, giving them a chance to demonstrate their potential capabilities. As a second step, we may try to gently guide the free play into desired directions, and it is only in a third step that the emerging behaviors could be valued and stored for later use in behavioral architectures following prescribed goals. The methods for autonomous development presented in this book are a first step towards this very practical goal and they might help in the future to overcome the enormous difficulties in the behavior generation of complex systems.

Given the practical orientation of the book, the reader is invited to enjoy the numerous videos at `http://playfulmachines.com` demonstrating the specific applications and/or doing experiments using the simulator that comes with the book. You will find nearly 30 experiments ranging from simple systems in low-dimensional sensorimotor loops to the robotic zoo with creatures such as snakes, dogs, and humanoids in a highly complex virtual, but physically realistic world. Hopefully this will also help to understand the theoretical considerations based on dynamical systems theory.

The book is intended for students and researchers interested in self-organization on both the practical and theoretical level. It should also be of interest for direct practical application since the controller with the homeokinetic learning algorithm can easily be connected to any real robot, helping it to a playful self-exploration of its behavioral variety. In combination with the experiments, the book also offers itself as an undergraduate or graduate course on self-organization and the dynamical systems approach to robotics.

Acknowledgments

Homeokinesis has come a long way and there are many companions we are grateful to for their enthusiasm and support. One of the early and constant companions is Michael Herrmann whom we like to thank for asking critical questions, digging out exotic papers, and collaborating with us for many years. In early phases the idea was supported by René Liebscher by creating and controlling hardware robots, providing a first proof of principle. Further thanks go to our colleagues Frank Hesse and Frank Güttler for constantly probing the ideas and assisting with soft- and hardware issues.

Big thanks go to Nihat Ay and Keyan Zahedi for inspiring discussions and important impulses in recent years. I (Ralf) would like to express my sincere thanks to Nihat Ay for hospitality in his group and an exciting, long-lasting cooperation on the mathematical foundations of embodied AI which was essential for writing this book. We are also grateful to Frank Pasemann, Manfred Hild, Mikhail Prokopenko, and Florentin Wörgötter for many discussions and sharing ideas.

Thanks go also to the researchers around the globe from many different disciplines for their ideas that have provided inspiration for our work. In particular, we thank Rolf Pfeifer for pioneering the exciting field of embodied artificial intelligence and for writing the foreword. We are grateful to the artist Julius Popp for inspiring us by his fascinating work and for the fruitful cooperation on spherical robots.

Without the support of Theo Geisel and Jürgen Jost this work would not have been possible—thank you for giving us space and time for conducting our research. We would also like to express our thanks for the financial support by the German Research Foundation (DFG) and the German Federal Ministry of Education and Research (BMBF).

I (Georg) thank all the contributors to the software development of LpzRobots, in particular Frank Güttler, Frank Hesse, Dominique Schneider, and Antonia Siegert.

Warm thanks to Joseph Lizier for resolving our frequent English language uncertainties. We also thank the team at Springer, in particular Olga Chiarcos and Federica Corradi dell Acqua for escorting us through the preparation of the manuscript and coordinating the details.

Last but not least, I (Ralf) would like to thank my wife Marianne and my sons Jakob, Uwe, and Holger for their love, understanding, and their continuing encouragement that was essential for completing this work. I (Georg) am deeply grateful to my wife Anya and my son Jan for their love, permanent support, and tolerance for long working days.

Contents

Chapter 1
Introduction

Robots and their relation to mankind have taken a long and diversified development. Starting from the romantic desire to have a workmate and/or playmate centuries ago, the modern history of robot control starts with the birth of artificial intelligence (AI) about 50 years ago. In the hype of that time, robots were considered as machines under total control of an artificial intelligence thought to understand the world and the physics of the body well enough in order to control the robot by a set of rules defining its behavior.

Subsequent developments have generated on the one hand the marvelous machines welding, mounting, and painting cars, but, on the other hand, also the bitter insight that the complexity of the world exceeds by far the scope of the internal models necessary for the AI approach. Even now, after 50 years, behavioral skills of the most advanced robots are far behind that of any insect. The rethinking began more than two decades ago, prompting a drastic change of paradigms in the control of autonomous robots [25, 26]. The new or embodied AI recognizes the role of the body as an equal partner in the control process. The exploitation of the specific properties of the body, sometimes called morphological computation [132], not only reduces the computational load on the controller in specific tasks like walking or swimming but also leads to smoother and more natural motions of the robot, see [131, 136] for a comprehensive and inspiring presentation of these ideas. The research has produced many successful and inspiring results. Of particular importance is the attempt to understand more about intelligence when using the embodiment approach.

Although quite successful, the method seems to be restricted to specific tasks and requires an inspired designer for the "morphological computer," instead of the intelligent programmer of the classical AI systems. Moreover, a general theoretical foundation is still missing. In our view, the achieved results suggest a next step on the way that gives machines more and more autonomy. Our ambition is to make the machines discover their behavioral options in a playful individual development in the first place and to look only afterwards for uses of the emerging behavioral options. Different from developmental robotics [97, 179], our approach is not focused so much on mental development but on the playful unfolding of the sensorimotor contingencies that form the basis for mental development.

R. Der and G. Martius: The Playful Machine, COSMOS 15, pp. 1–8.
springerlink.com

Why Activity

In order to make this idea more concrete, let us formulate it in a casual way. As introduced in the Preface, let us consider robots in a kind of paradise, instead of a world driven by needs. With unlimited access to resources and without any given task, goal, purpose, or other pressures why should such a creature do anything at all? Borrowing from psychology, we may call this the problem of self-actualization.[1]

Besides being of practical interest, self-actualization (at any level of the behavioral hierarchy) is also of principal interest in modern biology, artificial intelligence, and robotics. For instance, robotics and embodied AI both are in quest of features ranging from artificial curiosity to internal motivation to the very emergence of a machine-self. In biology this problem is related to the origin and realization of behavioral variability under identical circumstances (a question that is discussed much in connection with establishing free will as a biological trait, see [24] for a review).

There are some approaches known from the literature. One is making use of information theory. Considering the robot with its brain as an information processing system, the optimization of information flows might be an option. So far, different information measures have been shown to drive the robot to self-induced activity. Examples are empowerment [83, 87, 148] or predictive information [19] in the flow of sensor values the robot induces by its behavior [10, 162, 191]. While of high generality, the application of these principles is hindered by excessive sampling costs. Information theory also allows for the formalization of conceptual terms like autonomy [17, 18], which may help in developing truly autonomous systems. Other general paradigms like Autopoiesis [109], although very intellectually appealing and helpful in describing the fundamental nature of living systems, are not constructive enough so that they are difficult to operationalize.

More concrete approaches address the question of artificial curiosity or intrinsic motivation in reinforcement learning. Pioneering work has been done by Schmidhuber using the prediction error as a reward signal in order to make the robot curious for new experiences [150]. The approach has been further developed in a number of papers, see e. g. [151, 164]. Related ideas have been put forward in the so called playground experiment by Kaplan and Oudeyer [85, 117], using the learning progress as a reward signal. Steels [157] proposes the Autotelic Principle, i. e. the balance of skill and challenge of behavioral components as the motivation for open ended development. This are interesting and encouraging developments towards agents with an internal, self-determined drive for activity. However, so far the developments are rather application related and require a convenient pre-structuring of state-action spaces whenever the systems are getting more complex. Information theory, while being domain invariant, is restricted in applicability by its tremendous sampling costs.

---

[1] However, different from psychology we will use this term not at the mental but at the sensorimotor level, meaning the self-determined unfolding of sensorimotor contingencies instead of the strive for personal growth and fulfillment in the sense of Maslow´s “what a man can be, he must be” [108].

The Machines

Our approach to self-actualization is not philosophical but practical, aiming at the creation of a universal concept that can be operationalized and tested in complex robotic objects. Our approach is machine driven. So, before formulating the general principle let us have a closer look at the kind of machines we are interested in. The different strands of development in robotics and AI have produced very different kinds of robots. On the one side of the spectrum, there is the highly sophisticated machine for the reliable execution of motion plans like complicated dance moves, see Fig. 1.1(a).

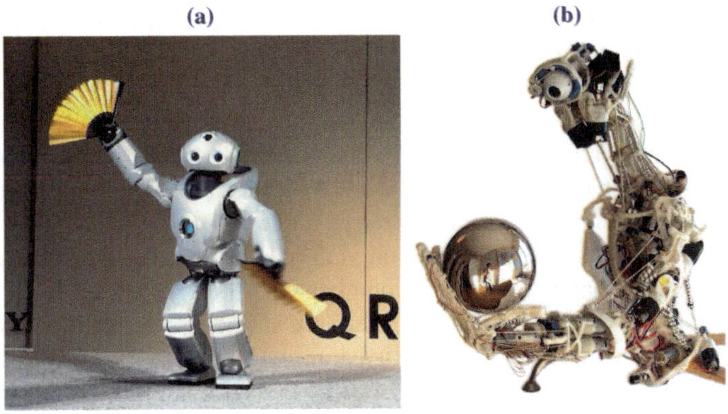

Fig. 1.1: Two different worlds of robots. **(a)** Sony robot QRIO dancing. These robots in an ideal manner realize the dream of having a machine that is authentic in executing a given motion plan. **(b)** Anthropomimetic robot ECCEROBOT. Instead of just mimicking the shape of the human body the compliantly designed robot tries to copy also the inner structure and mechanisms made up by bones, joints, muscles and tendons. In this way, it has the potential for human-like action and interaction with the world. This is considered the main prerequisite for the emergence of human-like intelligence. Images included with kind permission of (a) PC Watch [126] and (b) R. Knight [169].

On the other side, we find machines, like for instance the anthropomimetic ECCEROBOT in Fig. 1.1(b), designed to reflect the morphology of the human body as closely as possible. Copying the internal structure and mechanisms made up of bones, joints, muscles and tendons, these machines are compliant to the influence of external forces and intra-body couplings in a similar way as humans are. There are severe practical reasons for considering such machines. For instance in health care, service robots are to behave human-like and in particular should not react rigidly in encounters with humans. Moreover, they also have a high intellectual appeal, cumulating in the question whether the human-like body shapes cognitive processing into human-like dimensions.

It is in these compliant machines where the principles of embodied robotics find their real playground. In fact, robots of that kind are a nightmare to any classical control approach and there is no chance for anything like classical AI realization, based on planning and a concrete world model, under these circumstances. Instead, the controller is challenged to maximally exploit the physical peculiarities of the body in its interaction with the environment. However, even in the embodied AI approach, the control of such systems is still in its infancy.

Compliant machines are the target group of this book, using however an extended concept of compliance. Compliance can be extended to the way the robot is related to its environment. In fact, we will introduce the notion of an extended body so that we can have a "rigid" robot (like Sony's Qrio) surrounded virtually by a compliant sphere provided by its sensors.

The robots studied in this book are simpler than the ECCEROBOT but we claim that they share many features and fundamental challenges with the latter. Let us demonstrate the parallels by one of our machines, the SPHERICAL, or its simpler variant, the BARREL, both being driven by an internal mechanism for shifting the robot's center of gravity, see Fig. 1.2. Rolling is the most natural but by far not the only form of motion of this physical system. As demonstrated in experiments, while stable rolling modes can be excited by very simple closed loop controllers even on structured grounds, the execution of a motion plan, formulated in the space of the motor commands, may become quite complicated.

In this kind of machine, obviously there are specific patterns of behavior that, while singled out by a certain complexity, are achievable with minimal control. So, the idea is not to force the body into a specific behavior but to come into a kind of

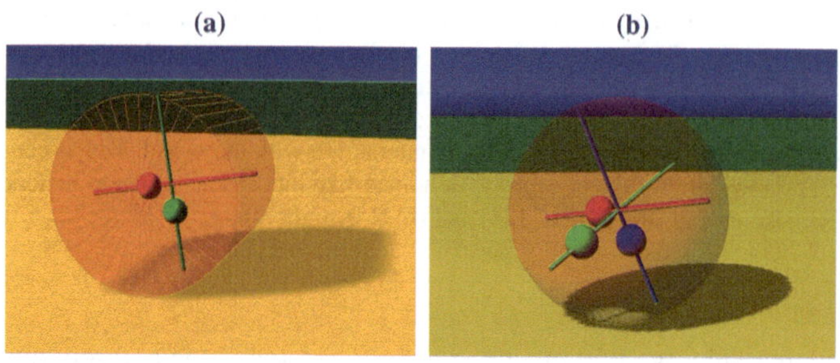

Fig. 1.2: Robots dominated by embodiment. Two robots arguably matching the conceptual level of the anthropomimetic robots. The robots locomote by shifting their internal masses defining the center of gravity. The only sensor values are the inclination of the axes. As with ECCEROBOT, there are no single actions with definite consequences, even in the probabilistic sense. Instead, each action, a motion of the internal weights, produces a whole body reaction that depends on the current physical state of the mechanical system. **(a)** The BARREL, a physical system with 8 degrees of freedom. If simply rolling, the number of degrees of freedom reduces by 4. **(b)** The SPHERICAL has only one additional degree of freedom, the third internal mass, but displays much more complex motion patterns due to its different physical properties.

functional resonance with its specific physical properties. This is the point where the different machines pose the same challenge: feeling the body instead of forcing it is a indispensable prerequisite for any successful control paradigm. Of course, there are differences in the level of complexity and in the scientific questions. In particular, when looking for the roots of human like intelligence, the SPHERICAL probably can not make a real contribution.

Facing the Unknown

Now we can return to our original question. Keeping to the example of the sphere, why should a controller try at all to start shifting the internal weights so that the robot is driven into moving? And what can a general principle look like that drives the brain to find out about the most natural modes of behavior and the ways it can excite them? This book presents a solution that is based on the dynamical systems approach, thereby working with continuous state-action spaces right from the outset. We realize the brain of the robot by two neural networks, one for control and the second one acting as an internal forward model, predicting the next sensor state based on the present sensor and motor values. However, in contrast to classical AI, we do not use the model for planning ahead. Instead, by its prediction error, the model just quantifies the ability of the brain to look ahead, separating in this way the knowable components of the sensorimotor dynamics from what we call the unknown.

In these terms, a self-determined and explorative behavior can be understood as ways of facing the unknown. A naive way of doing so would be to "get things under control," i. e. adapt behavior in order to reduce the influence of the unknown on the future evolution of the system. However, in our robot paradise, this principle generates in most cases systems that self-regulate into a "do nothing" behavior. This "lazy robot effect" is easily understood by the fact that the consequences of doing nothing are perfectly predictable, in a static environment at least.

The astonishingly simple solution to that problem is found in a new representation of the unknown. Technically this amounts to replacing the prediction error with its time inverted counterpart, the reconstruction or time-loop error.

Homeokinesis

Facing the unknown is now defined as a continual process of adaptation directed towards reducing the size of the time-loop error. This book will show that this method not only solves the lazy robot problem but also provides a systematic approach to the self-determined individual development of embodied and compliant robots. The minimization of the time-loop error is shown to generate a common kinetic regime, called "homeokinesis", jointly involving the physical, the neural, and the synaptic processes of the artificial brain-body system formed by the robot and its controlling unit.

We show both theoretically and in many applications that homeokinesis is a fruitful solution to the problem of self-actualization and realizes in a systematic way the playful self-exploration of complex robotic objects. In the example of the spherical robot we observe that the homeokinetic brain excites the most stable rolling modes if running on a level surface, changes between modes in a self-determined manner, and finds another adequate rolling mode if in a spherical basin. These results propagate through to other robots with always the same findings—emerging search and the playful exploration of complex motion patterns with high sensorimotor coordination. Surprise is one of the inherent features of homeokinesis. In many cases we were witnessing the emergence of quite unexpected behaviors, e. g. various wrestling scenarios if two humanoid robots come into closer contact.

The approach to a self-organized, playful exploration of robot behavior is, in its concrete realization, based on many novelties making the principle applicable to a broad variety of robots with only minor changes. In particular, we realize a real-time application to high dimensional systems of many active degrees of freedom.

Guided Self-Organization

Without any purpose or goal, the emerging motion patterns are contingent, meaning many and varied, and transient by nature. This is vital for a developing system but less attractive from the practical point of view. The book also introduces a new research field, called guided self-organization, showing how external influences can be integrated in order to guide the self-organization into given directions, like the emergence of specific locomotion patterns. We introduce and investigate in practical applications several guidance principles, giving guidance for instance by directly influencing the motor patterns, by phase relations inducing symmetry breaking, or by rewards.

How to Use the Book

Besides conveying conceptual and theoretical foundations for the world of playful machines the book aims at making the reader interested in doing their own experiments. Therefore, the book comes with a demonstration software using our fully fledged robot simulator called LpzRobots and numerous suggestions for experiments scattered throughout the book. The software can be downloaded for free from `http://www.playfulmachines.com`. Furthermore, we demonstrate our results by a large number of videos that can be watched at the same site. For illustration the videos are referenced in the book with a single image or a series of frames, however the captions are referring to the entire video clips.

The book is written on different levels of detail. Most chapters are organized such that they start with the more basic content and dive deeper and deeper into the details. So you may choose your own level by skipping later parts of the chapters. To better digest the theoretical content you can also at any time procrastinate your study by playing around with the experiments. They can run for hours while you read the book and you may come back to see what happened.

The mathematics may appear a bit hard at the beginning, but **don't panic**! Particularly hard sections are marked with a * sign and are intended for advanced study.

Organization of the Book

We start in Chap. 2 with a condensed introduction to self-organization phenomena in nature and machines as they are known from the literature. We have included this chapter in order to give the reader some background on the basic mechanisms of self-organization, since the latter are helpful for understanding the specific properties of homeokinetic systems.

Special to our approach is what we call the externalization of complexity. While using extremely simple structures for both internal model and controller, the behavioral patterns generated by our approach are of a highly complex nature. The very basis of that phenomenon is the use of closed loop control in a tight sensorimotor coupling. This is explained in detail in Chap. 3. Moreover, we use extremely fast learning procedures so that both model and controller are relearning if situations change, i. e. we replace internal complexity with flexibility.

Chapter 4 gives some theory and illustrative examples for the interplay between model and controller, demonstrating thereby the "lazy robot effect" that appears here in a kind of self-regulated stability. This phenomenon is considered in the historical context of homeostasis.

The subsequent Chap. 5 introduces homeokinesis on the basis of the time-loop error and its use for the self-organization of control. Homeokinetic learning by gradient descending the time-loop error is shown to generate a specific dynamical regime of the brain-body system that gives the synaptic dynamics a primary functional role in behavior generation.

In Chaps. 6 and 7 we give an analytical investigation of homeokinesis as a self-referential dynamical system. Chapter 8 gives first applications focusing on specific features like the role of symmetries, spontaneity, and the emergence of specific embodiment effects in both simulated and real robots with only two motors but many physical degrees of freedom. The often surprising effects may be interpreted as a kind of arousal of the most natural motion patterns latent in the physical system.

Chapter 9 is concerned with another unique feature of our approach, namely the bootstrapping of both model and controller from scratch, i. e. if starting from our general initialization of "do nothing" and "know nothing." Moreover, it introduces the cognitive deprivation effect and shows how systems can recover from the deprivation by the homeokinetic learning process.

The specific phenomena observed in the low-dimensional systems, the excitation of latent whole body modes in particular, find their counterpart also in the high-dimensional examples considered in Chap. 10. These complex robots are the actual target group of homeokinesis. Therefore, we try to present a broad spectrum of different robotic systems subject to homeokinetic control. In all examples we observe complex motion patterns with high sensorimotor coordination emerging, so to say, out of nothing, demonstrating the playfulness of these machines under the homeokinetic control paradigm.

This chapter is followed by an alternative approach to homeokinesis, based on a novel representation of the sensorimotor dynamics that we call the interaction representation. Besides giving an additional motivation for homeokinesis, this chapter will extend the considerations to the case of several time steps and will eventually consider infinite time horizons making contact with the global Lyapunov exponents and chaos theory. The chapter is a little more mathematically demanding, which is the reason for locating it after the concrete applications.

As mentioned above, a serious practical concern is to find ways for guiding self-organization towards specific goals. The results obtained so far in this novel direction of research are presented in Chap. 12 through Chap. 14, giving both the methodological ideas, concrete realizations, and a first analysis of concrete applications.

Details on the realization and several extensions of the algorithm are presented in Chap. 15. Although many of the algorithmic details are already referenced in earlier chapters, we have pushed this chapter towards the end of the book in order not to burden the reader with too much details for a first reading. The same is true for Chap. 16 devoted to a detailed description of the LPZROBOTS simulator that is the basis for all experiments in the virtual worlds proposed in the book.

Chapter 2
Self-Organization in Nature and Machines

Abstract: Self-organization in the sense used in natural sciences means the spontaneous creation of patterns in space and/or time in dissipative systems consisting of many individual components. Central in this context is the notion of emergence meaning the spontaneous creation of structures or functions that are not directly explainable from the interactions between the constituents of the system. This chapter presents at first several examples of prominent self-organizing systems in nature with the aim to identify the underlying mechanisms. While self-organization in natural systems shares a common scheme, self-organization in machines is more diversified. An exception is swarm robotics because of the similarity to a system of many constituents interacting via local laws as encountered in physics (particles), biology (insects), and technology (robots). This chapter aims at providing a common basis for a translation of self-organization effects to **single** robots considered as complex physical systems consisting of many constituents that are constraining each other in an intensive manner.

Self-organization is a ubiquitous phenomenon observed in many complex systems in the fields of physics, chemistry, computer science, economics, and biology. While synergetics provides a general theoretical framework for the wide field of such phenomena [60, 62, 182] there are many examples that are controversial and difficult to fit into a quantitative explanatory system [139]. This chapter will not try to shed new light onto this research field but instead work out by way of examples the most prominent features and underlying mechanisms of self organizing systems in nature and machines. Once identified, these mechanisms may serve as a guiding principle for autonomous robot development.

After identifying one cornerstone—self-amplification—by the example of spontaneous magnetization (Sect. 2.1.1), we are going to study systems as different as convection patterns (Sect. 2.1.2), reaction diffusion systems (Sect. 2.1.3), and Turing patterns (Sect. 2.1.2) in order to understand the role of symmetries and spontaneous symmetry breaking, the second corner stone of self-organization.

After giving examples from biology (Sect. 2.1.4) that may help to understand the comprehensive nature of the general scheme, we will consider swarm robotics (Sect. 2.2.1) because of its close relationship to biology. Further fields of robotic research are related to self-organization only in a more distant way. We discuss in particular the research in artificial evolution in Sect. 2.2.2.

R. Der and G. Martius: The Playful Machine, COSMOS 15, pp. 9–21.
springerlink.com

Let us start with the self-organization phenomena in physics since they are very clear cut.

2.1 Self-Organization — The Physical Perspective

There are common features of self-organization that are shared by many examples in physics. The most essential ingredient in a self-organization scenario is the effect of self-amplification of small perturbations. Let us work out this in a simple case.

2.1.1 Phase Transitions

Many self-organization phenomena happen at phase transitions. A prominent example is the spontaneous magnetization in ferromagnetic materials [30]. These materials consist of a field of spins, which may be consider as little magnets. At high temperatures their orientation is random due to thermal fluctuations. The system could be given an order from outside by applying an external magnetic field that forces the magnets to orient.

However, even without the external field, when lowering the temperature below a critical value T_c, there is a phase transition to a magnetic domain structure consisting of clusters of aligned magnets. How can this be without any external guidance? The explanation rests on then "willingness" of the magnets to align along a given field. The thermal fluctuations are still active also below T_c and may create small clusters of aligned magnets by chance. However, such a micro-cluster acts like a little magnet that may force neighboring magnets to align so that the cluster increases in size and influence on its neighborhood. In this way, an originally small region will increase by this self-amplification mechanism so that the new order can expand over macroscopically large regions. The competition between clusters growing at different places in the substance leads eventually to the mentioned magnetic domain structure.

The magnetization direction of each of the domains is essentially random since it is the result of a thermal fluctuation. The space, however, is invariant against rotations of the magnets in any direction so that we have a breaking of this spatial symmetry. By the described scenario the original symmetry is broken spontaneously. This is one of the central effects in self-organization.

2.1.2 Convection Patterns

The structures created by the above effect are quite irregular and thus display only one part of the self-organization phenomena we are interested in. Of more interest are the regular structures observed in open physical systems driven by external influences into a steady state far from equilibrium. Well known is for instance the emergence of convection patterns in fluids under a heat gradient. A famous example are the Bénard cells, named after the French physicist Henri Bénard who discovered the phenomenon in 1900, see Fig. 2.2 for a specific example. He studied thin layers of water on a homogeneous surface heated from below such that the heat gradient is constant throughout the liquid, see Fig. 2.1.

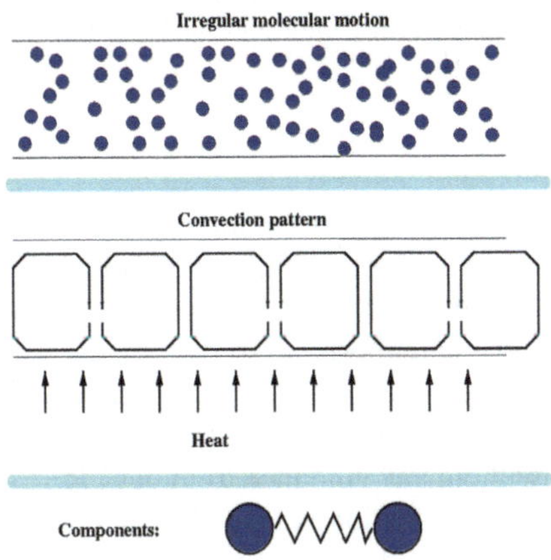

Fig. 2.1: Convection patterns. Heat transport in a layer of water with small (top) and large heat gradient (middle). The interaction between the particles depends on the distance alone (bottom) so that there is no relation to the global structure.

As long as the gradient is not too large, heat transport is realized by conduction, which is based on energy transfer by collisions between particles of different kinetic energy. This is a random process, which is stable against perturbations but not very effective. Instead, heat transport is realized more effectively by convection instead of conduction. An example is given by the radiator driving a convection pattern in a room since air is ascending over the heat source and descending after cooling at the ceiling and the walls. The reason for the emerging convection pattern is the different distribution of heat sources and sinks in the room.

We could organize the Bénard system to develop a convection pattern, too, by using an inhomogeneous heat source, but this is not necessary. Instead, once the gradient exceeds a certain critical value, a phase transition is taking place towards a surprisingly regular convection pattern, a hexagonal structure in the concrete case. The emergence of that pattern may be considered as self-organized, since no external help has been given. How can this be? The explanation has two steps. First, with a large gradient, the system becomes unstable against thermal fluctuations. If somewhere a higher concentration of hot molecules is emerging, this acts as a kind of mini-radiator causing a tiny upward convection with a subsequent counter-con-

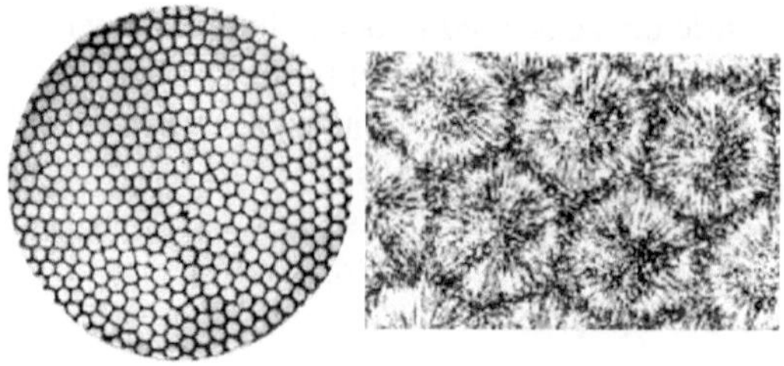

Fig. 2.2: Bénard cells. Macroscopic regular structures as a ubiquitous phenomenon of self-organization: Heated fluid forms regular hexagonal convection patterns. On the **right** is a small zoomed in part of the **left**. To illustrate the flow elongated particles have been added. © Herbert Oertel [181].

vection of cooler molecules in the surroundings. Such a mini-convection cell can induce further convection cells in the vicinity since the downward current will facilitate other upward currents close-by. In this way, one convection cell acts like a germ for further cells in the neighborhood so that, by the instability of the system, a small thermal fluctuation can self-amplify to a macroscopic pattern, which may spread over the whole system (or parts of it). In this way the original symmetry is broken spontaneously, without any guidance from outside.

So, the first step in the explanation is given by the ability of the system to amplify microscopic thermal fluctuations due to the emerging instability. But how can we explain the observed regularities. And why are the patterns hexagonal? Well, the concrete form of the patterns in such phenomena depends on circumstances but there is a "weak" rule of thumb. Coarsely speaking, one can say that spontaneous symmetry breaking is in a sense economical, i.e. the symmetry is broken in the least possible way. In other words, the emerging structures are as simple as possible. This is explained by the fact that the establishment of the ordered structure is a complicated process that involves the reorganization of the many constituents of the system, the re-organizational efforts being the smaller, the simpler the structure. Hexagonal patterns are preferred if the system (under equilibrium conditions) is isotropic and homogeneous since they preserve the original invariance against translation and rotation, in a restricted sense, at least. This principle of "least commitment" is a general phenomenon that gives a qualitative understanding of the observed patterns.

This is interesting and intriguing. But there is more. An essential point is the tremendous reduction in dimensionality. The system actually consists of roughly 10^{23} components whereas the observed pattern is a regular macroscopic structure that essentially can be described by a few so called order parameters. Moreover, the patterns reflect a new dynamical organization of the system which is neither inscribed explicitly into the microscopic dynamics nor into the external conditions. This is why we may say that the system created this organization "out of nothing" by itself.

2.1.3 Reaction-Diffusion Systems

Further generic examples for self-organization are reaction-diffusion systems. Let us take this as another example in order to work out the fundamental principles of self-organizing systems in nature. As we have seen in the Bénard example the emergence of a pattern is the result of two opposing drives, a constraining one, which tries to conserve the overall symmetry of the system and some self-amplification mechanism that tries to destroy symmetries by amplifying the effects of local fluctuations. In a reaction-diffusion system the tendency towards symmetry is realized by diffusion, which is destroying any local density fluctuations. The self-amplification in reaction-diffusion systems is caused by an autocatalytic chemical reaction like

$$A \to A + A$$

This generates the desired self-amplification. If, by a fluctuation, the concentration of A is locally enhanced, the concentration of A is increasing exponentially in this region. This process is counteracted by the diffusion and the presence of other processes, which are responsible for the supply of raw and the disposal of waste material. Both are indispensable, since the autocatalytic reaction can not exist by itself but needs a support system.

2.1.3.1 Turing Patterns

There are many different reaction-diffusion systems leading to different kinds of patterns by a skillful combination of the components of the process. Alan Turing [173] was probably the first to investigate systems of the above kind in order to understand morphogenesis as a pattern formation process guided by the interplay of reaction and diffusion. He was far ahead of his time and it took more than 40 years until the first reaction proposed by Turing could be realized. Nowadays Turing patterns are very popular since they can be easily simulated by cellular automata.

Reaction-diffusion processes play also an important role in modern biology. For instance, they are postulated to cause the pigmentation of animal coats during morphogenesis. An illustration is given in Fig. 2.3, where the fur of a jaguar and two different results of a reaction-diffusion system are displayed.

2.1.3.2 Self-Organization in Space and Time

Classical Turing patterns are stationary. Even more interesting are reaction-diffusion systems, which realize a self-organization not only in space but also in time. The most famous example is the Belousov-Zhabotinsky reaction [192]. The competing processes, diffusion and chemical reactions, need time in order to be executed. By a skillful combination of the time scales and the nature of the involved reactions, the interplay between diffusion and reaction can be designed such that the respective

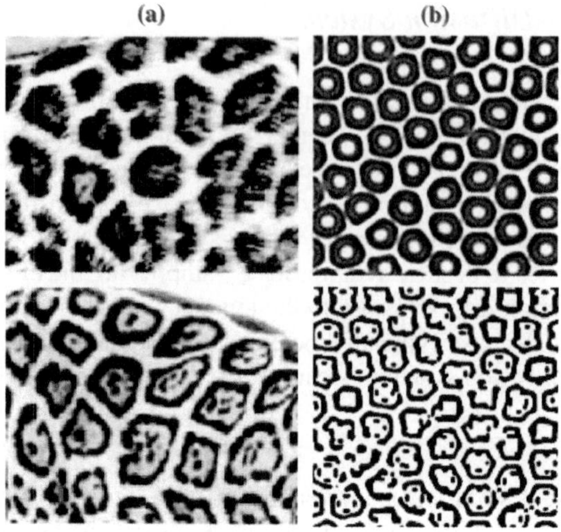

Fig. 2.3: Self-organized pattern formation. **(a)** Coat pattern of a jaguar at about 3 month **(top)** and in adulthood **(bottom)**. **(b)** Results of a 2D reaction diffusion system with changed parameters to induce the transient from the upper to the lower pattern. © American Physical Society and R. T. Lui et al 2006 [95].

phase shifts lead to an oscillatory behavior of the emerging patterns, see Fig. 2.4(a). There are several models which lead to oscillatory and more complicated time behaviors with many very beautiful patterns. Fig. 2.4(b) displays examples of more recent studies.

These patterns are more complicated than the stationary ones. Nevertheless they also represent a tremendous dimensionality reduction demonstrating how very high dimensional systems can find by themselves coherent and highly coordinated modes of behavior. This can be seen as a challenge for the creation of artificial systems, the topic we pursue in this book. We aim to understand and create behavior in highly complex artificial beings as low-dimensional spatiotemporal modes of a self-organizing system.

2.1.4 Examples from Biology

One of the most prominent examples of emerging collective behavior is seen in flocks of birds or schools of fish. Each individual has, via its senses, access only to signals from its immediate neighborhood. It can for example sense the directions of heading and distances of the neighboring individuals. With the appropriate alignment and integration into the local situation a global pattern forms as for instance depicted in Fig. 2.5(a).

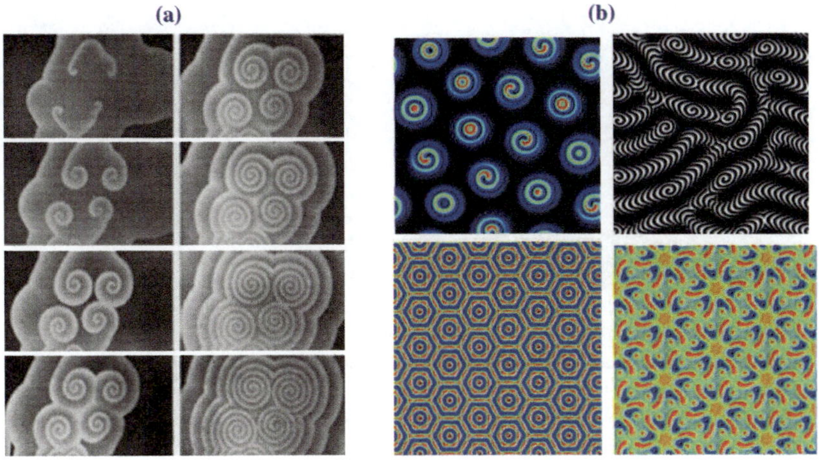

Fig. 2.4: Oscillatory patterns as a generic example of self-organization. **(a)** Time evolution of a Belousov-Zhabotinsky reaction in a petri dish [192] © Zhabotinsky and Zaikin 1971. **(b)** Patterns produced by two coupled layers of reaction-diffusion systems © American Physical Society and L. Yang et al 2003 [190].

Self-organization is also very well visible in the behavior of insect colonies. A typical example is how ants find their way to food sources. Experiments showed that only local interaction is required to make the ant colony choose the nearest available food source [37]. Each ant leaves a pheromone trace on its way, whether it searches for food or it returns to the nest. At crossings where several paths intersect they usually choose the direction that is most strongly marked. Ants from nearby food sources will return more quickly, such that those paths are more intensively marked. This in turn increases the probability of more ants following the shorter paths, see Fig. 2.5(b). This is a positive feedback mechanism providing the self-amplification that is common to most self-organizing systems.

Also the human brain self-organizes to a large extend. For example the primary visual cortex develops different cortical maps that arise from local interactions of the neurons early during development [49]. Similarly, the representation of the body parts in the somatosensory cortex is also self-organized [23]. During development a topological map of the body is formed, nearby neurons in the cortex getting sensitive to stimuli from nearby regions of the body. The relative size of the representation of the different body parts depends largely on the number of receptors and stimulations from this region, such that e. g. the representation of the tongue occupies about the same space as the legs and feet together.

This flexible representation is very powerful and gave rise to the development of computational models, the most prominent being the self-organizing map [89]. Let us briefly take a look at its function to see the parallels to the other systems. Each neuron in the network has initially a random receptive field. For a given stimulus the best fitting neuron is identified and optimizes its receptive field. Additionally the

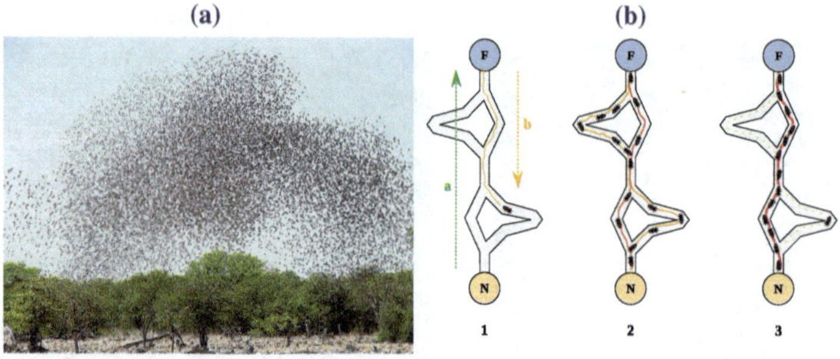

Fig. 2.5: Self-organization in collective behavior of animals. **(a)** Flocking of Red-billed Quelea © Alastair Rae [181]. **(b)** Shortest path finding by an ant colony. **1)** The first ant finds a food source (F), using some path, then it comes back to the nest (N), laying a pheromone trail. **2)** The ants follow one of the 4 possible paths, but the reinforcement of the trail makes the shortest path more appealing. **3)** The ants follow the shortest path, the pheromone trail of the longer ones evaporates © Johann Dréo [181].

neurons in its local neighborhood shift their receptive fields towards the presented stimulus. Altogether this acts as a self-amplification mechanism that eventually results in a self-organized topology preserving map.

The term self-organizing is also used in conjunction with the self-regulation of the control parameters of a dynamical system to a critical point. The phenomenon, introduced by Bak, Tang and Wiesenfeld [11], was called *self-organized criticality*. It is a general concept to explain the spatial and/or temporal scale-invariant characteristic of many natural systems [112]. For example the size of earthquakes, the size of homogeneously magnetized regions, the frequency of words in a text, and the size of neural burst activities in the brain [14, 94] all exhibit a power law statistics. These scale-invariant phenomenon are mostly observed at phase transitions and require a precise tuning of the control parameters. A "self-organized critical" system self-regulates the control parameters to the critical point. The term "self-organization" is used in a different meaning here, nevertheless, the bridge to our understanding of self-organization is that these systems self-regulate to a regime where self-organization is possible.

2.2 Self-Organization in Machines

The interrelation between biology and artificial worlds has been largely investigated by the Artificial Life community. The idea is to build artificial systems in order to capture the key aspects of living systems, see [3, 93]. For example, the mentioned reaction-diffusion systems have been used as operational models of pattern formation in biological systems (as observed for instance on shells or in nest structures of insect societies), see above and [27] for many examples. The spontaneous cooperation of activities in many component systems has crystallized indeed as a key concept from these studies thus providing a causal insight in how these phenomena arise. In robotics the closest relation to these natural systems is found in the field of swarm robotics, which deals with systems composed of many robots as we will discuss in the next section. Afterwards we will review self-organization of single robot behavior. There are of course also a large number of attempts to self-organize the structure of artificial systems. A popular example is the self-assembly of robotic structures based on simple but universal building blocks but we do not give any details here since we are interested in the self-organization of robot **behavior**.

2.2.1 Swarms and Multi-Robot Systems

When observing the beautiful patterns of a bird flock (see Fig. 2.5(a)), how they form and dynamically restructure when changing direction or floating around an obstacle—all without a central unit, then it is not surprising that researchers aim to understand these systems [144]. It is about the emergence of form and function in large communities of agents by simple local interactions, which has attracted the interest of researchers in artificial life and robotics, especially in the field of swarm robotics. We will only mention a few studies here, for comprehensive reviews on swarm robotics see [12, 15, 20].

One of the most influential works in this field was by Reynolds [144], who showed with computer simulations that it is sufficient to use a set of simple local interaction rules to achieve the dynamic formation of global flocks. The particles or agents in his simulation align to the orientation of neighboring agents, trying to keep a certain distance to their neighbors, and avoiding obstacles. An illustrative sequence of the evolution of a large flock is displayed in Fig. 2.6.

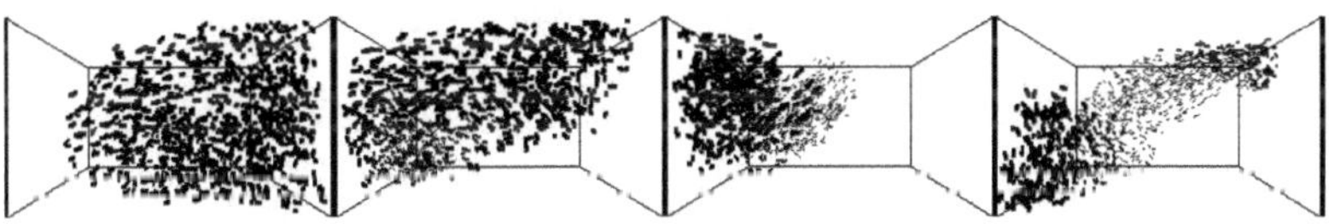

Fig. 2.6: Formation of a set of simulated particles self-organizing into a flock. Each particle has three behaviors: obstacle avoidance, keeping distance to particles in neighborhood and aligning to average alignment of neighborhood. Time progresses from left to right. Included with permission by K. Wiley [184].

Again we find the self-amplification mechanism in the alignment, similar to the magnetic dipoles (Sect. 2.1.1). Different global patterns emerge depending on the density, the noise and the parameters of interaction. Essentially, the system of hundreds of degrees of freedom organizes into a dynamics of a few dimensions that can be controlled qualitatively with a few parameters. One of the early demonstrations of this phenomenon was given by Deneubourg and colleagues in [38] where the different architectures of a wasp nest are explained by a change in the behavioral rules of the individuals.

Swarm robotics yields a collective behavior in a group of simple individuals that interact only locally, like the birds in the flock. The vision is that a set of comparably cheap robots can solve a complicated task. An example are rescue scenarios, where the fault tolerance and the speed of search are important aspects that cannot be achieved by a single (and large) robot. The following areas are of special interest to swarm robotics: collective decision making, path optimization and navigation, division of labor, self assembly, the role and emergence of communication. For example a self-organized path formation and homing behavior was demonstrated in [115, 172] using a set of mobile driving robots that have a camera and a circular ring of color LEDs at the outer boundary. Equipped with a set of rules describing how to react on the different colors of the neighboring robots they could organize into a path structure between the home location and distant location even in cluttered environments.

Another example are terrain covering robots [168] also inspired from ants. The robots mark their trail and at the same time sense the markings of other robots. With a combined obstacle- and trail-avoidance behavior the robots effectively explore an unknown environment without complicated localization methods or centralized control. In the two examples the robots used direct and stigmergic signaling for communication.

One of the major questions in the research on self-organizing collective behavior is about how much complexity is required at the individual level to generate the observed sophisticated pattern of biological systems [27] or to perform a desired task [38]. It is interesting that a certain cooperativity can also emerge without explicit communication, see [185]. That study also showed that decentralized control is more robust than a centralized control in achieving a collective behavior task, even though neither of them was explicitly optimized for robustness. It seems that robustness is an intrinsic property of the distributed organization as it can be vastly found in self-organizing biological systems as well [27].

2.2.2 Behavioral Self-Organization

The main objective of this book is the self-organization of the behavior of autonomous robots. In the literature, this question is often considered in connection with reinforcement learning or artificial evolution. Both methods offer the possibility to drive agents towards certain goals by very unspecific drives. As a consequence, the agent finds by itself a solution that may be called self-organized because of the large number of different potential solutions.

Let us consider a few examples from evolutionary robotics: by means of genetic algorithms a fitness function, encoding some desired goal, can be optimized in a high-dimensional search space. In this way the control network/program and/or the morphology of a robot are optimized to perform a desired behavior. This is achieved by generating a set of possible solutions (individuals), which are then iteratively recombined, mutated and selected until a satisfying solution is found. From generation to generation a certain structure arises such that specific and sometimes surprising behavior can emerge from a rather unspecific fitness function [113].

What is the role of self-organization in this scenario? If the fitness function is very specific, for example to optimize a gait of a quadruped robot for fast locomotion, e. g. [73], then there is only little room for self-organization. The dynamics is very much restricted by the goal. Nevertheless, the precise implementation of the behavior can be argued to self-organize. In natural evolution we encounter the opposite extreme, namely the goal is simply to survive. The emerging forms and functions can be considered as purely self-organized.

In evolutionary robotic it is unfeasible to work with such unspecific fitness functions, because this produces not enough feedback in a huge search space. Nevertheless there are some examples of largely self-organized behavior in evolutionary robotics. An illustrative example is the coevolution of two robots, a predator and a prey [52]. The predator is to catch the prey (high fitness for close distance) and the prey has to run away (high fitness for large distance to predator). As a byproduct a set of interesting behaviors emerged such as obstacle avoidance, straight navigation, visual tracking, object following and of course specific strategies optimized for the specific opponent, see Fig. 2.7. Also very interesting results have been obtained with the Polyworld simulation environment of Larry Yaeger [189].

There is also an excellent review [134] making the interplay between self-organization and embodiment more transparent. The review presents a variety of examples for the self-organization realized by emergent self-stabilization.

Fig. 2.7: Trajectories of coevolved predator and prey robots (simulation). Black circle stands for predator and empty circle for prey © Morgan Kaufmann Publishers and D. Floreano and S. Nolfi 1997 [52].

2.3 Exploiting Self-Organization for Robot Control

The aim of the present section is to extract the characteristic features of self-organizing systems so that these system can serve as a kind of prototype for the design of artificial systems. Let us therefore summarize the essential points and discuss the usefulness for robotic self-organization.

2.3.1 Mechanisms of Self-Organization

Although largely different in detail, there is a common principle driving self-organization. The respective systems are characterized by (i) observing global symmetries (like the homogeneity and isotropy of a fluid) or constraints and (ii) the presence of feedback mechanisms able of amplifying local, symmetry breaking fluctuations up to macroscopic scales. The emerging patterns are the result of a spontaneous symmetry breaking arising as the minimal compromise of violating versus conserving the global symmetries and/or constraints of the system.

Generic is the dependence on some few control parameters which regulate the transition from the "normal" to the self-organizing regimes. The control parameter for the emergence of Bénard cells is the temperature gradient, which measures so to say the stress exerted on the system from outside to organize heat transport as efficiently as possible. When increasing the gradient, transitions from the conduction regime to more and more complex convection patterns are known to occur. In the reaction-diffusion system the diffusion constant and the local reaction rates determine the kind of patterns observed. In the case of the ants the amount of pheromone an ant elicits can be considered as a control parameter. Since the pheromones evaporate in the course of time, a subcritical marking of the tracks does not lead to sufficient self-amplification such that no common path to the nearest food place would emerge.

The next point is the tremendous reduction in dimensionality. The systems are composed of many small elements interacting locally in a high dimensional space, but the emerging global patterns can be described by just a few order parameters. While the control parameters are given from outside, the order parameters are emerging with the patterns and are sufficient to describe the phenomenon entirely at the macroscopic level. Most importantly, the global structure is not at all inscribed into the local interactions as is most clearly demonstrated for instance in the case of the Bénard cells, see Fig. 2.1.

2.3.2 Self-Organization of Robot Behavior

The self-organization processes realized in machines so far are restricted to specific conditions and goals. How can the knowledge about self-organization in nature help in formulating a general and systematic approach? What is the relation between a convection pattern (like the Bénard cells) and the behavior of a robot. From a fundamental point of view, the robot is nothing else but a number of components, linked together by some joints that can be actuated by motor forces. Moreover, there is the interaction with the environment both by physical actions and sensorial responses. By sending electrical energy to the motors nearly everything is possible that can be done by this complex physical system.

If the robot follows closed loop control, the motor forces are a function of sensor responses so that we have a feedback system. The main ingredient of self-organization, the self-amplification of local fluctuations, seems to be reachable if only the feedback parameters are appropriately chosen. Such a system will also obey lots of symmetries and natural constraints, which may provide the confinement necessary to keep the self-amplification in bounds. Can we than expect that the system will develop spatiotemporal patterns – behaviors—as a compromise between these two opposing tendencies? And can we expect the same richness and beauty observed in the natural self-organization systems?

The answer in the first place is no. It will be possible only in exceptional cases to tune the parameters of the controller by hand so that the system has an appropriate feedback structure. This is in contrast to natural systems, where the interaction parameters are set, so that the system has the ability to self-organize (or not) and, if so, all one has to do is to choose the control parameters. Due to the dimensionality reduction phenomenon, the latter is a task of low complexity.

However, there is hope if we find adequate principles of self-regulation for the parameters of the system. This book presents homeokinesis as a general self-regulatory paradigm driving complex physical feedback systems towards a working regime where self-organization can happen. We will demonstrate both theoretically and by many robotic experiments that the new approach rests on the same principles: self-amplification of perturbations in balance with symmetry conserving and other confining effects that constrain the system from running into chaos. In this sense, homeokinesis is a valid realization of the general principles of self-organization in embodied robotic systems.

The concrete realization was guided by a set of premises that may serve as a kind of "acid test" that any principle claiming to engender self-organization should pass. These premises can be formulated in the following way. Any principle driving agents to self-organized behavior (i) should be domain invariant, i. e. be independent on any specificities of the physical system under control; it is (ii) to be operational in continuous state-action spaces; (iii) it must be simple in order to be as universal as possible; and eventually (iv) it should exclude any random generation of actions as they are used for instance in reinforcement learning. Instead, search has to be an emerging phenomenon that takes place also in the deterministic case. In other words: search is an element of behavior, any randomness (noise) must be of physical origin.

Chapter 3
The Sensorimotor Loop

Abstract: This chapter aims at providing a basic understanding of the sensorimotor loop as a feedback system. First we will give some insights into the richness of behavior resulting from simple closed loop control structures in a robotic system called the BARREL. This richness is a lesson we can learn from dynamical systems theory: even very simple systems can produce highly complicated behavior. Nearly everything is possible in such a feedback system that is provided with enough energy from outside. Surprisingly, this is accomplished even with extremely simple, fixed controllers, to which we will restrict ourselves here. In later chapters we will see how the homeokinetic principle makes theses systems adaptive and drives them towards specific working regimes of moderate complexity, loosely speaking somewhere between order and chaos.

The aim of this chapter is to make the reader familiar with the general structure and specific properties of tightly coupled sensorimotor loops. In these loops the motor commands are directly related to the sensor readings, so that the robot with its "brain" forms a feedback system. In this context the framework of dynamical systems, known from mathematics and physics, started to get increasing attention in the last two decades [13, 81, 120, 152]. It is a powerful method to analyze [75] and construct [39, 72, 78, 81, 152, 160] robot controllers, as it allows one to formulate the time evolution of the system, in a quantitative manner and to obtain both analytical and qualitative predictions. Dynamical system theory also led to the application of chaos control and coupled chaotic oscillators to robotics [91, 138, 159].

After a general introduction of closed loop control, using the framework of dynamical systems, we study a specific example, the BARREL, that is particularly interesting by its strong embodiment effects. The general settings are investigated subsequently in an elementary sensorimotor loop, a one-dimensional system controlled by a single neuron. Without noise, a pitchfork bifurcation occurs and a pronounced hysteresis behavior is established if the neuron has a bias. We introduce our concept of an effective bifurcation point to allow for noise effects. This concept defines a specific working regime, where robots can already take decisions while still being sensitive to perturbations by the environment. Eventually, we present briefly neural networks as universal tools for the realization of the control system. The investigations are based on dynamical systems theory but the mathematics will be kept

R. Der and G. Martius: The Playful Machine, COSMOS 15, pp. 23–58.
springerlink.com

simple and essentially self-contained so that no special knowledge of the latter is necessary.

The theoretical studies are underpinned by robotic experiments that can be executed with our simulation environment. Experiments are provided for studying closed loop control and in particular the concept of the effective bifurcation point under various conditions. The role of the embodiment can be investigated with wheeled robots. When interconnected to form a chain of robots emergent cooperativity is observed even though decentralized control is used. In this way, the present chapter makes first contact to the central idea of externalizing complexity, namely to control complex physical modes with very simple control structures and thus to source the complexity out to the interaction with the environment.

3.1 Sensorimotor Loop — The General Case

In a self-consistent approach to self-organizing robot behavior, the sensor values are the only source of information for the robot. This is also the *credo* of our approach. The communication between the "brain" and the body of the robot takes place at the discrete instants of time $t = 0, 1, 2, \ldots$. In practice, typical clock frequencies are ranging from 10 to 100 Hz, depending on the speed of the information processing. In each time step a vector of sensor values $x_t \in \mathbb{R}^n$ is reported. Let us illustrate this by two examples. Firstly we consider a wheeled robot with sensor vector

$$x = (v_l, v_r, s_1, \ldots, s_k)^\top \tag{3.1}$$

where v_l and v_r are the wheel velocities of the left and right wheel, respectively, as measured by the wheel counters, and s_i are the values of the infrared sensor i with $0 \le s_i \le 1$. The wheel velocities are examples for proprioceptive sensors. Infrared sensors are examples of exteroceptive sensors since they get information about the relation to the outside world. In many of the applications treated further below the robot has only proprioceptive sensors providing informative feedback on the state of its body, an example being our dog robot, see Fig. 3.1, where

$$x = (x_1, \ldots, x_n)^\top , \tag{3.2}$$

x_i are the joint angles.

3.1.1 The Controller

At each time step t, the controller sends a vector $y_t \in \mathbb{R}^m$ of target values to the motors of the robot. Closed loop control means that the controller is given by a function $K : \mathbb{R}^n \to \mathbb{R}^m$ mapping sensor values $x \in \mathbb{R}^n$ to motor values $y \in \mathbb{R}^m$. In the most simple case this is a function

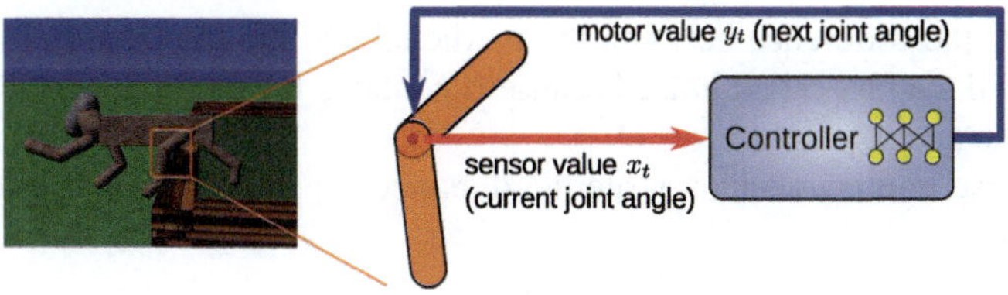

Fig. 3.1: Example of a sensorimotor loop with the joint angles as proprioceptive sensors. Only a single joint is depicted.

$$y = K(x) \tag{3.3}$$

depending in general on a vector of parameters p that can be adapted in order to realize a desired behavior. More generally, the map may depend on an internal state $s_t \in \mathbb{R}^k$, which is updated in each time step as well so that the controller is realized as

$$y_t = K(x_t, s_t) \tag{3.4}$$
$$s_t = O(x_t, s_{t-1}) \tag{3.5}$$

where now $K : \mathbb{R}^n \times \mathbb{R}^k \to \mathbb{R}^m$. Controllers with internal states are realized conveniently by recurrent neural networks as introduced later in Sect. 3.6.4. However, the main problem in using an internal state is to find the update rule (3.5) for the latter such that the system develops the desired behaviors.

We will use a very simple realization of the parameterized controller function $K(x)$, see Eq. (3.3), but complement it later with a dynamics for the parameters p, which will be a function of the state dynamics. We are free to consider the parameters p as internal state variables so that the new feature introduced by the homeokinetic principle is an explicit rule for the function $O : \mathbb{R}^n \times \mathbb{R}^k \to \mathbb{R}^k$ in Eq. (3.5).

3.1.2 Forward Model and Sensorimotor Dynamics

Let us stipulate that our robot has a certain ability for cognition. We understand here cognition in a minimalist sense as the ability of the robot to predict the consequences of its actions in the near future with a forward model. Formally, this is realized by a function $M : \mathbb{R}^n \times \mathbb{R}^m \to \mathbb{R}^n$ mapping the sensor values x and the actions y of the robot onto the new sensor values, i.e.

$$x_{t+1} = M(x_t, y_t) + \xi_{t+1} \tag{3.6}$$

where ξ_t is the difference between the predicted and the true sensor values. This quantity will also be called noise because it contains the unpredictable effects like sensor noise and so on.

With these notions we may write the dynamics of the sensorimotor loop in the closed form

$$x_{t+1} = \psi(x_t) + \xi_{t+1} \tag{3.7}$$

where

$$\psi(x) = M(x, K(x)) \ . \tag{3.8}$$

The function $\psi : \mathbb{R}^n \to \mathbb{R}^n$ is called the dynamics model and can be understood as a time series predictor for the time series of the sensor values x_t, see also Fig. 3.2.

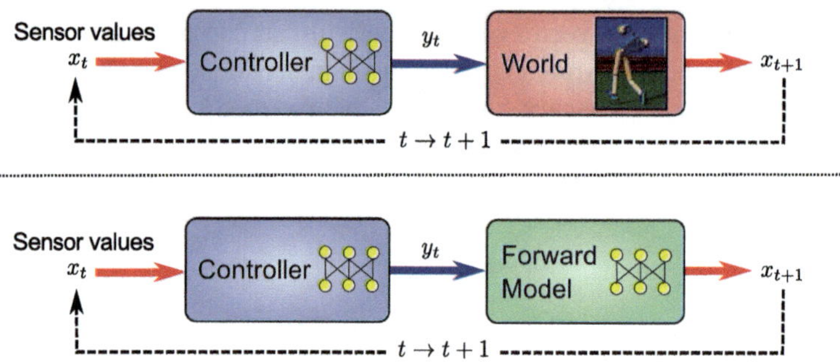

Fig. 3.2: The sensorimotor loop (top) and its model (bottom).

The prediction error, also called the model error, can be defined as (dropping the time indices)

$$E_{\text{pred}} = \xi^\top \xi \ . \tag{3.9}$$

Let the model M be realized by a parameterized function with parameters $a \in \mathbb{R}^M$. In order to minimize the prediction error (3.9) the parameters can be adapted by following the gradient of the error function in descending direction, see Fig. 3.3 for the signal flow. Let a be any of those parameters, then the learning step is defined as

$$\Delta a = -\varepsilon_A \frac{\partial E_{\text{pred}}}{\partial a} \ , \tag{3.10}$$

with a learning rate ε_A that is kept so large that fast parameter changes are possible in the following applications. The model and the learning dynamics can be realized for instance by a neural network as introduced in Sect. 3.6.

The structure of the model and the learning procedure define the passive cognitive abilities of the robot as will be worked out in more detail in the following.

We use the word *model* in two different contexts. There is the *forward model* M that predicts the outcome of the actions. For simplicity it is sometimes called

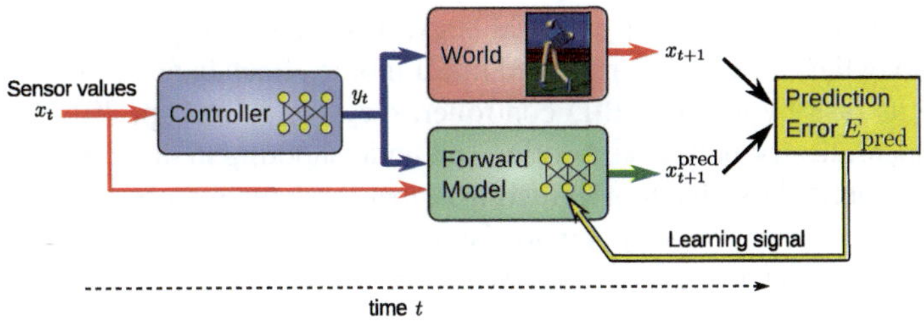

Fig. 3.3: The sensorimotor loop and the forward model of the robot. The controller receives the current sensor values and generates corresponding motor values, which are sent both to the robot and its forward model. The difference between the predicted new sensor values and the measured ones forms the prediction error E_{pred}. A learning signal for the improvement of the model is derived by gradient descending the error E_{pred}.

just the model. In its most simple realization it only receives the motor values (actions), in which case it can be considered as a self-model, especially if the sensors are essentially only the proprioceptive ones. The second context is the model of the entire sensorimotor dynamics $\psi(x)$, which we call the *dynamics model*. Quite generally, we will call the robot together with its controller and the forward model the *brain-body* system.

In order to illustrate the formalism of the sensorimotor loop and the merits of closed loop control we consider now an example using a robot with strong embodiment effects, i. e. a robot where the physical effects like inertia and centrifugal forces are heavily interfering with the effects of the actions taken. Afterwards we will return to a simpler case for analytical considerations.

3.2 Dominated by Embodiment: The BARREL

While our emphasis lies on controlling more complex robots we will in the current chapter restrict ourselves to a case of moderate complexity so that analytical considerations are still possible. The machine we are going to use is the BARREL (which is short for barrel robot), see Fig. 3.4.

Fig. 3.4: The BARREL. It consists of a cylindrical encasement with two weights sliding on two orthogonal axes perpendicular to the cylinder axis. Each of the weights is moved by a linear motor so that the system can shift its center of gravity. The sensor values measure only the inclination of the axes, i. e. $x_i = \sin(\theta_i)$. Thus, the true physical state of the system is largely unknown to the controller. Note that $x_1 < 0$ in the displayed situation.

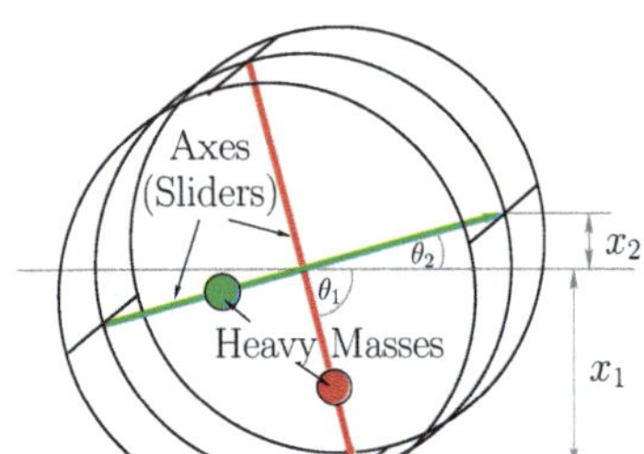

The robot has a cylindrically shaped body. Inside there are two weights that are moved along the two axes by simulated linear motors. Each bare motor is supported in doing its task by an extra PID controller, see Sect. 16.4.4 (p. 301) for details, that compensates for both overshooting and undershooting in the movement of the weights. Nevertheless, due to a limited maximal force the motors cannot move the weights with arbitrary velocity and precision. The movements of the weights induce a change of the center of gravity that causes the robot to roll in one direction or the other.

The BARREL can unfold many different kinds of motion, despite its simple construction, as will be seen below.

3.2.1 Properties

The reason why we have chosen the BARREL is because it is a body-dominated system. What we mean by this is that, similar to many systems used in robotics, the consequences of an action are largely dominated by the current physical state of the robot. By way of example the effect of shifting a weight on its axis will be very different if the system is at rest or in a rapidly rolling motion so that the sensor reaction on any action is highly ambiguous. The situation is even more complicated due to the physical properties of the robot which are simulated realistically by the ODE physics engine [154] embedded in our simulation software, see Chap. 16. The execution of an action (moving the weights) is largely influenced by the inertia of the weights, the Coriolis forces due to the motion of the weights on a rotating axis, centrifugal forces if the BARREL is rolling with high velocity, and others.

In the following we are going to discuss different control paradigms using the example of the BARREL. This allows us on the one hand to demonstrate some features of the embodied artificial intelligence approach in a simple and transparent manner and on the other hand to outline the perspectives of self-organization for extending this approach to a wider field of applications.

3.2.2 Open Loop Control

One way to control the BARREL is in the open loop setup where a sequence of motor signals is sent to the robot. This sequence may be for instance generated by a central pattern generator. Let us first assume that the BARREL is to roll with a fixed velocity. This may be achieved by shifting the internal weights periodically with a convenient frequency and a phase shift of $\pi/2$. The velocity of the BARREL is then determined by that frequency – one rotation of the BARREL corresponds to one period of the pendular oscillations.

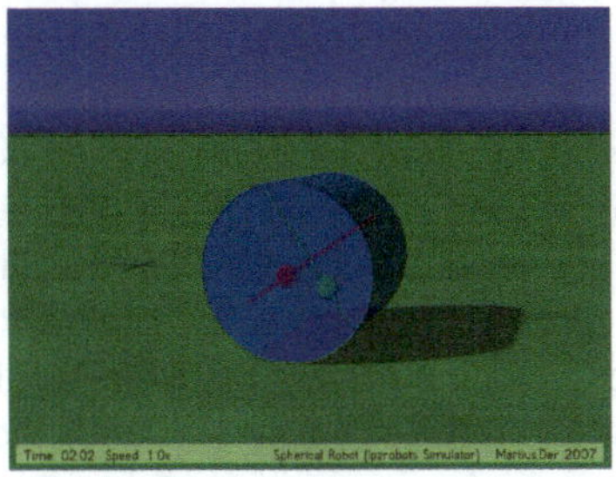

Video 3.1: Open loop control of the BARREL The robot is controlled by a periodic control signal driving the internal weights with a phase shift of $\pi/2$. Starting with a very low frequency of the controller signal, the frequency is doubled at time 2:15, 2:40, 3:05, and 3:30. At 3:45 the BARREL is accelerated by a force (red dot) but is seen to return rapidly to the original mode of behavior. The higher frequencies very clearly reveal the difficulties in realizing a fast motion under the open loop control paradigm. The video can be watched at `http://playfulmachines.com`.

When doing so we find that, at low frequencies and with some friction at least, the BARREL adapts its (average) rotational frequency to the frequency of the pendular oscillations indeed, see Video 3.1.

This behavior is stable against moderate perturbations. To understand this, imagine a stroboscopic mapping depicting the BARREL every moment the red weight, say, has maximal downward elongation. In an ideal and constant rolling mode (without friction) the red weight will be exactly below the cylinder axis since then there is no torque (the green weight is in the center due to the $\pi/2$ phase shift). If the BARREL is externally decelerated, the stroboscopic mapping will show the weight to rotate slowly against the rotational direction of the BARREL (since the maximal elongation is reached before the tip of the axis reaches ground). Hence a torque is exerted counteracting the slowing down. This stabilization mechanism works just as well if the BARREL is being accelerated from outside, as long as within certain bounds.

Let us try to make contact with the embodied AI paradigm [131] at this point. Its aim is to shift the computational load from the controller to the morphology and physical properties of the embodiment. In the above case this is realized due to the self-stabilization effect—a simple periodic signal generates a stable motion pattern in a complex physical object. If the aim was to realize nice harmonic motion one had two options, either to change the wave form of the control signal (to increase the control effort) or to modify the physical properties of the body.

The embodied AI approach takes the latter route. As the experiments show, for a BARREL with fixed physical properties, there are one or more preferred frequencies where the motion is most harmonic. This is also where minimal control efforts[1] are required. On the other hand, in the sense of the embodied AI, given the frequency, one may change the physical parameters like the PID settings, the forces and pen-

[1] Control efforts are understood in terms of the complexity of the required controller that can generate the behavior.

dular ranges, the mass of the cylinder and so forth in order to get the desired nice harmonic motion.

However, this only works as long as the frequency is not too high. At higher frequencies, the self-stabilization effect is lost and different modes of behavior are induced by the periodic signal. It is here that the physical effects, like the Coriolis force, start to dominate so that the BARREL is driven into irregular behaviors, see Video 3.1. Following the embodied AI paradigm, one would try to modify the physical properties appropriately to make the system obey the simple periodic control signal but there is no simple and/or straightforward solution to that problem.

The experiment can also be performed by the reader as described in Experiment 3.1.

Experiment 3.1: Open loop control of BARREL

Many of the experiments described in this book can be performed by the reader using our simulation software that is online at [103]. Here we will describe the handling of the software in more detail. Choose one of the options described at the website to get the software on your computer. You will get a folder with an `index.html`, which when opened in a browser shows all experiment descriptions and provides links to start them. For each simulation a terminal window and a graphical window opens. The latter shows the rendered scene, see Video 3.1. At the same time the terminal window shows a welcome text and the parameters that are used. While working with the simulator both windows are important. The terminal window allows to check and change parameters via a text-based console, which can be entered by pressing <Ctrl>+C in the terminal window. A prompt appears (>) and you can type
`>help`<Enter>
to obtain a list of possible commands, see Fig. 3.5(b). The graphical window allows to observe and possibly interact with the robots. For a list of keystrokes and mouse actions type `h` (make sure the focus is on this window).
Our first experiment is Open loop control of BARREL. The simulation is now running with the default parameters: `period=300` given in control steps (1/50 s) and `phaseshift=1` given in multiples of π.
Changing parameters of the robot or controller is done by using the pattern "Parameter=Value" on the command prompt. For instance, in order to decrease the period duration (increase frequency) of the sine signal type (after pressing <Ctrl>+C in the terminal window)
`>period=200`<Enter>
If done correctly, the actually set value is echoed, i. e. `period = 200.0000`,
otherwise the parameter name was probably misspelled. Hint: you can use the <Tab> key to do automatic completion.
In the default situation there is no rolling friction. The barrel does not move with a constant speed but oscillates instead. Switch on the friction by
`>friction=0.1`
Now, decrease the period further, try: `>period=100,50,10`. Note, that the robot cannot follow the periodic commands if too fast. You can also change to another control mode for instance by using a colored noise with the parameters
`>strength=`Strength
`>color=`Correlation time of the noise in 1/50 s
Try `>strength=1`, `>color=100` and disable the sine generator with
`>amplitude=0`
Most of the parameters of the simulation can be monitored by starting the GUILOGGER by pressing <Ctrl>+G in the graphical window. Then a new window appears where you may tick the boxes for the on-line display of sensor and/or motor values, see Fig. 3.5(a). Alternatively you can invoke the MATRIXVIZ with <Ctrl>+M, which is especially handy in highdimensional systems, see Fig. 3.5(c).

3.2.3 Closed Loop Control

Outside the self-stabilization region of a rolling mode, the BARREL does not obey the periodic motor signals any longer and realizes more or less chaotic behaviors.

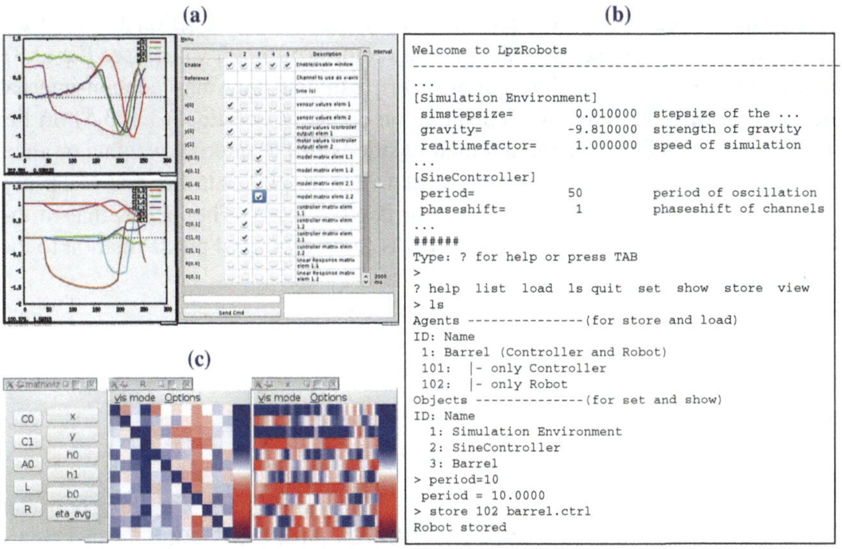

Fig. 3.5: User interface of the LPZROBOTS simulator. **(a)** GUILOGGER window with two controlled plotting windows. In the main window (**right**) a set of channels are selected. Their temporal evolution is shown in the subwindows (**left**), here sensor and motor values, and the parameters of the controller. **(b)** Terminal window with console interface. **(c)** MATRIXVIZ showing the state of the matrix R and the time evolution of x.

In some sense the body takes over the command and if the controller is to achieve a certain objective, it has to "watch" carefully what the body is doing and develop the right "tact" for it. However, this can not be done in the open loop control mode we used in the previous section, because the sensors have to be taken into account. In the special case of the BARREL, we have two sensors measuring the inclination of the axes, see Fig. 3.4.

As before, we assume that the sensor values at time t are comprised in the vector $x_t \in \mathbb{R}^n$. In the case of the BARREL we have $n - 2$ and we may normalize the sensor values so that $-1 \leq x_{ti} \leq 1$ for $i = 1, 2$. If the robot is rolling with a fixed velocity, the vector of sensor values x_t is rotated in each time step by a fixed angle α so that we have the very simple sensor dynamics

$$x_{t+1} = U(\alpha) x_t \,, \tag{3.11}$$

where U is the rotation matrix

$$U(\alpha) = \begin{pmatrix} \cos\alpha & -\sin\alpha \\ \sin\alpha & \cos\alpha \end{pmatrix}$$

rotating a vector by the angle α.

Under the closed loop control paradigm, the controller is a function $K : \mathbb{R}^n \to \mathbb{R}^m$ mapping the vector x_t of sensor values to the vector y_t of motor values, see Eq. (3.3). These controllers can be rather complex, in particular they may contain internal states that modify the mapping depending on contexts, see Eqs. (3.4, 3.5). In the BARREL case $y_t \in \mathbb{R}^2$ gives the nominal positions of the internal weights on their respective axes. We will use for the moment a very simple controller that will prove sufficient to produce a rolling motion with fixed velocity. The idea is that in the stable rolling mode, the vector of actions y_t is to be in a fixed phase relation to the vector of the sensor values x_t.

Let us therefore tentatively put the controller as

$$K(x) = Cx \tag{3.12}$$

where

$$C = cU(\phi) \tag{3.13}$$

with c defining the amplitude of the weight shifting. When using this controller we find stable rolling modes as expected. We may push the BARREL or reverse its velocity from outside but after a very short time the robot returns to its stable rolling mode with fixed velocity, to reproduce follow Experiment 3.2.

If C is a rotation matrix one observes stable rolling modes with a frequency defined by ϕ, see Eq. (3.13). However, it is observed that ϕ is in general larger than the true rotation angle of the sensor vector in one time step α (3.11). Instead one may choose ϕ rather large without increasing the velocity considerably. The phase difference is due to the specific physical properties of the robot described above and could be evaluated empirically or calculated by knowing the physics of the system in detail. However, this is not in the spirit of this book, which is devoted to the self-organization of specific modes in a physical system without knowing it in advance.

The linear controller, see Eq. (3.12), is appropriate in the BARREL case since the sensor values are not in a direct proportionality to the motor values. In general, we have to make sure the actions (motor values) are kept in bounds by introducing a smooth squashing function $g(u)$, that is applied componentwise putting

$$K(x) = g(Cx + h) \tag{3.14}$$

where $h \in \mathbb{R}^m$ is a bias introduced for greater generality. In this way, motor values are kept safely in a defined interval.

The nonlinearity in Eq. (3.14) does not qualitatively change the behavior if C is chosen as in Eq. (3.13). However, we are now free to chose the matrix elements of C arbitrarily. As the experiments show, depending on the parameters, one observes a variety of behavioral modes, which are sometimes surprising given the extremely simple nature of the controller. An example is the "lolloping" mode, see Video 3.2. The variety of modes is even larger if the bias values h are also chosen arbitrarily, see Experiment 3.2.

By varying the parameters in a systematic way, we obviously can find out about the scope of modes that are realizable by this type of controllers. In the BARREL

Experiment 3.2: Closed loop control of BARREL

Start the simulation Closed loop control of BARREL. The simulation is now running with the C matrix as $C_{11} = C_{22} = 1, C_{12} = -C_{21} = .1$ and $h_0 = h_1 = 0$. Change the parameters `coupling1` for the diagonal and `coupling2` for the non-diagonal matrix elements of C, e. g.
`>coupling1=-0.2`
Use the GUILOGGER (<Ctrl>+G) for watching the sensor values, which are a good indicative for the behavior of the BARREL. In order to select random control parameters in the interval $(-5,5)$ press `r` on the graphical window. The new parameters are printed on the terminal and you can use the GUILOGGER for monitoring all parameters (x,y,C,h). To randomize also the bias terms h in the interval $(-3,3)$ press `R` (<Shift>+R), which results in even more different behaviors. With `L` (Shift+L) you can set to $C_{11} = C_{12} = 2$, $C_{21} = C_{22} = -1$, which will cause the robot to roll and jump such that we call it the lolloping mode.

Video 3.2: BARREL performing a "lolloping" behavior. A fixed closed loop controller ($C_{11} = C_{12} = 2$, $C_{21} = C_{22} = -1$, $h_{1,2} = 0$) was used. The robot jumps by rapidly moving the internal weight while rolling from left ro right. The video can be watched at `http://playfulmachines.com`.

case the space of parameters is 6 dimensional so that the search space is already quite large. However, in the systems to be considered later, the parameter space is of several hundred dimensions so that an unbiased search is impossible. The new perspective self-organization can bring into this paradigm is to find out, by self-exploration, about the whole range of **low-complexity** modes the system is able to support by low-complexity controllers. These modes may then be viewed as the candidates for embodied AI realizations of more complex behavior architectures.

Before going to the more theoretical considerations let us remark a remote similarity of the BARREL to the famous passive walker [111] when driven by some periodic motor forces enabling it to walk on a horizontal plane [32]. This walker works by having its center of gravity a little ahead of the point of support, falling over being avoided by the leg swinging forward to support the body in the right moment. The BARREL has an easier life since it is always protected from falling over by the cylinder encasement. Nevertheless, in both cases there is a self-stabilizing and self-promoting effect due to the specific embodiment so that the amount of control can be kept small.

3.3 Analyzing the Loop

In order to treat essential features of the closed loop control analytically, we consider now an even simpler example, namely a one-dimensional sensorimotor loop. One special application will be the velocity control of wheeled robots. Actually, this seems a little strange, given that the realization of an externally determined wheel velocity is a standard task of classical robotics with many reliable solutions. However, our goal is different. We want to use the wheel velocity control as a transparent example for the more involved cases to be considered later and moreover we aim to underline the merits and the potential of closed loop control for embodied systems. Furthermore, we want to show how minimalist decentralized control can give rise to self-organization effects as a mere consequence of the specific physical properties of the embodied system. This will be exemplified by a chain of such robots with each wheel being controlled independently so that control is strictly decentralized. We will see that under certain conditions the robots spontaneously cooperate. This takes place even when moving in a maze with only narrow passages between obstacles. Thereby, the investigation of this special system gives us the opportunity to come into contact with basic notions and effects relevant for the self-organization of robot behavior. Moreover we will also argue that this kind of control opens new perspectives for robotic tasks in quite practical settings.

Let us consider a wheel of a robot rotating with velocity v_t. It is monitored by a wheel counter, which outputs a measured velocity x_t so that

$$x_t = Av_t + \kappa_t$$

where κ_t is the measurement (sensor) noise and A is a hardware constant assumed to be known for the moment. The controller of the robot outputs the target velocity y_t for the current time step. Assuming that the motor is able to realize in one time step the target wheel velocity we would have $v_{t+1} = y_t$ and $x_{t+1} = Ay_t + \kappa_t$. However, there are always perturbations so that we put

$$x_{t+1} = Ay_t + \xi_{t+1} \; . \tag{3.15}$$

We will use this equation also for the case that the robot is moving on the ground or embedded into the chain. In this case the "noise" term ξ has to account for all effects due to the coupling with other robots, the slip and friction on the ground, inertia effects, collisions, and the like. Fig. 3.6 gives an exemplification of this situation in an experiment with a realistically simulated robot.

In a closed loop approach without internal states, the controller is given in terms of the sensor value x_t as

$$y_t = K(x_t) \; . \tag{3.16}$$

In the simple case we use

$$K(x) = \tanh(Cx) \tag{3.17}$$

where the hyperbolic tangent (tanh) confines the motor values between -1 and 1, see Fig. 3.7, so that the wheel can rotate forward and backward. We may interpret this as a simple rate-coded neuron with tanh activation function and synaptic strength C, the neural network realization, see Sect. 3.6, of the controller consisting just of this one neuron.

Using Eq. (3.16) in Eq. (3.15) leads to the following dynamics of the sensorimotor loop

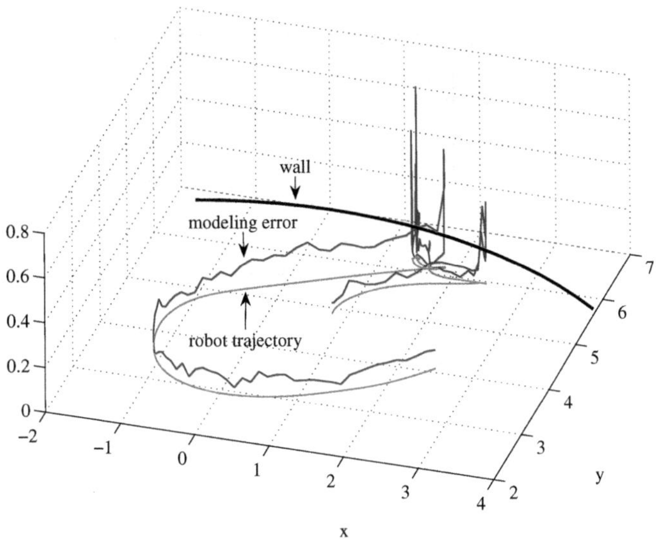

Fig. 3.6: Model error of a wheeled robot. A typical trajectory of the robot with the model errors ξ^2, see Eq. (3.18). While during the normal drive the error is essentially given by the measuring uncertainties, the error increases drastically at the repeated collisions with the wall since wheels get blocked or may also rotate freely if the robot gets jammed. Control is realized by a homeokinetic controller as introduced later. Graphics taken from [68]. © MDPI Publishing and Frank Hesse.

$$x_{t+1} = A\tanh(Cx_t) + \xi_{t+1} \tag{3.18}$$

so that the map ψ of Eq. (3.8) is

$$\psi(x) = A\tanh(Cx) \ .$$

It will prove convenient to use also the formulation of the dynamics in terms of the membrane potential $z_t = Cx_t$ of the neuron where $R = CA$

$$z_{t+1} = R\tanh(z_t) + \rho_{t+1} \tag{3.19}$$

and

$$\rho = C\xi \tag{3.20}$$

is a modified noise variable.

3.3.1 Feedback Strength

The sensorimotor loop is a feedback loop, the (linear) feedback strength being given by

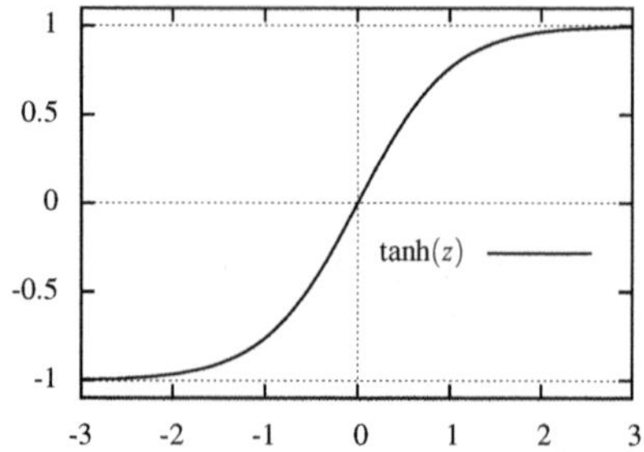

Fig. 3.7: Hyperbolic tangent squashing function. The range of $\tanh(z)$ is $(-1,1)$. Used as an activation function the neuron has a linear response around $z = 0$ with slope 1. For large absolute z the neuron is in its saturation region with little input dependence.

$$R = CA \; .$$

The effect of R can be seen by the following argument. We consider the case of small z where the approximation $\tanh(z) \approx z - \frac{z^3}{3}$ can be used. Using the linear term only, one gets (ignoring the noise for the moment) from either Eq. (3.18) or Eq. (3.19)

$$x_t = R^t x_0 \tag{3.21}$$

meaning that the wheel slows down for $0 < R < 1$ ($x_t \to 0$ as $t \to \infty$) and accelerates for $R > 1$ (x_t increases exponentially), but the velocity can not increase unrestrictedly since $|\tanh(z)| < 1$, i. e. the nonlinearities confine the further growth of x_t.

3.3.2 *Fixed Points*

The asymptotic value of x in the nonlinear case is given by the solution of the fixed point equation obtained from Eq. (3.19) as

$$z = R \tanh(z) \; , \tag{3.22}$$

which has always $z^* = 0$ as a fixed point, which is stable for $0 < R < 1$ and unstable for $R > 1$. For $R < 0$ the state changes signed every iteration and is thus not useful for our purposes. At $R = 1$ there is a pitchfork bifurcation with two stable fixed points emerging. The stable fixed points can be found for instance by graphically solving the transcendental equation as shown in Fig. 3.8(a).

Another way is to use the approximation $\tanh(z) \approx z - \frac{z^3}{3}$ in order to get results for the case that $|z|$ is not too large. This is supported by the well known fact from dynamical systems theory that a system given by Eq. (3.19) is topologically equivalent [165] to the more simple system

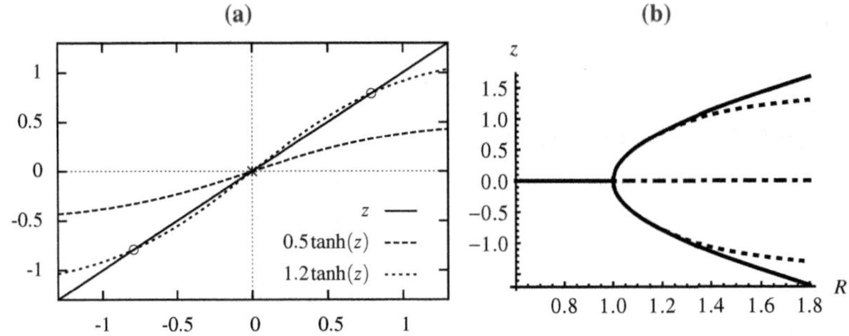

Fig. 3.8: Pitchfork bifurcation. (a) The graphical solution of $z = R\tanh(z)$ is exemplified for $R = 0.5$ with the stable fixed point at $z = 0$ and for $R = 1.2$ with two stable fixed points at $z = \pm 0.79$, the fixed point at $z = 0$ being unstable; **(b)** The bifurcation diagram is showing the fixed points in dependence on R. The **solid** line indicates the stable fixed points of the exact system and the **dashed** line of the approximation Eq. (3.23). The *dash-dotted* line marks the unstable fixed points.

$$z_{t+1} = R\left(z_t - \frac{z_t^3}{3}\right) + \rho_{t+1}.$$

In this way we may get a qualitative analysis of the system in simpler terms. This point will be discussed again in the gradient descent formulation of the dynamics, see Sect. 3.3.3.

Using this simplification, we obtain for $R = 1 + \delta$ with $0 < \delta \ll 1$ the two stable fixed points as (setting $\rho = 0$)

$$z = \pm\sqrt{3(R-1)} + O(\delta^2)\,. \tag{3.23}$$

The approximation is valid only for $0 < \delta \lll 1$. The expansion in next order of the noise strength is given in Sect. 3.A.2.2 (p. 57), see Eq. (3.50). Further analysis of such a system (self-coupled single neuron) without noise is given in [121]. In [122] it was shown that such a neuron can even exhibit chaotic dynamics, however, not for the parameters used here.

3.3.3 Dynamics as Gradient Descent

The above derived properties of the sensorimotor loop can be seen more simply if rewriting the system dynamics as a gradient descent. This is possible for any one-dimensional system. In the present case we use

$$\frac{\partial}{\partial z}\ln\cosh z = \tanh z$$

to introduce a potential

$$V(z) = -R \ln \cosh z + \frac{z^2}{2} \tag{3.24}$$

so that Eq. (3.19) can be rewritten as

$$\Delta z_t = -\frac{\partial}{\partial z_t} V(z_t) + \rho_{t+1} \, . \tag{3.25}$$

As usual the gradient dynamics of Eq. (3.25) may be visualized by that of a sphere sliding down on the walls of a vessel filled with a viscous fluid.

The potential picture allows the discussion of the properties in simpler terms. For instance, the potential

$$V(z) = \gamma z + \alpha z^2 + \beta z^4 \tag{3.26}$$

with $\alpha = \frac{1}{2}(1-R)$, $\beta = \frac{1}{12}R$, and $\gamma = 0$, corresponds to the third order approximation of the tanh function. It is easily seen that the potential differs from the full potential by a smooth deformation (without introducing new extremes) so that the qualitative properties of the gradient dynamics are preserved. In this way the notion of topological equivalence of dynamical systems gets a clear graphical demonstration, see Fig. 3.9. The above potential is well known from many branches of physics and dynamical systems theory.

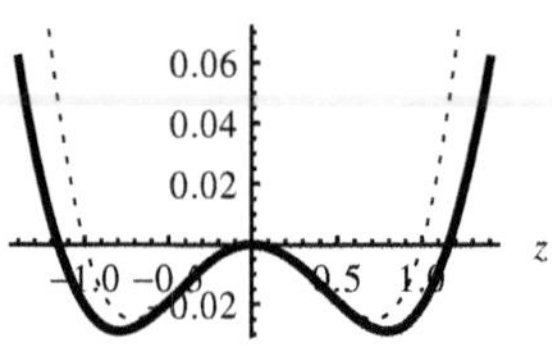

Fig. 3.9: Double well potential with approximation. Comparison of the exact potential, see Eq. (3.24) (**solid** curve), with its z^4 approximation Eq. (3.26), both for $R = 1.2$.

3.3.4 The Effective Bifurcation Point

If the noise is not zero it will not only modify the gradient step in size but may even change its sign. As a result, the system does not converge to the fixed point but instead fluctuates around the latter, provided the noise is weak and/or the fixed point has a high stability. In the bistable situation the noise may even drive the state over the potential barrier by a longer sequence of sign inverted gradient steps. Intuitively, the probability of such a transition will decrease exponentially with the length of an appropriate sequence which, at its hand, must be the longer the higher the potential wall and the weaker the noise.

Interestingly, the average time between such transitions is given [55] as a function of the strength D of the noise ρ Eq. (3.25) as

$$T_{\text{cross}} \propto e^{\frac{\Delta V}{D}} \tag{3.27}$$

where ΔV is the height of the potential barrier. The probability of crossing is close to one if $D \gg \Delta V$, i. e. if the potential barrier is low as compared to the fluctuations produced by the noise. On the other hand, if $\Delta V \gg D$, there is no barrier crossing at all in any physical time. There is a certain region where the system still feels the bistability but the barrier crossing is nevertheless substantial. In this region, the system, so to say, is already taking decisions (choosing one of the two fixed points) but is still very sensitively reacting to perturbations (e. g. by the environment) by revising decisions.

Right in that region we may define a point, called the *effective bifurcation point*,[2] where the number of barrier crossings becomes critical in a relevant physical time. In the robotics application this will be the time scale of the behavior so that barrier crossing events will directly influence the behavior of the robot.

The concrete position of that point is empirical and can be defined by specifying the crossing time. In view of the exponential behavior of the crossing time, see Eq. (3.27), we better define the position of the effective bifurcation point by specifying the quantity (which is the logarithm of the crossing time)

$$\chi = \frac{\Delta V}{D} \,. \tag{3.28}$$

Details may be found in Sect. 3.A.2. The idea is illustrated in Fig. 3.10, which demonstrates that the effective bifurcation point is rather sharply defined even if the noise is rather strong. This is a direct consequence of the exponential dependence on the noise strength, see Eq. (3.27). Moreover, Fig. 3.10 shows that the effective bifurcation point reveals itself rather sharply if looking at the time averages of the noisy trajectories.

We can even find an analytical expression for the effective bifurcation point. As shown in the Appendix, see Sect. 3.A.2 (p. 56) Eq. (3.52), the effective bifurcation point is for small noise strength D given by the expansion

$$R_{\text{eff}} = 1 + \frac{2}{\sqrt{3}}\sqrt{\chi D^*} + \frac{22}{15}\chi D^* + O\left(D^{\frac{3}{2}}\right) \tag{3.29}$$

where $D^* = D/A^2$. With infinitesimal noise, the effective bifurcation point deviates from the deterministic one (which is at $R = 1$) by a square root law in the variance D of the noise. However, since this is an expansion in terms of $\sqrt{D}$, the next order terms rapidly come into play as the noise increases. Figure 3.11 confirms Eq. (3.29) with $\chi = 3.5$ in a convincing way. The idea is also illustrated by Fig. 3.12 displaying the probability distributions of the system state with different values of R. At the bifurcation point of the noise-free system ($R = 1$) a broad unimodal distribution is observed. For larger values of R a clear bimodal structure appears, and eventually

[2] The notion should not be confused with the stochastic bifurcation point, an abstract mathematical concept introduced by Ludwig Arnold [5]. In contrast to our effective bifurcation point the stochastic bifurcation is obtained from asymptotic limits. In our pitchfork bifurcation setting the bifurcation structure would not be revealed. Moreover, asymptotic limits are not useful in robotics due to finite behavior timescales.

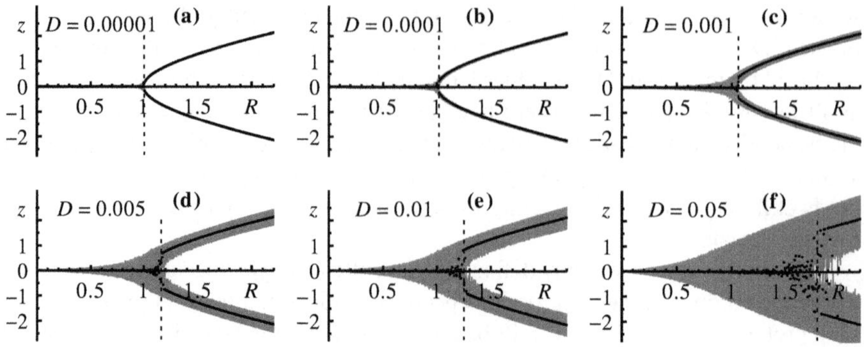

Fig. 3.10: Effective bifurcation diagrams for different noise strengths. The variance D of the uniform distributed noise is increased from panel (a) to (f). The **gray** area marks the hull of all trajectories (for each value of R 5000 iterations with positive and negative initial value). The **black** dots mark the mean value of 20 000 iterations for each value of R and the **dashed** line shows the effective bifurcation point determined by eye.

the probability for the state to be in the region around $z = 0$ decays exponentially to zero.

We will encounter many illustrative examples showing that the effective bifurcation point defines a working regime, where systems are particularly avid to self-organize. First examples will be given in the following.

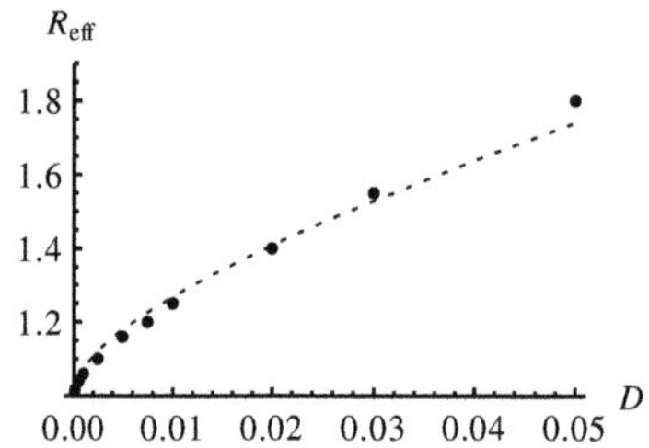

Fig. 3.11: Effective bifurcation point in dependence on the noise strength. The **dashed** line shows Eq. (3.29) with $\chi = 3.5$ and the **black** points show the data taken form the simulations (partly displayed in Fig. 3.10).

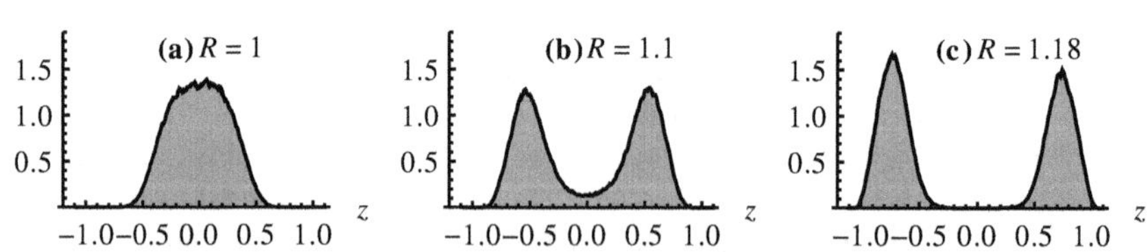

Fig. 3.12: Stationary probability distributions of the states (z) for different values of R. In all panels $D = 0.005$, see Fig. 3.10(d). **(a)** $R = 1$ (bifurcation point of the noise-free system); **(b)** $R = 1.1$ (R_{eff} for $\chi = 1.24$); **(c)** $R = 1.18$ (R_{eff} for $\chi = 3.5$).

3.3.5 Effective Bifurcation Point and Explorative Behavior

The effective bifurcation point is a novel concept with many interesting and far reaching consequences for the closed loop control of embodied robots. Let us consider a few examples, starting with a two-wheeled robot with each wheel being controlled individually and the sensor measuring the actual wheel velocity. We use in all examples the above controller with parameter C defining the feedback strength $R = AC$, using $A = \mathbb{I}$ for simplicity in the following. In the applications, the noise may be very strong and of varying nature, thinking for instance of a collision with an obstacle as in Fig. 3.6, so that the above requirements for the validity of Eq. (3.29) are not guaranteed. Nevertheless the concept of the effective bifurcation point remains valid. In fact, numerical experiments support that there is a relatively sharp value C_{eff} defining a favorable working regime in the sense explained above. You may convince yourself by doing Experiment 3.3.

Experiment 3.3: Closed loop control of wheeled robots.

For this experiment you have two choices: (a) a single wheeled robot in a square arena and (b) a chain of robots. The simulations start with the coupling set to $C = 1.2$ and the noise strength set to `noise=0.1` ($D = 0.01$). Decrease the coupling constants by entering:
`>coupling=value`
until the behavior starts to change frequently, which should happen around $C_{\text{eff}} = 1.05$. This is especially visible in simulation (b), where below C_{eff} the chain barely moves. In order to exert external forces to the robots use either <Ctrl>+<left Mouse button> or <Ctrl>+<right Mouse button>. You can add/remove random obstacles by pressing `o` or `O` in all simulations.

The behavior of the robot is in good agreement with the message from the effective bifurcation diagrams of Fig. 3.10 and Fig. 3.12. With values $C \ll C_{\text{eff}}$ the wheel velocity is essentially fluctuating around zero, so that the robot executes a very slow random walk. The amplitude of the fluctuations is increasing rapidly if C approaches C_{eff}. With $C \gg C_{\text{eff}}$ the state is caught essentially in one of the fixed points so that the wheel velocity is fluctuating around this fixed point. Then, the robot is in one of four states, moving either forward or backward or rotating on site clockwise or counterclockwise.

A more favorable behavior occurs if C is close to C_{eff} since then there are occasional random transitions between these four states so that the robot realizes a rather effective exploration of the space. The nature of this exploration sensitively depends on the value of C since it defines the frequency of the switching between the states.

3.3.5.1 Robot in Cluttered Environments

The benefits of this closed loop control mode reveal themselves most clearly if the robot is colliding with an obstacle. In collision events, the wheels can get blocked

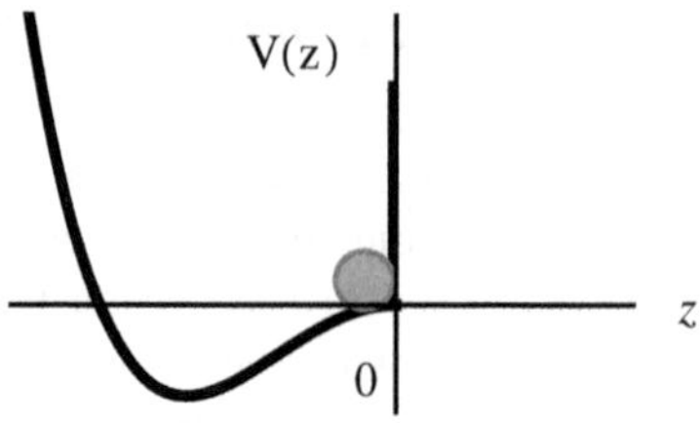

Fig. 3.13: The collision with an obstacle in the potential picture of the dynamics. The robot was moving forward, i.e. the state was at the r.h.s ($z > 0$) minimum of the double well potential. The impenetrable object corresponds to an infinitely steep rise in the potential so that the state is bound to move to the left minimum and the robot starts moving backward.

or at least slowed down.[3] In either case the **effective** feedback strength of the loop decreases so that z and with it the output y of the neuron decays. The only activity in the loop then comes from noise events, either by the measurement noise or the actuator noise (there are many different possibilities for a jerkily motion of the wheels, see the experiment). These fluctuations will be amplified if they are of the right sign, i. e. if the robot is momentarily moving away form the obstacle since then the wheels get free and the velocity can grow due to the still positive feedback. Hence, after a (short) sequence of collision events the state of the system goes to the other fixed point so that the robot is moving away from the obstacle, see Fig. 3.13 and Fig. 3.14. We may say that in this elementary sense the robot 'feels' the obstacles and generates a behavior away from them after a collision. There are no bumper sensors necessary for this to happen. Instead the behavior is an emerging property of our specific control mode close to the effective bifurcation point. You can verify this by doing Experiment 3.3 with a single robot.

The experiments reveal also another important property, namely that, once the initial hard collision has taken place, the probing of the obstacle by collisions happens in a quite gentle way so that the risk of damages is reduced. This is also an immediate consequence of the dynamics under the effective bifurcation regime, see Fig. 3.14.

As a consequence of this 'feeling' for the body, the robot will explore its environment in an effective way also in the case of a heavily cluttered arena. In Sect. 3.5 we will discuss how the robot can get this 'feeling' of the body without collisions by a virtual extension of the body using proximity sensors.

3.3.5.2 Emerging Cooperativity in a Chain of Robots

The sensitivity towards external influences produces an interesting effect if we consider a chain of passively connected robots (via ball-joints) with each wheel being controlled individually by a controller of the above kind. The behavior depends in an even more sensitive way on the value of the feedback strength in the loops. The region around the effective bifurcation point is again that of the most interesting behaviors. There the wheel velocities can become already quite large (decisions are taken) but can easily be switched by (i) noise events caused for instance by collisions

[3] The wheel velocities may also change sign due to an elastic collision, in which case the robot will automatically move away from the obstacle.

with obstacles or (ii) by the influence of the forces exerted by the other robots in the chain. In fact, due to the high sensitivity, the wheel velocity will have a tendency to switch sign if a torque in the opposite direction is exerted by the other robots in the chain. By switching the velocity, the wheel is now acting in the direction of the force exerted on it. This self-amplification effect is essentially what is necessary for a self-organized synchronization of the wheel velocities.

Video 3.3 demonstrates quite clearly the strength of this self-organized synchronization effect, which not only makes the robot chain move into one direction but also keeps it still explorative in the sense that, after some time, it spontaneously inverts its direction of motion. Moreover, when colliding with an obstacle, the chain of robots often will change velocity in an integrated manner. Finally and most importantly, it will also effectively explore the spatial extensions of a maze, see Video 3.4 and/or do Experiment 3.3 with a chain of robots.

In order to demonstrate this point in a more general context, we have carried out a series of experiments with a more complex controller. Instead of using a controller for each wheel, each robot has now a controller with two motor neurons each of, which receiving the inputs from the two wheel sensors. Thus, there is a 2×2 coupling matrix C that we parameterize as

$$C = \begin{pmatrix} c & b \\ b & c \end{pmatrix}$$

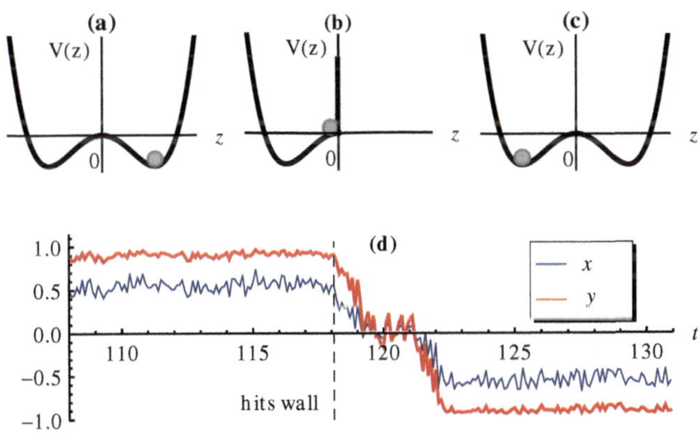

Fig. 3.14: Robot at a wall. (d) Neuron output y and wheel velocity sensor x of the closed loop control system (with feedback strength $R > R_c$) before and after a collision with a wall as obtained from an experiment with a real Khepera robot. Note that the wheels of the robot may slip, such that the activity in the loop is slowly decaying at the wall contact. Panels **(a-c)** display the corresponding potential $V(z)$ the sphere marking a stationary state (fixed point) of the system. Before the collision the system is in the fixed point with positive sign (robot is driving forward). During the time 118 to 122 the robot is kept at the unstable fixed point $z = 0$. At around 121.5 sec the robot starts moving backward because of the noise amplification effect.

Video 3.3: Emerging cooperativity in a chain of TwoWheeleds. The arena has no obstacles. Control is completely decentralized, but the individual wheels spontaneously cooperate in the working regime close to the effective bifurcation point. The video can be watched at `http://playfulmachines.com`.

Video 3.4: The chain of robots with decentralized control in a regular maze. Spontaneous cooperativity and sensitive reactions to collisions helps the chain to navigate in the maze without any proximity sensors. The video can be watched at `http://playfulmachines.com`.

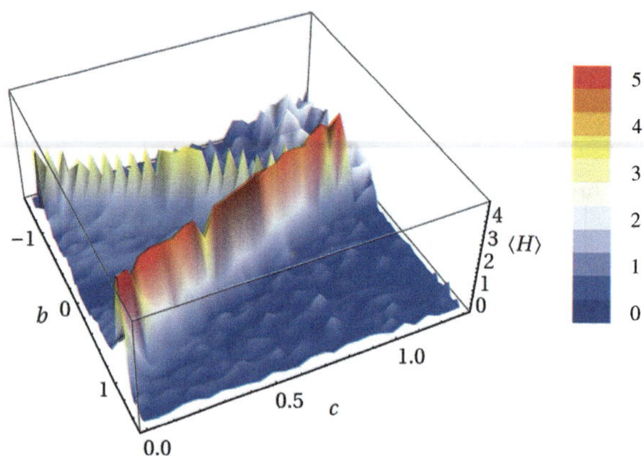

Fig. 3.15: Exploration of a maze by the chain quantified by the spacial entropy. The average spacial entropy $\langle H \rangle$ is depicted in dependence on the coupling parameters c and b. It is maximal if all sites are visited by the robot with equal probability and it is zero if the robot remains in its starting position. Interestingly, the coverage of the maze is best for the decentralized controller setting with $b = 0$, see [40] for details.

in accordance with the symmetries of the robot. The most essential point is that the robot chain explores the space effectively only for very subtle combinations of the parameters b and c. This is reflected by the spacial entropy (entropy of distribution of visited sites), see Fig. 3.15. It is only with those parameter values that the 10 wheels of the chain coordinate their activities so that the chain can act as a whole. In the two-dimensional parameter space, these combinations define a line of effective bifurcation points, see [40], which again shows the importance of that concept. Even more interestingly, the best exploration and hence the best spontaneous co-operation between the robots is obtained if $b = 0$, which means that the channels decouple or that each wheel has its individual controller. This is an interesting result in the direction of embodied robotics meaning that, due to the embodiment, higher coordination is achieved with less control, see [40, 191] for more details and the Videos 3.3 and 3.4.

3.4 Extending the Parameter Space

The analysis and results given so far have demonstrated that, under the closed loop setting, very rich behavior is observed even if the controller neuron has only a single parameter. This can be advanced further by including an additional parameter that acts as a bias for the neuron. As we will see, this extension of the parameter space creates a new behavioral feature—the system shows hysteresis. These properties are investigated as before by first analyzing the fixed point nature of the deterministic system.

3.4.1 Bifurcation Scenario.

When including a bias h into the controller, i. e.

$$K(x) = \tanh(Cx + h)$$

the fixed point equation of the deterministic system is, written in terms of the membrane potential, given by

$$z = R \tanh(z) + h \,. \tag{3.30}$$

Using the approximation of $\tanh(z)$ (Sect. 3.3.2), one gets from Eq. (3.30)

$$z = R\left(z - \frac{z^3}{3}\right) + h \,. \tag{3.31}$$

The stability and existence of the fixed points is well known so that we can draw upon the results obtained for such systems, see e. g. [121]. The bias leads to a cusp bifurcation where we have always one stable fixed point and additionally we observe

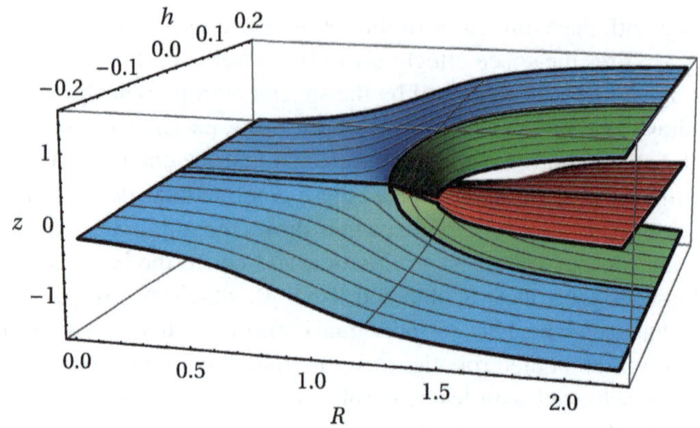

Fig. 3.16: Bifurcation diagram for the closed loop with bias. The diagram depicts the solution of $z = R\tanh(z) + h$, see Eq. (3.30), as a surface over R and h, that is called cusp bifurcation. Colors: *blue, green* are stable fixed points and *red* is unstable.

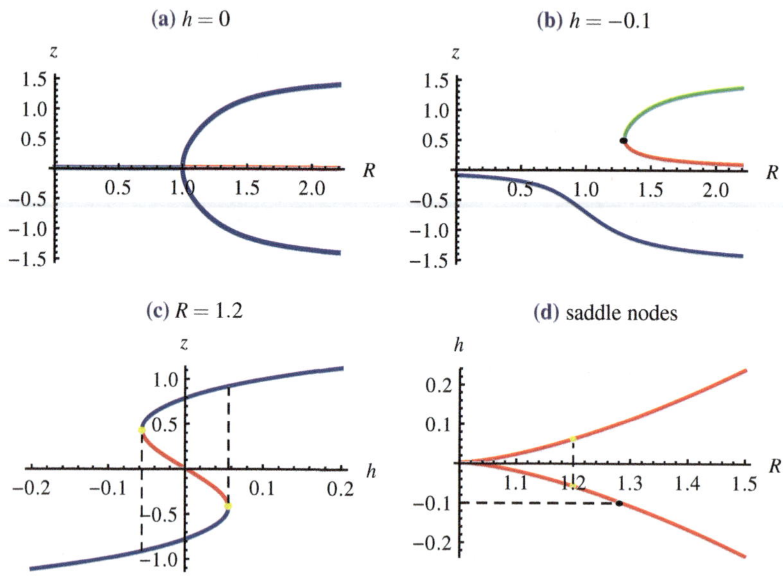

Fig. 3.17: Bifurcation diagrams for $z = R\tanh(z) + h$**.** The diagram depicts different cuts through the surface in Fig. 3.16. **(a)** Section at $h = 0$; **(b)** Catastrophic bifurcation at $h = -0.1$; **(c)** The hysteresis in dependence of h in the supercritical regime at $R = 1.2$, also indicated in Fig. 3.16. **(d)** Typical cusp wedge showing the saddle nodes in dependence of R and h. The colored points are correspondingly marked in **(b,c)**. Colors: *blue, green* are stable fixed points and *red* marks unstable fixed points.

the emergence of a so-called saddle-node bifurcation point, sometimes also called a catastrophic bifurcation because of the sudden appearance or disappearance of a pair of fixed points, one stable the other one unstable. A saddle-node bifurcation point occurs if the solution of the cubic fixed point equation (3.31) has three real solutions of which two are coinciding. The solution, is plotted as a bifurcation diagram for the parameters R and h in Fig. 3.16 showing a cusp bifurcation [2]. The saddle-nodes are located at the line where the red and green surfaces are intersecting.

Figure 3.17 shows different sections of the fixed point structure for better illustration. The positions of the saddle nodes form the typical cusp wedge, see Fig. 3.17(d). With $R < 1$, the system has only one stable fixed point. We call the system subcritical, since the fixed point is at small z and thus little activity occurs in the sensorimotor loop. For $h = 0$ we observe the same pitchfork bifurcation as before, see Fig. 3.17(a). For $h \neq 0$ two disconnected branches emerge, see Fig. 3.17(b), and the bifurcation becomes catastrophic, since the system state can undergo a drastic transition for small changes in parameter values. To illustrate the consequences, Fig. 3.17(c) displays the bifurcation diagram for the supercritical parameter $R = 1.2$ in dependence of h revealing a clear hysteresis effect. This means that the system resides in its fixed point when the parameter h is slowly decreased or increased until the fixed point disappears and the state jumps to the fixed point with the opposite sign (dashed lines). If the parameter h is changed in the other direction the same behavior is observed. Hence, for one parameter configuration the system can be in two possible states and h can be used to force a transition.

The visualization of the dynamics can again be done in the gradient descent picture, see Fig. 3.18.

3.4.2 Application: Biasing the Behavior

Hysteresis systems have interesting properties that have been exploited in many branches of science and technology. In our case, if we set the feedback strength a little above the critical point, a periodic signal of the bias with small amplitude is largely amplified by the dynamics of the sensorimotor loop. In the homeokinesis control mode to be discussed further below, the bias will be found to self-regulate. In that scenario, the amplification of the bias signals plays an essential role in the emergence of nontrivial behaviors.

The properties of the bias can immediately be used for biasing the behavior into a desired direction. For instance, in the simple two-wheeled robot experiment we can force the robot to move preferentially into the forward direction by choosing the same positive value of the bias for both the neurons. By choosing a bias of moderate size, we can preserve the reactions of our two-wheeled robot to collisions with some obstacle and nevertheless retain its forward preference, reducing the on-site rotation probability so that the exploration of the space will be enhanced.

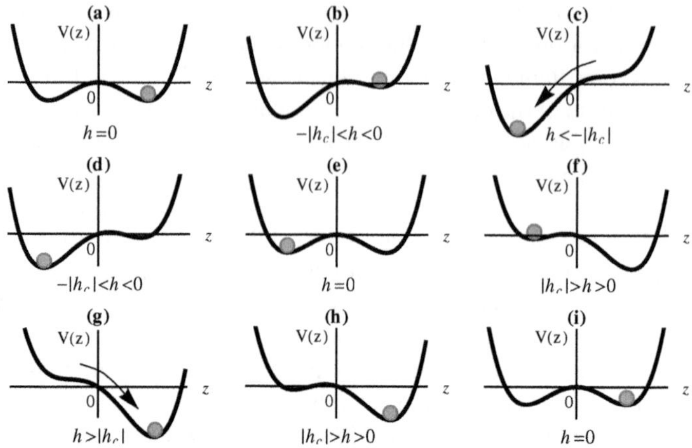

Fig. 3.18: The hysteresis cycle in the gradient picture. The diagrams show the stages of one hysteresis cycle starting from $h = 0$ **(a)** with the state at $z > 0$ as represented by the sphere. Decreasing h leads to a deepening of the left minimum, while the right minimum gets more flat, but the state remains at the minimum at $z > 0$, see **(b)**. If $h = -h_c$ the saddle-node bifurcation happens, i. e. both the maximum at z=0 and the right minimum disappear so that the system shifts to the left minimum of the potential **(c)**. Increasing h until $h = 0$ brings us back to the initial situation with the difference that the system is now at the fixed point with negative sign, see **(d,e)**. The diagrams **(f)** and **(g)** show the switching from the minimum at $z < 0$ to the minimum at $z > 0$ by increasing h. By decreasing h until $h = 0$ the hysteresis cycle is finished, see **(h,i)**.

3.5 Expanding the Body

In the scenarios considered so far, the robot can feel its environment only by collisions or other direct physical influences. This is of course annoying in practical applications where collisions might be dangerous. One way to help this is to virtually expand the body of the robot by exteroceptive sensors so that the collision can be felt before it takes place physically. A convenient choice are infrared sensors, which gradually change their sensor values as a function of the distance to an obstacle. Let us assume we equip the robot with six such sensors as depicted in Fig. 3.19. The output y_i of the controller for wheel $i \in \{\text{left, right}\}$ now is chosen as (dropping time indices, same C for both wheels)

$$y_i = \tanh(Cx_i + h_i)$$

where h_i is a bias given by a linear combination of the infrared sensor values s_j

$$h_i = \sum_j B_{ij} s_j \tag{3.32}$$

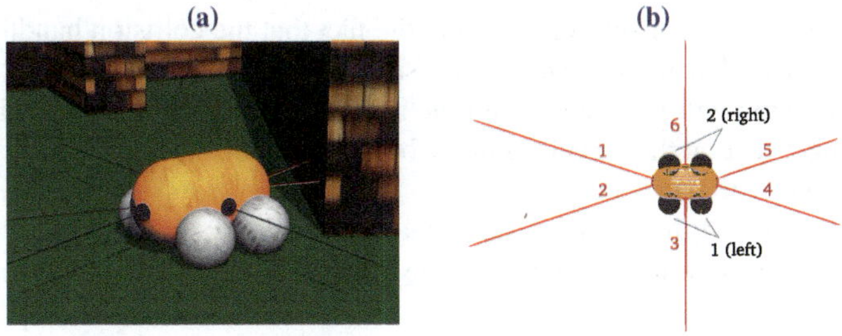

Fig. 3.19: Virtual expansion of the body. Infrared sensors, usually seen as exteroceptive sensors measuring distances to an obstacle, are considered here as defining a soft expansion of the body that is used as elastic collision sphere. The robot is the FOURWHEELED of the LPZROBOTS simulator that can be operated in a two-wheeled mode, meaning that the wheels at one side receive the same motor signal. The wheel velocity sensor reports the mean of both wheel sensors. **(a)** Screenshot taken from a simulation. The infrared (IR) sensors are drawn for illustration in *black* if they measure no obstacle and in *red* if they do; **(b)** Wire-frame view with IR sensor rays.

with a conveniently chosen set of parameters B_{ij} (typically in $[-1,1]$) that determine the reactions of the robot to the proximity sensors.

The biasing effect can be understood in terms of the unbiased loop by introducing an effective sensor value x_{eff}, defined via $Cx_{\text{eff}} = Cx + h$. Changing h according to Eq. (3.32) acts like a variation of the sensor values x_{eff} just like in a collision. The effects of collisions can be mimicked if the B_{ij} are chosen such that a slowing down or switching of the wheel velocity is generated when approaching an obstacle. Moreover, by choosing C appropriately, the reactions of the robot can be made more smooth or more robust against the switching signals h_i, so that, by choosing the couplings B and C, a large number of behavioral patterns can be achieved. Experiment 3.4 shows that the robot will again explore its environment effectively, however now without colliding with the obstacles.

We see here some parallel to the famous Braitenberg vehicles [22], in that we can by simple means generate complex behaviors of the robot moving in an environment. It is not difficult to imitate some of the vehicles proposed by Braitenberg but we do not go into the details here. Altogether, the parallel is also in the general attitude of externalizing complexity from the internal control structure to the interaction with the environment.

Experiment 3.4: Expanding the body.

Start the simulation Expanding the Body (FourWheeled). Some obstacles are placed in the environment, you may add more with `o` and remove them with `O`. Try different coupling strengths for the IR sensors by entering:

```
>alpha=value
```

The default value is 1.0. The coupling matrix set to $B_{ij} = \alpha \begin{pmatrix} -1 & 0 & 1 & 0 & 1 & -1 \\ 0 & -1 & -1 & 1 & 0 & 1 \end{pmatrix}$, which leads to an obstacle avoidance behavior. You can also change each individual entry with the parameters `B11-B26` to get a different behavior. Alternatively you can try Expanding the Body (LongVehicle) where only four infrared sensors are used.

One advantage of our setting is seen in the fact that the collision handling works even in the case that the obstacle is not seen by the IR sensors. In that case the physical collision will take place with the switching of the wheel velocities due to the blocking effect on the wheels as described above. Thus there is always a kind of minimal strategy for survival.

The remainder of this chapter is devoted to the introduction of neural networks that are used for the implementation of the control systems. It may be skipped and looked up as needed.

3.6 Realization by Neural Networks

The implementation of the parameterized functions for controllers and forward models are realized in practice in many different ways. A standard approach is to use artificial neural networks because they are universal, easy to implement, and have well structured algorithms for the adaptation of the parameters. Since we want to derive explicit expressions later on, we will introduce very briefly the main points of that approach in the following.

We will consider a network of rate-coding neurons, where the activity of each neuron at discrete time steps is represented by a single number, which can be interpreted as a mean firing rate. Actually, biological neurons have a very complicated internal dynamics, but in a simplified view we can think of a membrane potential that is driven by incoming spikes from other neurons. Once the membrane potential has reached a certain threshold, a single spike is fired and the membrane potential is reset. In this picture the time distribution of the spikes may carry information (time coding). This is known to play a role for instance in correlation learning via spike timing dependent plasticity (STDP) [36]. However, in many cases it was shown that neurons transmit information mainly via their firing rate (mean activity), which is especially prominent in coding of sensory stimuli. This is the so called rate-coding paradigm, which is used in the artificial neural networks (ANNs) we are dealing with. Viewed as information processing systems, ANNs are much closer to the brain than to a classical computer program. In particular the ANNs are prominent due to their ability to learn.

There are many excellent textbooks, e. g. [114, 167], dealing with the theory and application of artificial neural networks, especially in robotics, so that we will sketch here only the facts most relevant to our work.

3.6.1 Rate-Coding Neuron Model

Under the rate-coding paradigm, the neuron can be represented by a simple mathematical function. The neuron is considered as an input-output device. The input is represented by a vector $x \in \mathbb{R}^n$ and the output of a neuron i is written as

$$o_i = g(z_i)$$

where $g : \mathbb{R}^1 \to \mathbb{R}^1$ is the so called activation (or transfer) function and z_i is the post synaptic potential given by the weighted sum over the inputs as

$$z_i = h_i + \sum_{j=1}^{n} W_{ij} x_j$$

the W_{ij} being the synaptic efficacies of the i-th neuron connected to a neuron or sensor input j and h_i is a threshold or biasing term. We denote by $W_i \in \mathbb{R}^{n_i}$ the weight vector of neuron i. The function $g : \mathbb{R}^1 \to \mathbb{R}^1$ is typically a sigmoid function like

$$g(z) = \text{sigmoid}(z) = \frac{1}{1+e^{-z}},$$

which monotonously increases from 0 to 1 as z increases from large negative to large positive values. We mainly use

$$g(z) = \tanh(z) = \frac{e^z - e^{-z}}{e^{-z} + e^z} \tag{3.33}$$

mapping the input to the interval -1 to 1, see Fig. 3.20.

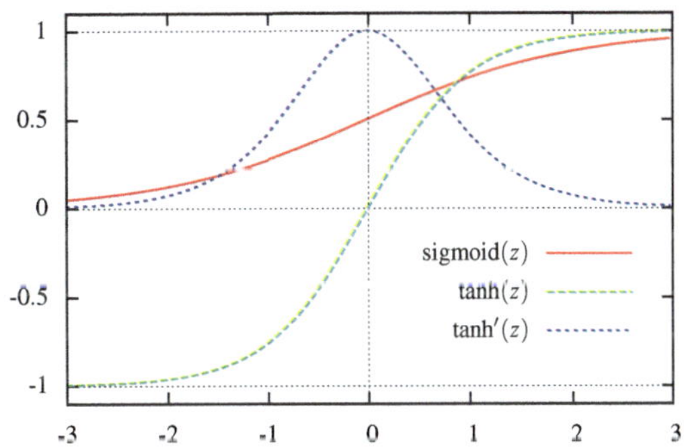

Fig. 3.20: Neuron activation functions. Note the range of sigmoid(z) is $(0,1)$ whereas $\tanh(z)$ maps to $(-1,1)$. The derivative of the hyperbolic tangent is given by $\tanh'(z) = 1 - \tanh^2(z)$.

3.6.2 Supervised Learning

A neuron i from above maps the inputs x to output o_i. The map is parameterized by the weight vector W_i. The weights can be adapted in the following learning scenario: given a set of input-output mappings (provided by a trainer) find W_i such that the neuron reproduces these pairs as close as possible. In an on-line learning scenario, the neuron sees an input x together with the target output o_i^{teach}, which is provided by the trainer, in each time step. The error of the neuron is

$$E_i = \frac{1}{2}\left(o_i - o_i^{\text{teach}}\right)^2 = \frac{1}{2}\delta_i^2 \tag{3.34}$$

where $o_i = g(z_i)$ is the current output of the neuron and $\delta_i = o_i - o_i^{\text{teach}}$ is the mismatch between current output and target output. Gradient descent on this error yields the update rule, also known delta-rule, for both the synaptic weights and the thresholds as

$$\Delta W_{ij} = -\varepsilon d_i x_j \tag{3.35}$$

$$\Delta h_i = -\varepsilon d_i \tag{3.36}$$

where

$$d_i = g'(z_i)\left(o_i - o_i^{\text{teach}}\right) = g'(z_i)\,\delta_i\,, \tag{3.37}$$

which can be considered as an error signal present at the output of the synapse W_{ij}. The structure of d_i is generic since it can be considered as the result of propagating the error signal δ_i at the output of the neuron through the activation function, thereby producing a factor $g'(z_i)$, backwards to the output of the synapse.

The update of W_{ij} is given by the product of this output signal multiplied by x_j, which is the activity at the input of the synapse. This is reminiscent of the general rule introduced by D. Hebb [66] who postulated that the synaptic strength increases if activities on the pre- and postsynaptic side are both high, often stated as "fire together—wire together." However the delta-rule is still different from pure Hebbian learning. The difference of Eqs. (3.35–3.37) to Hebb's rule is (i) that the postsynaptic activity is not that of the neuron but the error signal propagated back from the output of the neuron and (ii) that the update can be both positive and negative. Nevertheless we will find this parallel to Hebb's rule very helpful also with our more complicated learning rules driving the self-organization process.

3.6.3 Feed-Forward Networks

Neurons can be combined to neural networks in order to represent more complicated input output mappings. In a typical feed-forward architecture the net realizes a function $f : \mathbb{R}^n \to \mathbb{R}^m$, i.e. it maps inputs $x \in \mathbb{R}^n$ to outputs $o \in \mathbb{R}^m$. The network is organized as a sequence of layers, the neuron outputs of a layer forming the inputs to the neurons of the subsequent layer. A network with N layers consists of the input layer where x is fed in, a number of hidden layers, and an output layer of m neurons. Numbering the layers from the input layer ($k = 0$) to the output layer ($k = N$) the rule for feeding the input information through the network is given iteratively by

$$o_i^{(k)} = g\left(h_i^{(k)} + \sum_j W_{ij}^{(k)} o_j^{(k-1)}\right), \quad \text{for } k = 1, 2, \ldots, N \tag{3.38}$$

(identical activation functions for all neurons) starting with $o^{(0)} = x$. The output vector of the network is given by $o^{(N)}$.

There is well known theorem stating the universality of neural networks as a function approximator: Already a network with only one hidden layer (with a sufficient number of neurons) is able to represents any function $F : \mathbb{R}^n \to \mathbb{R}^m$ with arbitrary accuracy see [33]. An intuitive explanation is that the neurons of the hidden layer can be organized into groups so that the collective receptive field of a group covers regions in input space mapping to about the same value of the target function f. Then, all the output layer has to do is to weight the contributions of the groups in order to produce the correct output $o^{(\text{teach})} = f(x)$ to any input x.

This function approximation can be learned by gradient descending the error

$$E = \left(o - o^{\text{teach}}\right)^2 . \tag{3.39}$$

This can be transformed into a systematic procedure for the individual updates of the synaptic weights

$$\Delta W_{ij}^{(k)} = -\varepsilon d_i^{(k)} o_j^{(k-1)} , \tag{3.40}$$

which is again Hebb like. The error activities are obtained in an iterative way

$$d_i^{(k)} = g'\left(z_i^{(k)}\right) \sum_l d_l^{(k+1)} W_{li}^{(k+1)}, \quad \text{for } k = 1,\ldots,N-1 \tag{3.41}$$

starting with $d_i^{(N)} = g'\left(z_i^{(N)}\right)\left(o_i - o_i^{\text{teach}}\right)$ as in the single layer case. The error signal for neuron i in layer $k < N$ is produced by the weighted sum of the error signals in the layer $k+1$ multiplied by the derivative of the gain function. This is the famous error backpropagation rule [180]. In matrix notation we write

$$d^{(k)} - G'\left(z^{(k)}\right)\left(W^{(k+1)}\right)^\top d^{(k+1)}, \quad \text{for } k = 1,\ldots,N-1 \tag{3.42}$$

where G' is the diagonal matrix given by $G'_{ij}\left(z^{(k)}\right) = \delta_{ij} g'\left(z_i^{(k)}\right)$ with δ_{ij} being the Kronecker delta and $W^\top$ being the transpose of a matrix W.

3.6.4 Recurrent Networks

The feed forward networks serve well the purpose of parametrized function approximation. However in many applications, in time series prediction e.g., the output of the network is depending on previously seen inputs. Another example is the realization of a controller with internal states, see Eq. (3.4). This can be achieved with additional inputs into the network generated from earlier events. A more elegant way consists in recurrent neural networks (RNN), which internally built up a memory of the past by time delayed recurrences.

A simple RNN, introduced by Elman [48], is obtained by equipping a feed-forward network with feedback loops providing a time delayed copy of the outputs of all neurons of the net, or a group thereof, back to the neurons itself. As before the inputs are presented to the network at times $t = 0, 1, \ldots$. The iterative rule (3.38) for evaluating the output of the network is now (sum over repeated indices)

$$o_i^{(k)}(t) = g\left(W_{ij}^{(k)} o_j^{(k-1)}(t) + V_{ij}^{(k)} o_j(t-1) + h_i^{(k)}\right), \quad \text{for } k = 1, 2, \ldots, N \quad (3.43)$$

where $o^{(0)}(t) = x_t$ is the current input into the network, $o(t-1)$ the vector of previous activations of the ensemble of neurons in the net, and V is a weight matrix controlling the feeding back of these activations. We are not going into any details here since we are introducing recurrences in a different way by the principle of homeokinesis, see Chap. 5.

The network is now a discrete-time dynamical system driven by the inputs x_t. Depending on the weights, the network can develop the full scope of dynamical behaviors ranging from fixed point attractors over limit cycles to chaos. Again the important point is that the weights can be learned such that the network reproduces a target dynamics. The learning rules are a bit more complicated but are systematic and well investigated. It is this property which makes the recurrent networks valuable tools for time series prediction, system identification, signal processing, and last but not least robot control.

3.7 Summary

To summarize, given a simple closed loop system with a single nonlinear neuron we find non-trivial fixed points and hysteresis depending on the control parameters. This type of dynamics is found in many systems, e. g. in statistical mechanical models of magnets, see [165]. We introduce the effective bifurcation point, since we consider the system with noisy perturbations. For robot control it seems suitable to keep the feedback strength R at the effective bifurcation point, slightly above the bifurcation point of the deterministic system, in order to have two stable fixed points allowing the system to switch either by changing the bias h or by external influences. In the case of multiple sensors and motors we get of course a much larger number of attractors, Neimark-Sacker bifurcations, limit cycles and so forth, which will be work out in Chap. 7 and demonstrated by applications to robotic systems later in this book.

Appendix 3.A Mathematical Details

We give here the mathematical details for the stability analysis of the fixed points without bias and the evaluation for the position of the effective bifurcation point.

3.A.1 *Stability Analysis*

The properties of the fixed points are most easily obtained by a linear stability analysis. We put $z_t = z^* + u_t$ where u is small. The state dynamics $z_{t+1} = Rg(z_t)$ is linearized as

$$z^* + u_{t+1} = Rg(z^* + u_t)$$

and by using Taylor expansion one gets

$$u_{t+1} = L(z^*)\,u_t$$

where

$$L = Rg'(z^*)$$

is the derivative. One has only to show that $0 < L < 1$ in order to prove that $u_t \to 0$ as $t \to \infty$. Using the approximation Eq. (3.23) we get

$$L = Rg'(z^*) = R\left(1 - \frac{3(R-1)}{R}\right) = 3 - 2R$$

so $L < 1$ if $1.5 > R > 1$, provided $\tanh(z) - z - \frac{z^3}{3}$ is valid, i.e as long as R close to 1. However if $R \gg 1$ then $|z|$ is very large and one gets from the fixed point equation ($z = R\tanh(z)$) that $z^* \simeq R$, so $L = R(1 - \tanh^2(R))$ or approximately $L = 4R\exp(-2R)$, since at large R we may write approximately

$$\tanh R = \frac{e^R - e^{-R}}{e^R + e^{-R}} = \frac{1 - e^{-2R}}{1 + e^{-2R}} \approx 1 - 2e^{-2R}$$

Hence L goes exponentially to 0 as $R \to \infty$ so that the stability of the fixed point increases with increasing R.

These results have been derived in some detail since L is the central quantity of the time-loop error, as we will see later. We have seen here that it is directly related to the stability of the fixed points of the state dynamics. In the higher dimensional cases L is the Jacobian matrix of the dynamical system the eigenvalues of which define the stability of the system.

3.A.2 *Determining the Effective Bifurcation Point*

The aim now is to find an explicit expression for the position of the effective bifurcation point (EBP) in the case of the pitchfork bifurcation for the dynamics given by Eq. (3.19), i. e.

$$z_{t+1} = Rg\left(z_t\right) + \rho_{t+1} = -\frac{\partial}{\partial z_t} V\left(z_t\right) + \rho_{t+1}$$

The position of the EBP is defined by the empirical constant χ defining the relation between the barrier height and the variance of the noise as

$$\Delta V = \chi \left\langle \rho^2 \right\rangle \tag{3.44}$$

$\left\langle \rho^2 \right\rangle$ being the variance of the noise in the z dynamics.

3.A.2.1 Series Expansion of the EBP in the z^4 Case

In order to find ΔV we consider the z^4 potential case first, see Eq. (3.26)

$$V\left(z\right) = -\frac{R-1}{2} z^2 + \frac{1}{12} R z^4 ,$$

where the maximum is given by $V\left(0\right) = 0$ and the minimum by $V\left(z^*\right)$ with

$$z^* = \pm \sqrt{3 \frac{(R-1)}{R}}$$

being either one of the stable fixed points so that the barrier height is

$$\Delta V = V\left(0\right) - V\left(z^*\right) = \frac{3}{4} \frac{(R-1)^2}{R} .$$

Eq. (3.44) reads now

$$\frac{3}{4} \frac{(R-1)^2}{R} = q . \tag{3.45}$$

where $q = \chi \left\langle \rho^2 \right\rangle$. The solution and its power series expansion (Taylor) is

$$R = 1 + \frac{2}{3} q + \frac{2}{3} \sqrt{3q + q^2} = 1 + \frac{2}{\sqrt{3}} 3 \sqrt{q} + \frac{2}{3} q + O\left(q^{\frac{3}{2}}\right)$$

In order to get an expression in terms of the original noise strength $D = \left\langle \xi^2 \right\rangle$, i. e. the one featuring in the x dynamics, we use $\left\langle \rho^2 \right\rangle = \left\langle \xi^2 \right\rangle C^2 = DR^2/A^2 =: D^* R^2$ so that now $q = \chi D^*$ and we have to solve

$$\Delta V = q R^2 . \tag{3.46}$$

This equation is of third order in R so that it can be solved explicitly. The solution however is quite complex so that we better aim at establishing a series expansion in terms of q since we are mainly interested in the low noise case anyway. We try a power series expansion for R in terms of $\sqrt{q}$ as

$$R = 1 + m\sqrt{q} + nq + pq^{\frac{3}{2}} + O\left(q^2\right) \tag{3.47}$$

and expand the condition given by Eq. (3.46) in terms of q as

$$\frac{3}{4}\frac{(R-1)^2}{R} - qR^2 = \left(\frac{3}{4}m^2 - 1\right)q + \left(\frac{3}{2}mn - \frac{3}{4}m^3 - 2m\right)q^{\frac{3}{2}} + O\left(q^2\right) \tag{3.48}$$

where now $q = \chi D^*$. If the expansion Eq. (3.47) is to be a solution of Eq. (3.46), the coefficients in Eq. (3.48) must be zero in all orders that are relevant for the expansion of the solution. Up to order $q^{\frac{3}{2}}$ the coefficients are equal to zero if $m = 2/\sqrt{3}$ and $n = 2$ so that we get the result

$$R = 1 + \frac{2}{\sqrt{3}}\sqrt{q} + 2q + O\left(q^{\frac{3}{2}}\right).$$

Putting the results together, the low noise expansion of the EBP in the z^4 case is

$$R_{\text{eff}} = 1 + \frac{2}{\sqrt{3}}\sqrt{\chi D^*} + 2\chi D^* + O\left(D^{\frac{3}{2}}\right) \tag{3.49}$$

with $D^* = \frac{D}{A^2}$.

3.A.2.2 Series Expansion of the EBP in the tanh Case

The expression of the fixed point (FP) in the case of infinitesimal noise has been given already in Eq. (3.23). In order to get an expansion for the EBP comprising the next order we have to find first the corresponding expansion of the FP. We put $R = 1 + u$ with $0 < u$ and try $z = m\sqrt{u} + nu + pu^{\frac{3}{2}} + O\left(u^2\right)$ in the expansion of the FP condition $z = R\tanh z$ obtaining

$$R\tanh(z) - z = \alpha u^{\frac{3}{2}} + \beta u^2 + \gamma u^{\frac{5}{2}} + O\left(u^3\right)$$

where $\alpha = \left(m - \frac{m^3}{3}\right)$, $\beta = \left(n - m^2 n\right)$, and $\gamma = \frac{2}{15}m^5 - \left(\frac{m^3}{3}\right) - mn^2 + p - m^2 p$. The coefficients α, β, and γ are shown to be zero if $m = \sqrt{3}$, $n = 0$, and $p = \sqrt{3}/10$ so that the two stable FPs have the expansion

$$z^* = \pm\left(\sqrt{3u} + \frac{\sqrt{3}}{10}u^{\frac{3}{2}}\right) + O\left(u^{\frac{5}{2}}\right). \tag{3.50}$$

The potential now is, see Eq. (3.24),

$$V(z) = \frac{1}{2}z^2 - R\ln(\cosh z)$$

with the height of the barrier obtained as

$$\Delta V = 0 - V(z^*) = \frac{3}{4}u^2 - \frac{3}{20}u^3 + O\left(u^4\right)$$

The EBP is found by using for u the expansion $u = m\sqrt{q} + nq + sq^{\frac{3}{2}} + O\left(q^2\right)$ and $R = 1 + u$ in the condition $\Delta V = \chi D^* R^2$ so that

$$\frac{3}{4}u^2 - \frac{3}{20}u^3 - qR^2 = \left(\frac{3}{4}m^2 - 1\right)q + \left(\frac{3}{2}mn - 2m - \frac{3}{20}m^3\right)q^{\frac{3}{2}} + O\left(q^2\right) \quad (3.51)$$

the first two coefficients being zero if $m = \frac{2}{3}\sqrt{3}$ and $n = \frac{22}{15}$ so that for small noise strengths we obtain the solution of Eq. (3.45) for R_{eff} by the expansion

$$R_{\text{eff}} = 1 + \frac{2}{\sqrt{3}}\sqrt{\chi D^*} + \frac{22}{15}\chi D^* + O\left(D^{\frac{3}{2}}\right) \quad (3.52)$$

with the leading square root dependence for very small D as in the case of the z^4 potential.

Chapter 4
Principles of Self-Regulation — Homeostasis

Abstract: We have seen in Chap. 3 that there are specific working regimes in closed sensorimotor loops where the agents exhibit interesting behaviors. The challenge is now to develop general principles so that the agent finds these regions by itself. One essential point at this level of autonomy is the ability to survive in hostile situations, which, as a first prerequisite, requires a certain stability against external perturbations. An example of this is homeostasis, one of the prominent self-regulation scenarios in living beings. This chapter introduces Ashby´s homeostat as a concrete example from cybernetics and develops a general principle of self-regulation as a first step towards a general basis for the self-organization of behavior.

As a general principle of self-regulation, homeostasis would deserve a broad appreciation in the context of this book, even more so as it has recently found revived interest not only as a guiding principle for the resilience of the brain [50, 50, 54, 171, 174–176] but also as a general concept in the robotics community, see [80, 118, 119, 193]. Nevertheless, in order to focus on the developments to be presented later, we will only select one specific trait, Ashby's homeostat, of the earlier history of homeostasis. We select the homeostat since we feel very much intrigued by the audacious initiative of William Ross Ashby to built a machine in order to prove a general principle, homeostasis, of living beings. This is very close to our own ambitions with the main difference that we hypothesize the principle, homeokinesis, first, and try it in various machines to understand its consequences. Different from the homeostat, these machines do not realize the goal of ultrastability in challenging environments but establish a playful self-exploration of potential motion patterns, thereby reaching a degree of autonomy that goes much beyond the objective of pure self-stabilization.

This chapter will, after the short excursion into the history, present a general principle of self-regulation based on minimizing the prediction error of the adaptive forward model. The general setting is exemplified by experiments with our BARREL, demonstrating the emergence of different behavior modes. In most cases, the principle is shown to drive the system towards maximum stability, but in exceptional cases we observe also oscillatory modes depending on the embodiment of the system. Despite of these interesting findings, the principle is argued to lack an

R. Der and G. Martius: The Playful Machine, COSMOS 15, pp. 59–74.
springerlink.com

intrinsic drive for innovation. This drive will emerge with homeokinetic learning to be presented in the next chapter.

4.1 History and Development of Homeostasis

Originally homeostasis was conceived as an approach to complex control mechanisms in living beings [28]. For instance, blood pressure depends on many local variables, which are in turn affected by many hormonal, nervous and chemical processes, but it is stabilized globally without a central control unit. In the cybernetic era the notion of homeostasis has served as a kind of metaphorical explanation of various phenomena ranging from physiological control to social relations and formed an important basis for the intended development of self-organizing machines.

4.1.1 Ashby's Approach

William Ross Ashby, displayed in Fig. 4.1, studied the underlying properties of such mechanisms in living beings from a theoretical point of view and tried to transfer this principle to the design of control architectures in machines [8]. He found it natural in this context to consider dynamical systems involving switching processes, intermittency, memory, and dynamic relations with the environment. In particular

Fig. 4.1: Portrait of Ashby aged 45 in 1948. Image included with the permission of *The Estate of W. Ross Ashby.*

the notion of ultrastability appeared to him of central importance for the mechanisms underlying life.

Ashby claimed that in complex dynamical systems the notion of linear stability is not sufficient. Such systems should be able to cope with large deviations from a target state or even to actively produce such deviations in order to achieve structural reorganization in their course. Ultrastability refers to systems that in addition to their state dynamics also possess a parameter dynamics, which is supposed to modulate the system dynamics such that eventually a target state is approached even in presence of heavy structural deformations.

4.1.2 The Homeostat

Ashby's homeostat [8], see Fig. 4.2, was invented and actually built more than half a century ago. It was intended as a working illustration of the principle of homeostasis, a term coined by Cannon [28] in the 1930s. The machine built by Ashby consisted of four rotatable magnets whose deviations from their target orientation gave rise to currents in a number of coils that also influenced the orientation of the magnets. The strengths of the interactions among the magnets are subject to switching processes realized by resistor values produced by pseudo-random multi-selectors.

(a) (b)

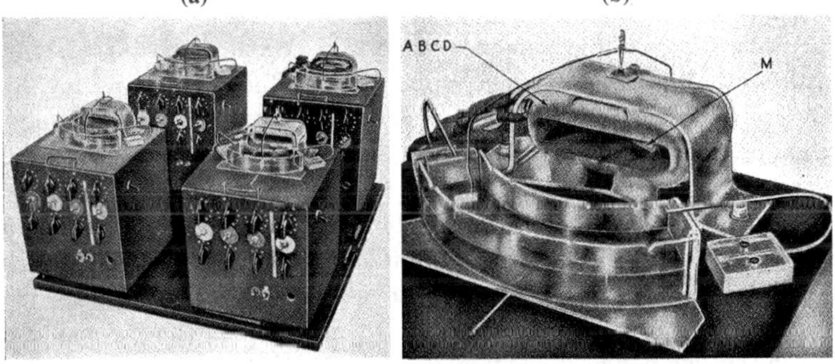

Fig. 4.2: Pictures of the Homeostat. (**a**) The four units, each one reacting on all the others. (**b**) Close view of one unit. The coil ABCD encircles the magnet M, which is suspended by the needle pivot. The suspending wire extends forward ending in a semicircular plastic trough filled with water, which has electrodes at each end. The potential is measured at the pivot socket. Images from [7].

4.1.2.1 Theory

The dynamics of the system can be described mathematically, see [6, 8] by a dynamical system in $n = 4$ dimensions

$$\dot{x}_i = \sum_{j=1}^{n} a_{ij} x_j \tag{4.1}$$

the x_i measuring the angles of the magnets. The diagonal self-couplings a_{ii} are adjusted by hand to negative values initially, such that the system has a bias towards stability. The coefficients a_{ij} for $i \neq j$ are subject to the switching dynamics and are automatically chosen from a set of 9 possible values, which are fractions as 0.48, 0.73, 0.89 and 1.0 of some maximal value, the corresponding negative values or zero so that actually there are 12^9 different dynamical behaviors. The switching dynamics introduces a state dependence into the parameters so that the Eq. (4.1) is a nonlinear dynamical system.

In a run, the system state x is allowed to evolve for some time. In the course of time it will either approach some attractor inside the critical surface or eventually reach the critical surface: If $|x_i| \geq \frac{\pi}{4}$ a new set of random variables a_{ij} for all off-diagonal couplings towards unit i is drawn such that the system is changed each time a potential instability is encountered. Since in the hardware implementation the critical surface is simply given by a mechanical constraint, the state is assumed not to evolve further once the critical surface, defined by the terminal positions of the magnets, is reached.

In the stable region of the combined parameter and state space, the system tolerates small noise. But by an external impact that is sufficiently strong, e.g. the loss of a physical connection between node i and j or by reversing the sign of an a_{ij}, the system returns, via (repeatedly) hitting the critical surface, to a parameter setting with the desired dynamical properties. For random couplings, negativity of all eigenvalues is realized only with probability 2^{-n}, i.e. homeostats with large n will converge, in general, only after very long transients. Further, oscillatory behavior or slow dynamics may occur inside the critical region such that the time between switching events may be prohibitively long. If the eigenvalues have a small negative real part, the system will be prone to noise.

The homeostat provided an early example of how an adaptive control system may solve complex problems even under heavy structural perturbations. However, as the theoretical considerations made clear, the approach does not scale. The complexity increases exponentially with the number of degrees of freedom so that Ashby never achieved the general purpose, brain-like devices, he initially was searching for. There have been later developments of the homeostat, but without significant results. Nonetheless, the homeostat over the years continued to influence our understanding as to how the large analogue, adaptive networks in biological brains, might achieve stability. Recently, as mentioned above, we have been witnessing a revived interest in the concept of homeostasis in various fields ranging from synaptic plasticity to the control of autonomous robots.

4.1.3 Learning Homeostat

The main idea of the homeostat can be reformulated in terms of continuous adaptive dynamical systems by replacing the hard boundaries with smooth nonlinearities, see [67]. If the critical surface is specified by soft thresholds, the state dynamics may be controlled to avoid the boundaries and to stay near the target state by following the gradients of the threshold functions. This can be translated into a learning algorithm for the parameters of the system. The resulting system could be treated analytically, so that explicit solutions were obtained at least in a number of special cases. It could be shown [67] that an equivalent of Ashby's parameter switching is obtained by the learning dynamics with much better scaling properties so that one of the main disadvantages of Ashby's homeostat can possibly be circumvented in that way, but the topic has not been followed up so far.

This way of merging state and parameter dynamics can be considered as a modern version of Ashby's original thinking. The productive interplay between learning and behaving is also one of the central topics of this book.

4.1.4 Relating Time Scales

One of the most remarkable features of the homeostat is the interplay between the system and the parameter dynamics. In order to make this more explicit let us compare Ashby's machine with nowadays realizations of adaptive systems. The latter are characterized usually by the separation of time scales for the learning and the behavior of the system. The reason is that most of the known learning algorithms are prone to instabilities if learning and system dynamics are not protected against each other—much in difference to Ashby's thinking where the system dynamics, when reaching the critical boundary, triggers a random change of the parameters, which leads to a largely different system dynamics and so on. This is an interplay between the two subsystems, which is remarkably stable and keeps the complete system both sensitive to external influences and stable against strong perturbations or even damages done to the system.

By the way, the human brain masters the same problem. Neuronal and synaptic dynamics are running concomitantly most of the time so that there must be a special organization preventing catastrophic interrelations between the two subsystems. Thus we feel challenged both by the results of modern brain research and the audacious thinking of Ashby and the early cyberneticists to tackle the problem of self-regulation by establishing a productive interplay of the neuronal and synaptic dynamics.

4.2 Self-Regulation

The capability to self-regulate under challenging environmental conditions is the essential feature of homeostasis. This has been achieved in the case of Ashby's homeostat by specifically engineering the system—the installation of the parameter switching dynamics—and in the learning homeostat by defining the learning dynamics and the set point. Instead of specifying a certain set point, we are looking now for a principle that is domain independent and self-contained so that we can extend the self-regulation paradigm to more general situations.

4.2.1 First Ideas

Self-organization and self-determined development require a certain amount of self-awareness of the robot. In particular, the robot should be able to predict the consequences of its actions in the near future. This can be achieved by equipping the robot with an adaptive internal model. We may restrict ourselves in the present setting to the most simple case where the model is to predict the sensor values measured in the next time step, see Fig. 3.3, so that we may use the forward model as introduced in Sect. 3.1.2.

This kind of model plays an important role in our approach so that we are going to go a bit deeper into detail here. Let us consider by way of example the BARREL as introduced in Sect. 3.2, in order to get an understanding of what such a forward model is about. Actually, the BARREL with its internal weights driven by restricted motor forces is quite a complicated physical object, with 8 degrees of freedom (three translational, three for the orientation in space and two for the internal weights), in general interacting with other physical objects like the floor (which may be structured and/or highly elastic) and other obstacles in an intricate way.

The BARREL can only be modeled in full generality by using its equations of motion in some convenient representation, like the Lagrange formalism of classical mechanics, together with the details of the interactions. Given the relevant information, including the coordinates of the BARREL and the structures of both the ground and the obstacles, these equations can be solved and this is exactly what the simulation software does. However, we neither have this information (we have only the inclination of the two internal axes via the sensor values) nor do we want to rely on them. Self-organization, at least as we know it from nature, means the emergence of dimensionally reduced modes in complex physical systems. The rolling mode on even ground is the best example since it can be described by a largely reduced dynamical system.

The reduced dynamical complexity of these "natural" modes is reflected in the complexity requirements for the model. Quite surprisingly we found in numerous applications that a linear or pseudo-linear model is all we need for our purpose of driving the system into a self-organization phase. In detail, we use in Eq. (3.6) in the most simple setting

$$M(x,y) = Ay + b \tag{4.2}$$

where the motor values are given by $y = K(x)$ with

$$K(x) = \tanh(Cx + h) \tag{4.3}$$

so that the dynamics model (3.8) is $\psi(x) = AK(x) + b$. The sensorimotor dynamics can be written as

$$x_{t+1} = Ay_t + b + \xi_{t+1} \tag{4.4}$$

where ξ gives the modeling error, b is a bias vector, and the $m \times n$ matrix A represents the response of the sensors to the motor values. Both quantities are subject to a fast and never ending learning dynamics, which is the essential compensation for the simplicity of the models.

The forward model can be trained by minimizing the prediction error $E_{\text{pred}} = \xi^\top \xi$ (3.9) by gradient descent as described in Eq. 3.10 (p. 26). The parameters to be adapted are the elements of both the matrix A and the bias vector b. In an online learning scenario we obtain in each time step a pair $(x_{t+1}; y_t)$ and try to adapt the parameters such that the modeling error $\xi^T \xi$ is reduced. In the concrete case, the gradient descent according to Eq. (3.10) yields the update rule (omitting the time indices)

$$\Delta A_{ij} = \varepsilon_A \xi_i y_j \tag{4.5}$$

$$\Delta b_i = \varepsilon_A \xi_i \tag{4.6}$$

with a conveniently chosen learning rate $\varepsilon_A > 0$. The index i is running over all sensors, i. e. $i = 1, 2, \ldots, n$ and j over the motors as $j = 1, 2, \ldots, m$. In practice, the updates should be complemented with a decay term so that the model has a chance to forget outdated contents. The update rule, essentially the famous Delta rule [183], has a simple interpretation in the context of neural networks introduced in Sect. 3.6. For instance, ΔA_{ij} may be interpreted as the product of the input y_j into a synapse A_{ij} times the error activity ξ_i at the output of the latter. This is reminiscent of Hebb's rule of learning in the brain indeed.

One important point is the choice of time scales. In our simple BARREL case the time scale for the parameter dynamics Eq. (4.5) is just a few seconds real time by choosing ε_A appropriately, see Fig. 4.3. This unusual fast learning is feasible in the context of self-regulation not only for the pure rolling modes, as we will see later. Learning in this sense does not mean that the parameters converge so that the prediction error is going to zero. Instead it turned out to be sufficient for the self-organization scenario developed in this book that relative information is provided in order to discriminate the good (in the sense of natural modes that can be modeled well) from the less interesting regions of the behavior space.

This is good news which, however, leaves the essential question open. If we want such modes to self-organize, how can a model help in this emergence process if it actually "understands" only the product and not its emergence, i. e. how to get there. The controller on the other hand urgently needs the help of the forward model

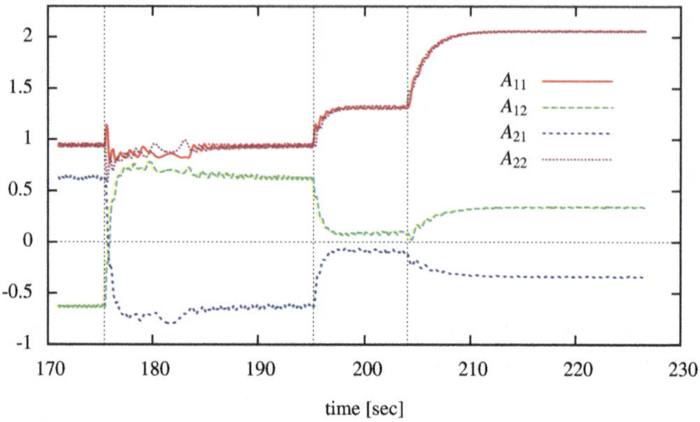

Fig. 4.3: The matrix elements of the model matrix *A* in a relearning process. The behavior of the robot is modified three times by changing the parameters of the controller (**dashed vertical lines**). The model is seen to adapt in a very short time to the new situation. In the final situation we had $C_{11} = C_{22} = 0.5$ and $C_{12} = -C_{21} = 0.1$. The relearning times are largely determined by the inertia of the BARREL since the direction of rolling has to change as a consequence of the parameter switching.

in order to find its specific control mode. This is a central problem of autonomous development, namely that controller learning and model learning have to bootstrap each other.

4.2.2 Learning the Controller from Specialized Models

We have seen above, that, given a controller, the model can learn easily the correlations between the actions and the induced sensor values. This is not a surprise given that we accept larger modeling errors and reduced competencies of the forward model (specialized model). In order to understand more about the mutual bootstrapping of model and controller, we have to consider the inverse problem. Let us assume we have learned a model for a given behavior generated by the parameters C and h of the controller. Can we reobtain the behavior from knowing the model? In principle this should be possible, since the model error can not only be minimized by changing the model for a given behavior, but the error of a given adaptive model, even if well trained, depends just as well on the behavior.

In fact, the error will be small if the system is in one of its more natural modes (like the rolling mode of the BARREL), it will be rather large if the system is in more complex modes (like the "lolloping" mode in Video 3.2), and very large if the system is in an erratic motion. Can the controller learn the natural modes at least from a specialized model, i. e. one that has learned to represent a specific mode?

The question can not be answered in general but the practical experiences show that such models are well able to attract the controller towards the pertinent behavior.

Learning the controller from the prediction error of the model is schematized in Fig. 4.4. In the specific setting given by Eqs. (4.3, 4.4) above, we find the learning rules as (omitting time indices, derivation in the Appendix, Sect. 4.A (p. 71))

$$\Delta C_{ij} = \varepsilon_c \eta_i x_j \,, \tag{4.7}$$

$$\Delta h_i = \varepsilon_c \eta_i \,, \tag{4.8}$$

where

$$\eta_i = \sum_j A_{ji} g'(z_j)\, \xi_i \,, \tag{4.9}$$

with $z_j = \sum_k C_{jk} x_k + h_j$ and $g'(z) = \tanh'(z) = 1 - \tanh^2(z)$. η is the result of back propagating ξ through the forward model and the output function of the controller. Similar to the case of model learning we may add a small decay term to the learning rule. The matrix A and the column vector b defining the forward model are assumed to be known for the moment.

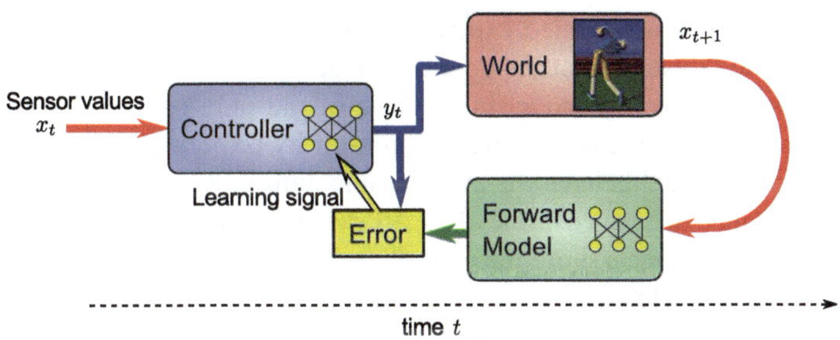

Fig. 4.4: Learning the controller from the forward model. The task is now inverted. Given the forward model, how can the controller be adapted so that the behavior of the robot can be well predicted by the model.

Experiment 4.1: Learning controller from forward model

Start the simulation Learning controller from model.It is now running with ε_c=`epsC=0`, ε_A=`epsA=0.1`. Check the convergence of the matrix A using the GUILOGGER (<Ctrl>+G). After convergence stop learning of the model by entering

`>epsA=0`

Change the behavior by pressing 's' or 'S' (in the graphical window) multiple times which scatters the parameters of C by random values in $[-0.2, 0.2]$ or $[-0.5, 0.5]$ respectively. Switch on learning of the controller:

`>epsC=0.1`

Watch whether the previous behavior is reestablished. Try more drastic changes by pressing 'r' or 'R' to reinitialize C randomly (see Experiment 3.2). Eventually you may wish to start anew with switching to exclusive model learning using `epsC=0`, `epsA=0.1`.

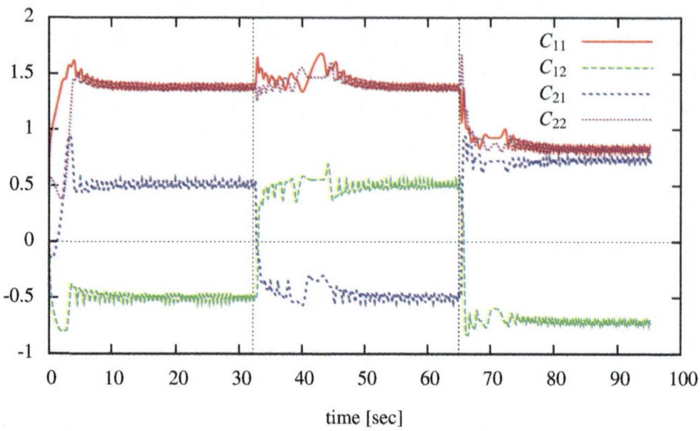

Fig. 4.5: Behavior of the controller parameters when learning the behavior from pre-specified models (rotation matrices). At seconds 32 and 65 the model was externally changed. Between the switches the controller matrix C clearly has the rotation structure. The relearning of the behavior takes only a few seconds.

The idea can be corroborated experimentally, see Experiment 4.1 or Fig. 4.5, which clearly demonstrate that the controller learns in just a few seconds the new behavior prescribed by the model in the case of the pure rolling modes. On the other hand, as the reader easily convinces himself by Experiment 4.1, learning the controller from a model behavior is also feasible with more complex modes of behavior. For instance, we may use one of the above learned models for the "lolloping" modes. As the experiments show, the controller can be recovered from such a model in nearly the same reliable way as in the rotational mode. This puts some credibility into the postulate that behavior, in the natural modes at least, can be learned from having a **simple** forward model. In other words, these extremely simple forward models can act as a kind of "role model" for quite complicated behaviors.

Note that even though the learning process may leave a non-vanishing residual error, the procedure does not fail because mainly the relative errors matter.

4.2.3 Homeostasis: Self-Regulated Stability

Before putting the pieces together, let us summarize the results achieved so far. We have seen that simple controllers can command complicated behaviors if they excite nontrivial physical objects in the closed loop control mode. On the other hand, we have seen that forward models with restricted competencies can host very complicated behaviors like rocking, jumping or kind of lolloping in our BARREL case. Moreover, we have seen that these two units can teach each other very effectively. We can imagine this as a kind of mutual, body mediated attraction between these two units. However, this mutual attraction phenomenon was possible only if one of the two was pre-specified from outside or learned in an earlier episode to realize a specific mode of behavior. What will happen if we let the two teach each other simultaneously? Obviously the attraction will now drive both model and controller towards each other. Without any goal or purpose given from outside there are more questions than answers. Will the system go into a specific mode of behavior and which one will it be? Can we expect any systematics from such a procedure and what is the role of the embodiment in such a process?

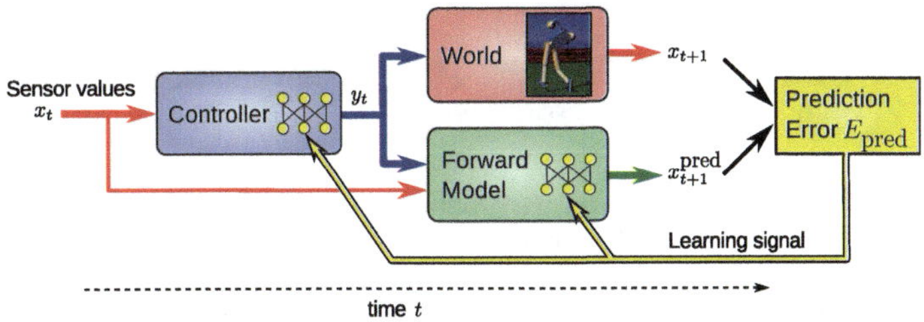

Fig. 4.6: How the controller and the model teach each other. The prediction error provides a learning signal for both the model and the controller.

The general setting is sketched in Fig. 4.6. In the above realization for both the model and the controller we simply have to learn both model and controller simultaneously so that we now have a combined dynamics consisting of the physics, the model parameters, and the controller parameters, altogether a dynamical system in a space of 20 dimensions with comparable time scales of the subsystems.

The idea can easily be checked experimentally, see Experiment 4.2. Considering many different initializations for the C matrix reveals that the concomitant learning of controller and model most of the time runs into a fixed point of the combined dynamics where the robot is at rest, see Fig. 4.7. Actually, this is not surprising since in a physically stable situation the model can easily learn the motor-sensor correlations so that the prediction error has a minimum.

Experiment 4.2: Homeostatic control of the BARREL

The simulation Homeostatic control of BARREL starts with the robot at rest and will not leave this state by itself. The parameters are ε_c=`epsC=0.1`, ε_A=`epsA=0.1`. Exert external forces to the BARREL by pressing 'x' or 'X' (in the graphical window) or use <Ctrl>+<left Mouse button> or <Ctrl>+<right Mouse button>.You can speed up/slow down the simulation by pressing '+'/'-'. Surprisingly the robot does not always come to rest but enters a periodic forward/backward locomotion behavior. Try different learning rates when the robot is in such a mode for instance:
`>epsA=0.005`
The interval of forward and backward motion becomes longer.
Try `>epsC=0.5` and `>epsA=0.5`. and observe the behavior. Eventually the robot will stabilizes into the "do nothing" regime, since the model can learn the behavior quick enough.
You may reinitialize both the parameters of the controller and the forward model randomly as described in Experiment 3.2. For large learning rates the system comes to rest in almost any case. For low learning rates we observe often oscillatory behavior.

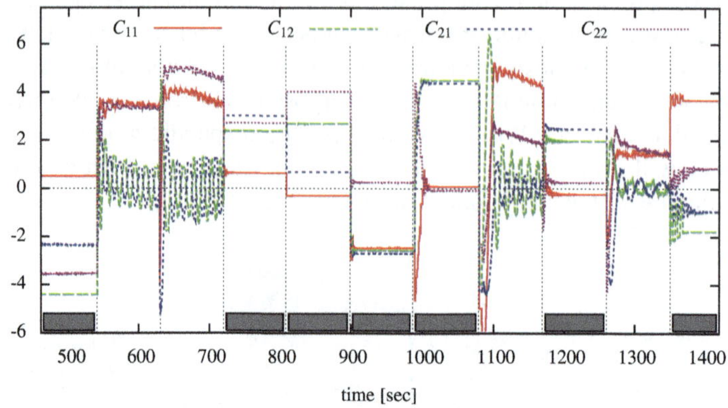

Fig. 4.7: A long run of the BARREL with random reinitialization of the C matrix every 90 sec. The reinitialization of the controller matrix is followed by a relearning of both the model and the controller. In most cases (marked by gray bars) there is a rapid convergence towards a stable situation. In other cases the learning dynamics remains active, which corresponds to a richer dynamics of the system in the sweeping modes. Parameters: $\varepsilon_c = 0.5, \varepsilon_A = 0.1$, no friction.

On the other hand, when starting the controller or the model with a rotation matrix or if the robot is kicked from outside into a rolling motion, something interesting happens: the BARREL starts rolling, but it will not end up in a fixed rotation mode. Instead it increases velocity then decelerates, reverses velocity and starts the same in the opposite direction. This slowly oscillating mode will then continue forever. So, instead of realizing just one behavior (rolling mode with fixed velocity) the robot sweeps through the behavior space of the possible rolling modes. This is not only seen at the level of behaviors but is clearly reflected by the structure of both the C and A matrices, which sweep (more or less periodically) through the space of rotation matrices. The phenomenon is discussed in more detail in Sect. 4.B (p. 73).

The reason for this behavior is seen in the mutual teaching scenario of model and controller. Both components are trying to imitate each other but since the learning needs time one of the two is always lagging behind the other with a change of roles after some time. Most importantly the model is lagging behind the physical mode. This is also supported by the experiment: for large learning rates the system comes to rest most of the time—the lag between model, controller and physical mode is small. However if the learning rates are small this lag is significant and we observe the periodic forward/backward behavior. In Chap. 5 we will trace back this behavior to a very general phenomenon, the spontaneous symmetry breaking, which is ubiquitous in self-organizing systems.

The preceding investigations show that, depending on the starting conditions and/or external perturbations, the mutual attraction mechanism between model and controller provides a self-regulation into body related behaviors, which may be oscillatory or at a stable fixed point attractor. The latter correspond to stable physical states where the body is at rest. The system can be switched from one to another by strong external mechanical perturbations, see Experiment 4.2.

Nevertheless, so far there is no intrinsic drive for innovations that might lead the system out of these states. The most obvious reason for the lack of initiative is seen in the fact that physically stable situations are candidates for small prediction errors, since the models can most easily adapt to such a situation. In exceptional cases, this may produce unexpected results, like the sweeping mode, if we have an exceptionally high degree of symmetry in the system. In other applications, like with a humanoid robot, one finds that the landscape of behaviors is mainly characterized by stable attractors of the dynamics. In a way, the method described so far realizes a version of Ashby's homeostasis [8] or even ultrastability in a self-referential way.

4.2.4 Conclusions and Outlook

The self-regulation towards the regimes of high stability is a self-supporting effect that destroys any drive for development. In fact, in order that the agents feel challenged to do something new, the controller should provide new experiences instead of stabilizing motion patterns established already in the model. The ultimate reason for that behavior is contained directly in our objective function, given that the prediction error is minimal if the robot is stalling so that there are no changes in the sensor values at all. In Chap. 5 we will define a new objective function, which introduces the desired drive for innovation in a most natural way.

Appendix 4.A Learning a Behavior from its Forward Model

Forward model learning, as introduced above, is a supervised learning task with the target values of the next state being given, in the perspective of the model, from outside. In a typical supervised learning scenario, the model learns the prediction task after a sufficiently long learning period with maximum precision. Still, in nearly all practical cases, there is a residual error the quantity and quality of which depends largely on the complexity of the predictor, the details of the learning process (annealing strategies for the learning rate, to give an example), possible strategies for avoiding overfitting, and the like. Given the learning process, the residual error reflects how well the model is able to represent the systematic features of the process.

The learning rule for learning the controller from a given model is obtained by considering x_{t+1} as a target value given from outside and trying to get the **behavior** (as described by the dynamics model) more in concordance with this value![1] Thus, the derivative w. r. t. any parameter p of the controller is taken as

[1] In other words, the new state x_{t+1}, is considered obscure despite our attempt to build a dynamical model by our function ψ of the time step. This attitude is a little "fatalistic" and all we can do then is to try to adapt our controller such that the dynamics model the true world match.

$$\frac{\partial}{\partial p}\xi^\top \xi = 2\xi^\top \frac{\partial}{\partial p}(x_{t+1} - \psi(x)) = -2\xi^\top \frac{\partial}{\partial p}\psi(x) . \tag{4.10}$$

In the specific setting $\psi(x) = Ag(Cx+h)+b$, the parameters are the elements of C and h.

Let us consider

$$\frac{1}{2}\frac{\partial}{\partial C_{ij}}E = \xi^\top \frac{\partial}{\partial C_{ij}}\psi(x) = \sum_k \xi_k \frac{\partial}{\partial C_{ij}}\psi_k(x) = \sum_k \xi_k \frac{\partial}{\partial C_{ij}}(Ag(z))_k$$
$$= \sum_k \xi_k A_{kl} \frac{\partial}{\partial C_{ij}} g_l(z_l)$$

where $g : R^m \to R^m$ and $z \in R^m$ with $z = Cx+h$. Using that

$$\frac{\partial}{\partial C_{ij}} g_l(z_l) = \delta_{il} g_i'(z_i) x_j$$

we find

$$\frac{1}{2}\frac{\partial}{\partial C_{ij}}E = \sum_k \xi_k A_{ki} g_i' x_j = \eta_i x_j$$

where the vector η is

$$\eta = G' A^\top \xi$$

and G' is the diagonal matrix with matrix elements $G'_{rs} = \delta_{rs} g_r'(z_r)$. In matrix representation we write the result as

$$\frac{1}{2}\frac{\partial}{\partial C}E = \eta x^\top .$$

Correspondingly we obtain for the bias using

$$\frac{\partial}{\partial h_i} g_l(z_l) = \delta_{il} g_i'(z_i)$$

that

$$\frac{1}{2}\frac{\partial}{\partial h}E = \eta .$$

With these results, the learning rules read

$$\Delta C_{ij} = \varepsilon \eta_i x_j \tag{4.11}$$
$$\Delta h_i = \varepsilon \eta_i . \tag{4.12}$$

These are the rules, see Eqs. (4.7, 4.8) used in the main text. In the neural network language, $\eta = G' A^\top \xi$ is considered as the result of propagating the prediction error ξ_{t+1} back to the internal state of the neuron so that the rule has a Hebbian like structure in the sense considered in Eq. (3.40).

Appendix 4.B A Bootstrapping Scenario

In order to make bootstrapping of specific modes more explicit, let us consider the case that x_t is a periodic motion as for instance with the BARREL being tossed by an external force. In the specific cases of the freely rolling BARREL, e. g. , the state dynamics is even a harmonic oscillation described by

$$x_{t+1} = O(\phi)\, x_t \tag{4.13}$$

with $O(\phi)$ a rotation matrix rotating the vector $x_t \in \mathbb{R}^2$ by the angle ϕ.

In general, we may assume that the state vector essentially is following this law plus perturbations due to the actions of the controller, nonlinearities, and several physical effects. Whatever the true motion is, there will always be a **misfit** of the model, which contains a component of the periodic motion (since it is present in both x_{t+1} and its model representation Ay_t, which contains the non-linearities), i.e. we can write tentatively

$$\xi_{t+1} = x_{t+1} - Ay_t \propto O(\phi)\, x_t + \dots . \tag{4.14}$$

Using this in Eq. (4.11) yields[2]

$$\Delta C \propto O(\phi)\, x_t x_t^{\top} + \dots$$

Consider, by way of example, a two-dimensional system with state vector $x_t \propto (\sin\omega t, \cos\omega t)^{\top}$. While the non-diagonal terms of the matrix $x_t x_t^{\top}$ average out to zero, the diagonal ones are always positive so that

$$\left\langle x_t x_t^{\top} \right\rangle_T \approx \frac{1}{2}\alpha^2 \mathbb{I} + \dots$$

where $\langle \dots \rangle_T$ means the averaging over a time window of length $T \gg 1/\omega$ and α is proportional to the average amplitude of x_t. We obtain

$$\Delta C \propto O(\phi) + \dots . \tag{4.15}$$

These crude considerations reveal an important tendency of the learning procedure: if the state vector is rotating with a certain frequency, the update of the C matrix is given by the corresponding rotation matrix so that, depending on the sign, the latent oscillation is either enhanced or suppressed by the learning procedure. In the former case, this is a bootstrapping of a specific oscillatory mode in the system. Based on these considerations, it is not a surprise that in the case of the BARREL the oscillatory modes are emerging if the robot is driven into a rolling motion by external impacts.

---

[2] We consider the case that A is proportional to the unit matrix and the g' are close to 1 corresponding to a weak excitation of the neurons.

Actually, the situation is even more involved. On the one hand, the neglected terms in Eq. (4.15) may dominate the periodic one and produce a more complicated behavior than the simple rotation by the angle ϕ. In particular, with comparable time scales of model and controller learning, the model can not follow fast enough to become bias free and sufficiently accurate. This introduces spurious correlations across time between the error ξ and the state x leading to additional updates for the parameters C and h. Moreover, by the above arguments, the h dynamics is approximately

$$\Delta h = \varepsilon O(\phi) x_t + \ldots \tag{4.16}$$

leading to a phase shifted oscillation of the bias, which influences strongly the dynamics of Eq. (4.13). Probably, these feedback effects are the reasons for the sweeping phenomenon observed in Sect. 4.2.3 above, see also Experiment 4.2.

The arguments are rather crude but give an impression how latent oscillatory modes may self-amplify by this homeostatic learning mechanism. As it turns out, in practice the bootstrapping of stable modes is a common phenomenon in these learning scenarios. The concrete mechanisms may vary but, as a general scheme, the learning improves the agent's ability to reenact the transition from the current state x_t to the next state x_{t+1} it has induced itself by its control action. The update of the controller is done by means of the agent's forward model, to the best of its knowledge, so to say. The true transition dynamics may largely differ from the predicted one but, as the above considerations have shown, as long as the model is qualitatively right, it still reflects the physical response of the system which is all the bootstrapping needs.

This is also an argument in favor of the simplicity of the models used in our practical applications. Although only qualitatively right, the forward model may be sufficiently helpful so that the controller is adapted into the right direction of getting better in harmony with the responses by the physical system.

Chapter 5
A General Approach to Self-Organization — Homeokinesis

Abstract: In this chapter we will introduce the concept of homeokinesis, formulate it in mathematical terms, and develop a first understanding of its functionality. The preceding chapter on homeostasis made clear that the objective of "keeping things under control" cannot lead to a system which has a drive of its own to explore its behavioral options in a self-determined manner. This is not surprising since the homeostatic objective drives the controller to minimize the future effects of unpredictable perturbations. This chapter uses a different objective, the so called time-loop error, derives learning rules by gradient descending that error and discusses first consequences of the new approach. Minimizing the time-loop error is shown to generate a dynamical entanglement between state and parameter dynamics that has been termed homeokinesis since it realizes a dynamical regime jointly involving the physical, the neural, and the synaptic dynamics of the brain-body system.

While homeostasis is seen as a general paradigm for the self-regulation towards stability, homeokinesis is intended as its dynamical counterpart—a general principle that drives agents to activity and innovation. Homeokinesis is realized as a supervised learning procedure, however with an objective that is completely internal to the agent so that the brain-body system can be modeled by a self-referential dynamical system. This chapter introduces first the new objective, the so-called time-loop error (TLE) that was introduced originally as the basis of homeokinesis in [42]. In Sect. 5.2 we will derive learning rules for the parameters of the controller and show that these learning rules realize an irreducible entanglement between parameter and state dynamics of the sensorimotor system. If the controller and the forward model of the agent are realized by neural networks, this entanglement involves also the synaptic dynamics of the neurons so that we have a system with fast synaptic plasticity, the latter playing a constitutive role in the generation of the behavior. We call such systems self-referential dynamical systems.

Section 5.3 represents, in the sense of a first study, key features of homeokinetic systems. In Sect. 5.3.1 we give the general structure as a self-referential dynamical system and describe in Sect. 5.3.2 one of their most essential properties: minimizing the TLE leads to a progressive destabilization of the system so that a general drive for activity and innovation is emerging. Section 5.3.3 is devoted to the central working mechanism of homeokinetic learning, the vivid interplay between state and parameter dynamics.

R. Der and G. Martius: The Playful Machine, COSMOS 15, pp. 75–106.
springerlink.com

Sects. (5.3.4–5.3.6) describe subsequently further key features of homeokinesis by demonstrating how the general trend for destabilization of the system is counteracted by the nonlinearities, the symmetries, and the evasive nature of true chaotic behavior. The compromise between the conservative (or confining) effects and the general drive for destabilization forms the basis for a behavioral variability that balances in complexity somewhere between order and chaos.

Further details together with an alternative derivation of the homeokinesis principle may be found in Chap. 11 where we introduce the so-called interaction representation. This new representation provides further theoretical insight into homeokinetic systems. In particular, by extending the interaction representation to infinite time horizons, we may understand homeokinesis as a flow of the **global** Lyapunov exponents. This theoretical analysis reveals a direct connection to chaos theory, which may be helpful for future research, given the wealth of results and knowledge available there.

5.1 Homeokinesis — Introduction

We have discussed in Chap. 4 how self-regulation can be realized based completely on the internal perspective of the agent. The idea was to make the agent control its behavior by letting forward model and controller teach each other, solely driven by the objective to keep the prediction error low. In the example of the BARREL we observed the emergence of different control strategies generating specific, body related behaviors. This principle is appealing because it brings the embodiment of the robot into play in a natural way. In fact, since we do not give any specific goals or external cues, it is the embodiment that decides where the system tends to develop. In the experiment the controller drives the system into physical states which it can stabilize best—in accordance with the principle since it is there where the prediction error is minimized.

5.1.1 Time-Loop Error

However, in this way one does not get the curious and self-exploratory robot we want to create. Instead of realizing a general stasis in the system we want brain, body, and environment to get into a joint kinetic regime. So, instead of homeostasis we want a homeokinetic system. The central question is now: how can we find a general drive for activity? Or phrased differently: how can we get out of the stable attractors? This would be simple if we could invert the arrow of time as indicated in Fig. 5.1.

Of course, the physics of the system can surely not run backward in time. However, what we can do is to let the controller learn to stabilize the behavior backward instead of forward in time as before. This means we must consider *reconstruction*

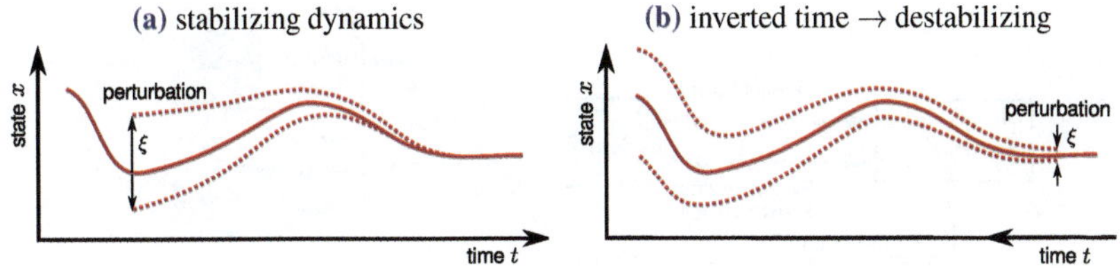

Fig. 5.1: Inversion of time converts stability into instability. **(a)** Schematic evolution of a state and its perturbations in a stabilizing dynamics, approaching an attractor. **(b)** If the arrow of time is inverted, the attractor is converted into a repeller and small perturbations are amplified. In homeokinesis, a controller learns to stabilize a dynamics backward in time. If applied to the real system (that is running forward in time), the controller aims at destabilizing the dynamics.

(in time) instead of prediction and reiterate accordingly our earlier considerations. Let us start from our general formulation of the sensorimotor dynamics as given by

$$x_{t+1} = \psi(x_t) + \xi_{t+1} \,. \tag{5.1}$$

A single step in the sensorimotor dynamics can be considered as the transformation of the input x_t into the output x_{t+1}, the systematic part of that transformation being given by the dynamics model represented by the map $\psi : \mathbb{R}^n \to \mathbb{R}^n$. The reconstruction of the true input x_t from the observed output x_{t+1} by means of the dynamics model corresponds to finding the input shift v that generates best the observed output shift ξ, i.e. find v_t so that

$$x_{t+1} = \psi(x_t) + \xi_{t+1} = \psi(x_t + v_t) \,. \tag{5.2}$$

This is a nonlinear inverse problem that can be solved explicitly if the inverse ψ^{-1} of the function ψ exists so that the reconstructed sensor vector $\hat{x}_t$ is

$$\hat{x}_t = x_t + v_t = \psi^{-1}(x_{t+1}) \,. \tag{5.3}$$

In the general case we define v_t such that $\|x_{t+1} - \psi(x_t + v_t)\|$ is minimal[1]

$$\|x_{t+1} - \psi(x_t + v_t)\| \stackrel{!}{=} \min \tag{5.4}$$

with solution written as

$$v_t = \arg\min_v \|x_{t+1} - \psi(x_t + v)\| \,. \tag{5.5}$$

We call either v_t itself or its square norm

$$E = v_t^\top v_t$$

[1] This may have multiple solutions but we do not want to go into mathematical subtleties here.

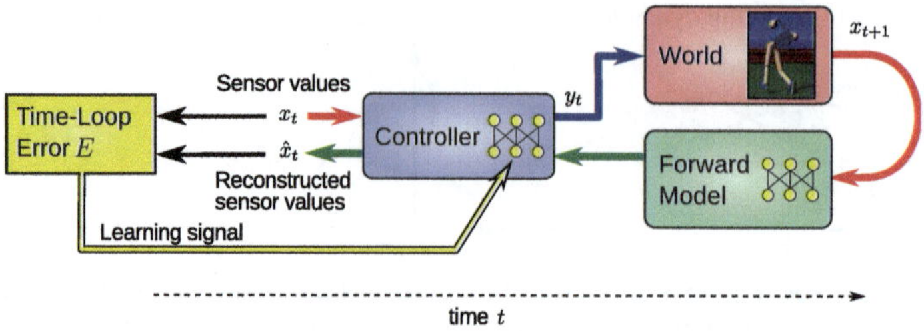

Fig. 5.2: Time loop and time-loop error. Starting at time t with sensor vector x, the signal flow in the time loop runs through the controller and the world (robot) producing the new sensor values at time $t+1$, which are projected backward through the forward model and the controller to yield the reconstructed sensor values $\hat{x}_t$ (5.3). The difference between reconstructed and true sensor values forms the time-loop error or reconstruction error, which is used as a learning signal for the controller.

the *reconstruction error*. The sequence of steps from x_t to x_{t+1} and back through the dynamics model ψ to time t is forming a time loop so that E is also called the *time-loop error* (TLE), see Fig. 5.2 for an illustration. The aim of homeokinetic learning consists in changing the behavior so that the TLE is being minimized.

The mathematical definition (5.5) is too complicated for practical use. A simplification is obtained in a natural way if the prediction error (also called noise) ξ is small so that we can use a Taylor expansion given by

$$\psi(x+v) = \psi(x) + L(x)\, v + O\left(\|v\|^2\right)$$

where the Jacobian matrix L is given in terms of its matrix elements by

$$L_{ij}(x) = \frac{\partial \psi_i(x)}{\partial x_j}$$

or more compactly

$$L(x) = \frac{\partial \psi(x)}{\partial x}\,. \tag{5.6}$$

Assuming that the inverse L^{-1} exists, we obtain

$$v = L^{-1}(x)\,\xi + O\left(\|v\|^2\right)\,. \tag{5.7}$$

In Sect. 5.B (p. 100) we discuss a method for making the higher order terms to disappear by replacing x in $L(x)$ by a so called effective state. In this way we can restrict ourselves to the lowest order term without loosing anything. In practice, the effective states may be difficult to obtained so that we will restrict ourselves to use the **definition**

$$v = L^{-1}(x)\,\xi \tag{5.8}$$

as the basis of our developments. Now, we can express the TLE explicitly as[2]

$$E = v^\top v = \xi^\top \frac{1}{LL^\top} \xi \,. \tag{5.9}$$

The step from Eq. (5.5) to Eq. (5.8) may seem an oversimplification so that we would ask the reader at this point to remember the goal of our project. We are looking for a general principle that drives agents to activities in agreement with their specific embodiment, thereby giving them a general scheme for self-exploration. Taking the reconstruction error as the basis of that intent is just a guess (an educated one, as we believe). We will see that this guess has far reaching consequences into the desired direction even if we replace the original with a further guess, given by the definition of Eq. (5.8). As we will discuss later, see in particular Chap. 11, the use of the simplified form of the TLE corresponds to treating our system on the basis of the local Lyapunov exponents[3] which measure the stability of the system at the level of a **linear** stability analysis. But this seems to be sufficient given that our dynamics model, the map ψ, is rather crude anyway and that increasing the local Lyapunov exponents is already all we need for getting the desired destabilization necessary for a self-induced activity.

This is a good point to mention that, beside the time-loop error or the prediction error, there are also other approaches to autonomous robot development. As mentioned in Chap. 1 one of the pioneering work first work was done by Schmidhuber concerning artificial curiosity or intrinsic motivation in reinforcement learning. The prediction error was used as a reward signal (large error is desired to learn new things) in order to make the robot curious for new experiences [150] and further developed in [151, 164]. Related ideas have been put forward in the so called playground experiment by Kaplan and Oudeyer [85, 117], by using the learning progress as a reward signal. Another idea is the Autotelic Principle proposed by Steels [157] using the balance of skill and challenge of behavioral components as the motivation for open ended development. Several promising options are also provided by information theory as discussed in [135] with its relation to embodiment. Preliminary work has shown a great similarity between using the time-loop error and the maximization of the so called predictive information of the sensor data, see for instance [9, 10, 40, 191].

[2] We write the inverse of a matrix M as $\frac{1}{M}$.

[3] The local Lyapunov exponents λ_i of a dynamics described by the map ψ are related by $2\lambda_i = \ln \rho_i$ to the eigenvalues ρ_i of $LL^\top$. The (local) Lyapunov exponents of the sensorimotor loop (described by its model ψ) are negative for a stabilizing and positive for a destabilizing dynamics (in the direction of the respective eigenvector). The global Lyapunov exponents and their relation to homeokinetic learning is introduced in Chap. 11.

5.1.2 Discussion

Equation (5.9) already contains the main features of homeokinesis in a nutshell. Coarsely speaking, the TLE is small if, on the one hand, the prediction error ξ is small and, on the other hand, the eigenvalues of $LL^\top$ are **large** (the matrix is in the denominator). The former condition means that the behavior of the system is well predictable, i. e. the system behaves in a more or less systematic way and is stable against perturbations. The second condition is best understood by noting that the eigenvalues of $LL^\top$ measure the local stability of the system against perturbations. Thus, the TLE is small if the eigenvalues are large meaning that the dynamics is destabilized.

The resulting behavior is a compromise between predictability (keeping things under control) and chaoticity (trying new solutions due to the inherent instability) and it will be seen that this compromise creates a high behavioral variety which is body related and adapted to the environment.

5.1.3 Standard Setting

In most applications it has proven sufficient to use a rather simple setting for both the forward model and the controller. This is (i) possible because of the strong interplay between state and parameter dynamics, the main effect of homeokinetic learning, that replaces complex internal states of the controller; and this is (ii) desirable since it allows the externalization of complexity (see Chap. 3) to the physics of the robot in its environment. These points will be elucidated in detail in the later developments.

In our standard setting we use a one-layer neural network for the controller $K : \mathbb{R}^n \to \mathbb{R}^m$ given by

$$K(x) = g(Cx + h)$$

with the $m \times n$ synaptic matrix C, the bias vector $h \in \mathbb{R}^m$, and the activation function $g : \mathbb{R}^m \to \mathbb{R}^m$. In most applications g is realized by the tanh function (taken componentwise) introduced in Eq. 3.33 (p. 51). With the forward model providing a prediction as

$$x_{t+1}^{\text{pred}} = Ay_t + b \tag{5.10}$$

the dynamics model of the loop, the map $\psi : \mathbb{R}^m \to \mathbb{R}^m$, is

$$\psi(x) = Ag(Cx + h) + b \tag{5.11}$$

as already discussed in Sect. 3.3 and in Chap. 4, see Eqs. (4.2, 4.3). The Jacobian matrix, see Eq. (5.6), is now explicitly obtained as

$$L(x) = AG'(z)C \tag{5.12}$$

where $G'(z)$ is the diagonal matrix defined as $G'_{ij} = \delta_{ij} g'(z_i) = \delta_{ij} g'_i(z)$ with the Kronecker delta δ_{ij} and

$$z = Cx + h$$

may be interpreted as the vector of the membrane potentials of the motor neurons. The learning of the forward model remains as given by Eqs. (4.5, 4.6), further details can be found in Chap. 9.

5.1.4 Landscape of the Time-Loop Error

Homeokinetic learning is realized as a gradient descent on the TLE. Some essential features of homeokinesis can already be seen by considering the landscape of the TLE in a simple example related to the one-dimensional system analyzed in Sect. 3.3. As a concrete realization we take our TwoWheeled (see Sect. 3.3.5), give both wheels the same motor command so that the robot can drive only straight, and control the wheels with a single motor neuron. Let us consider the system without bias h for a moment. As we know from Sect. 3.3.2 (p. 36) this is a dynamical system with a pitchfork bifurcation at a critical value for the (only) parameter C.

The landscape of the time-loop error is characterized by singularities acting as repeller (infinitely repulsive regions) for the gradient flow, see Fig. 5.3. These repulsive regions are easily identified. In the one-dimensional case, all matrices featuring

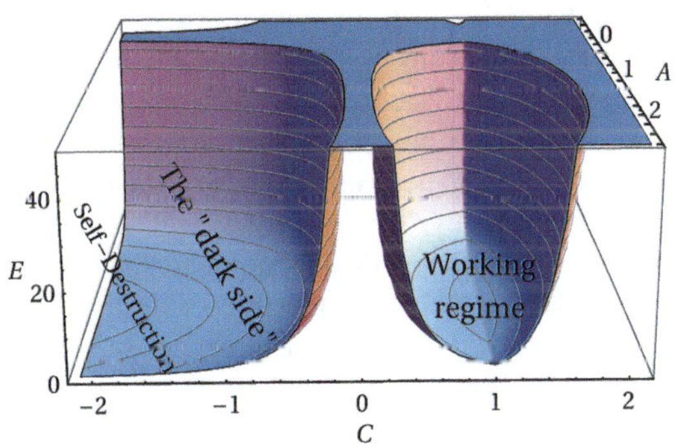

Fig. 5.3: Sketching the landscape of the time-loop error E. The landscape of a 1D system is structured by the singularities of E resulting from the zeros of the 1×1 Jacobian matrix $L = Ag'C$. The singularities act as repeller for the gradient flow of homeokinetic learning. The working regime is surrounded by singularities defined by $g' = 0$ (outer right, $C \to \infty$) and by $C = 0$ corresponding, respectively, to the saturation region of the neuron (minimal sensitivity) and the "do nothing" behavior of the robot. It was assumed that $A = 1$ is the correct value for the model matrix producing another repeller by the divergence of the prediction error ξ^2 if $A \to \infty$. The "dark side" region is reached when starting the gradient flow in the corresponding basin. As explained in the text, this side corresponds to a behavior of the robot leading potentially to self-destruction in the course of learning.

in the Jacobian matrix are 1×1 so that they are treated as scalars. The repellers are given simply by the zeros of L, i. e. by the zeros of A, C, and g'. The latter term is small in the saturation region of the motor neuron so that the TLE has a repeller there. As a consequence, the gradient flow will drive the system away from the saturation. This is in accordance with the expected sensitization since in the saturation region the neuron is insensitive to its inputs.

A second repeller is situated at $A = 0$. That one will be considered in detail in Chap. 9 so that we will not discuss it here.

A third repulsive region is given by $C = 0$. The role of that singularity is most immediately seen if the robot is initialized in the *tabula rasa* situation ($C = 0$) so that the controller does not react to its sensor values at all. This is the "do nothing" behavior (the robot stalls). However, now $C = 0$ is an instable fixed point for the gradient flow (due to the singularity) so that the slightest perturbation will quickly drive the parameters of the controller away from this "do nothing" region towards activity.

As Fig. 5.3 shows that, sandwiched between these two repelling regions, there is an attractive region, which we call the working regime since it is there where the robot is active with a high degree of sensitivity against the feedback from its body and the interactions with the environment. This point is illustrated by an experiment in Sect. 5.2.3 further below, a detailed analysis being given in Sect. 6.1 (p. 108).

Figure 5.3 also shows that, beside the "sound" region to the right hand side of the "do nothing" singularity, there is another region, called the "dark side." When initializing the robot in that region, the gradient flow will drive the system further down with C values becoming more and more negative. What does this mean for the robot behavior? In the present setting, with negative C values a positive sensor value (robot driving forward) is converted to a negative motor command, requiring the robot to drive backward and vice versa. This sign inversion pressure, encountered in each time step (e. g. 50 times per second), becomes the stronger the more negative C so that the motors or mechanical parts of the robot may become overloaded and break down eventually. This is why we call it the region of (potential) self-destruction. You may wish to test this phenomenon by doing Experiment 5.1.

Avoiding the "dark side" region is a question of an appropriate initialization, which turns out to be not too demanding in practice, see our cooking recipe in Sect. 15.2. Nevertheless, in high dimensional systems there are scenarios so that the system can switch to the wrong side, which often needs a restart for remedy.

The discussion of the landscape could easily be generalized to the multidimensional case by associating the zeros of L with the null spaces of the respective matrices C, A, and G'. Many more details on the landscape of the TLE will be revealed by both theoretical analysis and concrete examples in the following chapters.

5.2 Homeokinetic Learning

In the preceding Chap. 4 we have seen that minimal prediction errors are achieved best by a "do nothing" behavior. As discussed above, with the new objective there is an additional drive rendering the "do nothing" option instable. In the generic case, learning of the controller is realized by gradient descending the TLE E defined in Eq. 5.9 (p. 79). Writing $\Delta p_t = p_{t+1} - p_t$ for any parameter $p_t \in \mathbb{R}^1$ the learning step is obtained as (dropping time indices)

$$\Delta p = -\varepsilon_c \frac{\partial E}{\partial p} \tag{5.13}$$

with ε_c being the learning rate defining the width of the gradient step. Although this looks formally equivalent to common learning scenarios, it will become clear in the following that there are many novel features. Most prominently, the objective E is not given from outside but is a function of the state dynamics of the system alone and is therefore totally intrinsic to the agent. Moreover, state and parameter dynamics run simultaneously, there is no freezing or the like of the learning rate in order to get a convergence towards a behavioral goal. To the contrary, the parameter dynamics defined by Eq. (5.13) can be interpreted as the internal state dynamics in the sense of Eqs. (3.4, 3.5). This increases the complexity of the actually very simple controller structures we are using in the applications.

Let us consider the very general setting of a sensorimotor loop with a map ψ representing the systematic part of the dynamics. The map consists of the controller, mapping sensor states x_t to motor values y_t, and a forward model that predicts the next sensor state x_{t+1} on the basis of x_t and y_t as explained in detail in Sect. 3.1.2. The TLE depends explicitly, via L, on the parameters of the controller so that we can give generally valid expressions in terms of the derivatives of L. The derivation of the learning rules is simplified if we write the general expression of the TLE, see Eqs. (5.8, 5.9), as the overlap between the prediction error ξ and an auxiliary vector χ as[4]

$$E = \chi^\top \xi \tag{5.14}$$

where

$$\chi = \frac{1}{L^\top} v, \quad v = \frac{1}{L} \xi\,. \tag{5.15}$$

Note that ξ is at time $t+1$ all other quantities carrying time index t. To cope with singularities the inverses are understood as pseudoinverses, see Sect. 5.G. Let us concentrate in the following on the explicit dependence of E on the parameters via the Jacobian matrix L. Possible dependencies of ξ on the parameters p will be included later.

[4] Use that $a^\top M b = q^\top b$ with $q = M^\top a$.

5.2.1 Canonical Learning Rule

The gradient step is given in terms of the derivative of the Jacobian matrix w. r. t. any parameter p of the controller as, see Sect. 5.E (p. 103) for details,

$$\Delta p = \varepsilon_c \chi^\top \frac{\partial L}{\partial p} v \,, \tag{5.16}$$

(dropping time indices). **This is the central rule for the parameter dynamics generating the self-determined development of robot behaviors.**

The derivation of the rule is straightforward for the case that the matrix L is regular. A simple trick to avoid singularities would be to add a very small random matrix to L. This preserves the structure of the rule and shows that the singularities show up only in the vectors χ and v. Otherwise, one may use pseudoinverses, see Sect. 5.G or generalizations thereof as introduced in Sect. 15.4.

The general learning rule can be rewritten in many different forms that will be helpful in further developments. Equation (5.16) is written in a more symmetric fashion as

$$\Delta p = \varepsilon_c \chi^\top \frac{\partial Q}{\partial p} \chi \tag{5.17}$$

where

$$Q = LL^\top \quad \text{and} \quad \chi = Q^{-1} \xi \,.$$

Introducing a matrix scalar product between any two matrices A and B as

$$\langle A, B \rangle = Tr\left(A^\top B\right) = \sum_{ij} A_{ij} B_{ij} \tag{5.18}$$

we may also write the learning rule as

$$\Delta p = -\varepsilon_c \left\langle \frac{\partial E}{\partial L}, \frac{\partial L}{\partial p} \right\rangle \tag{5.19}$$

where

$$\frac{\partial E}{\partial L} = -2\chi v^\top \tag{5.20}$$

is called the *primary gradient*. This gradient drives L directly (without taking its parameter dependence into account) by $\Delta L = -\varepsilon_c \frac{\partial E}{\partial L} = 2\varepsilon_c \chi v^\top$ showing explicitly the role of χ and v in the full update step. Equation (5.19) defines the update step as the overlap between two matrices, the primary gradient and the parameter dependent term $\frac{\partial L}{\partial p}$ that will be discussed in the next section.

Equations (5.16) and (5.19) express quite clearly the essentials of the homeokinetic learning approach at a formal level. The parameter dynamics that changes the behavior of the agent is depending on (i) the driving term (5.20) containing the primary gradient of E that augments sensitivity and with it the amplification of perturbations in the physical system; and (ii) on the derivative $\partial L / \partial p$, which is related to the dynamical properties of the agent, in particular to the nonlinearities of both controller and forward model. It is this term that introduces the confinement against the unrestricted growth of the local Lyapunov exponents and that is responsible for the tight coupling between the state and the parameter dynamics, i. e. between the external and the internal world of the brain-body system including the synaptic level (parameter dynamics).

5.2.2 Explicit Learning Rules

Let us now consider the special case of our standard setting as defined in Sect. 5.1.3 with the Jacobian matrix given by Eq. (5.12). In order to formulate the learning rules, we introduce the auxiliary vectors ζ and μ defined as

$$\mu = G'A^\top \chi, \ \zeta = Cv \tag{5.21}$$

with the vectors χ and v given in Eq. (5.15). The learning rules as derived in Sect. 5.F are

$$\Delta C_{ij} = \varepsilon_c \mu_i v_j - \varepsilon_i y_i x_j \tag{5.22}$$
$$\Delta h_i = -\varepsilon_i y_i \tag{5.23}$$

where

$$y = g(z) = g(Cx + h)$$

(with $y \in \mathbb{R}^m$ and $z \in \mathbb{R}^m$) is the output vector of the controller and $\varepsilon_c > 0$ its overall learning rate. Indices are running over the m motors and the n sensors as $i = 1,2,\ldots,m$ and $j = 1,2,\ldots,n$. In practice it is sometimes appropriate to introduce additional damping terms into the rule. Remember, that all quantities carry time index t although v and χ depend on ξ_{t+1} (5.15).

We introduced the channel specific learning rates

$$\varepsilon_l = 2\alpha\varepsilon_c \mu_l \zeta_l \,. \tag{5.24}$$

valid for the tanh activation function, for the general case see Sect. 15.C. We introduced in the derivation given in Sect. 5.F (p. 104) the empirical factor α (default value 1) that is called `sense` in our algorithms, see for instance the `sox` algorithm in Sect. 15.1. By means of α one can define the sensitivity of the robot against variations in the sensor values. In particular, by choosing α one can drive the system into a desired working regime like the one given by the effective bifurcation point. Details may be found in Sect. 6.2 (p. 108), hints for the best choice of α can be found in different applications later in the book. However, in most applications we use the default value $\alpha = 1$ so that we may renounce from a more detailed discussion here.

In a more compact notation, we may write these rules also as

$$\Delta C = \varepsilon_c \mu v^\top - \epsilon y x^\top \tag{5.25}$$
$$\Delta h = -\epsilon y \tag{5.26}$$

the diagonal matrix ϵ containing the channel depending learning rates given by

$$\epsilon_{ij} = \delta_{ij} \varepsilon_i \tag{5.27}$$

with the Kronecker delta δ_{ij}.

With the regularization procedures given in Sects. 5.G and 15.4, these learning rules are valid for any combination of n sensors with m motors. The rules form a concrete example of homeokinetic learning and will be applied in this form (with some modifications) to a large number of different robotic systems. Note that the learning rates are not annealed or kept small in order to enforce convergence. Instead, the deeper sense of homeokinetic learning is in the vivid interplay between the system and the parameter dynamics, both running on comparable time scales.

In Sect. 5.C (p. 101) we provide further expressions of the learning rules derived for the case that the gradient is averaged over the noise.

5.2.3 Self-Actualization, Adaptivity, and Sensitivity — Example

Let us consider as a first illustration the behavior of the TwoWheeled as introduced in Sect. 5.1.4 with homeokinetic learning. Let us use the learning rule Eq. (5.22) for driving the parameter C, initialize C with a subcritical value so that the robot is stalling, stabilizing itself against small fluctuations in the sensor value (the wheel velocity). As seen in Fig. 5.4 or by doing Experiment 5.1, the learning dynamics is rapidly increasing the value of C. When exceeding its critical value, the robot starts to move into one of its possible directions since the parameter is driven towards a region of the then bistable system where the robot is taking decisions, driving either forward or backward (both directions are equally probable so that we actually observe an example of spontaneous symmetry breaking). We may interpret this initial phase as a self-regulation towards activity, calling it (sloppily) a kind of self-actualization.

The interesting situation happens if the robot collides with the wall where it is seen to invert velocity immediately. This rapid switching between the two fixed points of the sensorimotor dynamics is a result of the physical properties. The robot undergoes an elastic collision so that it is reflected from the wall with the effect that the wheels are also inverting their velocity, due to friction effects, at least partly, see also Fig. 3.6.

By this heavy perturbation, the bistable system is being switched, the robot driving to the other wall where the switching repeats. As Fig. 5.4 shows, this working regime of the robot is stable and would last forever, if the conditions were not changed. This specific working regime the robot has self-regulated into is characterized by its ability to take decisions, still remaining sufficiently sensitive to revert

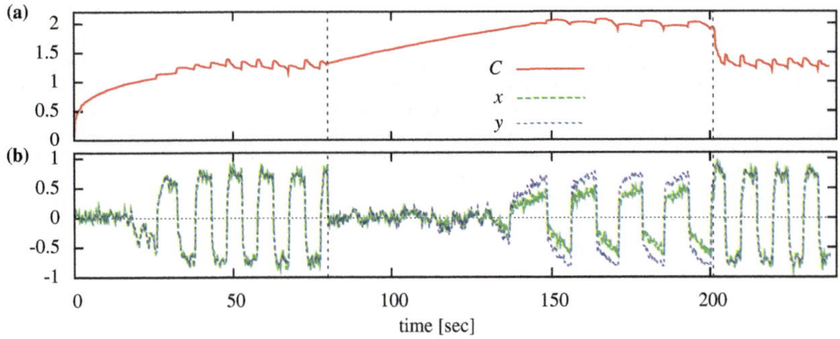

Fig. 5.4: Self-actualization and adaptation by homeokinetic learning. Experiment with TwoWheeled with connected wheels (1 DOF). **(a)** Evolution of the controller parameter C. **(b)** Sensor values x and motor values y over time. After 80 sec the trailer was attached (dashed lines). The robot comes to a rest temporarily but progressively increases its sensitivity until it recovers back to the unloaded behavior. After removing the trailer at time 120 sec the system self-regulates by the homeokinetic learning back to the unperturbed situation very rapidly. Settings: $\varepsilon_c = 1, \varepsilon_A = 0.25$.

decisions if perturbed by interactions with the environment or by changing physical responses of the body.

What happens if the physical properties of the system change suddenly. We model such a situation by attaching a trailer to the robot so that there is a higher friction. As the experiment shows, the sudden additional load has the expected result of lowering the effective feedback in the sensorimotor loop so that the robot stalls again. However, the learning dynamics quite rapidly increases the value of C so that after some time the robot recovers and the original motion pattern is reestablished also in the new physical situation. Moreover, after removing the trailer, the

Video 5.1: Sensitivity and adaptation to environmental changes. The robot starts with a unit initialization, such that it moves only slowly. After a short time the feedback strength has risen sufficiently such that the robot starts moving steadily forward and backward, bouncing at the walls. Then a heavy trailer is connected to the robot. Initially the robot hardly moves anymore, but the feedback strength is adapted so that stronger and stronger actions are performed, see Fig. 5.4. Settings: $\varepsilon_c = 1, \varepsilon_A = 0.25$. The video can be watched at `http://playfulmachines.com`.

Experiment 5.1: Sensitivity and adaptation to environmental changes.

Start the simulation Wheeled robot (1D-case) with trailer. At the beginning there are no obstacles except the walls of the arena. Watch how, after initialization, the robot just fluctuates until the critical value of the feedback strength is reached when the robot starts driving into one direction, eventually bouncing at the wall. Note that the h dynamics is disabled. If the robot is away from the walls press `b` to add a trailer box, which is connected automatically to the robot. Observe how the actions become more and more committed until the behavior without the trailer is being reestablished. You may remove the trailer with B. Use the GUILOGGER to watch evolution of the sensor values, motor values and parameter dynamics of C, and A. You can change the friction of the trailer with the ground by adjusting the parameter `friction` (default=1.8). To reproduce the parameter divergence on the "dark side" you can set C to a negative value by for instance `coupling=-0.5`.

learning dynamics very rapidly self-regulates the system back to the original working regime. This demonstrates the high degree of adaptivity realized by the homeokinetic learning dynamics.

5.2.4 The Homeokinetic BARREL

Before going to discuss the specific properties of the homeokinetic learning rules in detail by both theoretical analysis and practical applications, let us consider a second experiment using the BARREL in order to demonstrate the essential difference to the homeostatic learning considered in Chap. 4, in particular Sect. 4.2.3 (p. 68).

The BARREL, as introduced in Sect. 3.2 and already studied in Sect. 4.2.1, has two motors shifting the internal weights and two sensors measuring the inclination of the internal axes. Hence we put $m = n = 2$ in the above Eqs. (5.22, 5.23) and determine the learning rates empirically to make learning sufficiently fast. When connecting the homeokinetic "brain" to the robot the controller can be initialized such that it stabilizes the robot in a resting position. Then, if the learning is switched on, the feedback strength in the loop is steadily increasing so that the ensuing noise amplification effect creates a kind of restlessness of the robot and after a few seconds it starts rolling. Depending on the initial condition (and the prediction error/noise) the robot develops different motion patterns, in most cases it enters the sweeping mode, which we already observed in the homeostatic setting (Sect. 4.2.3).

The observed destabilization of the stalling mode, clearly demonstrated by Video 5.2, is the novel and important property, which makes the system much richer in the spectrum of behaviors. In the experiments one observes a lot of further active modes beside the sweeping mode. Many of them are long lived transients but sometimes changing between modes must be helped by either re-initializing the parameters or by shaking the robot by external force (see Experiment 7.2). By the mechanical influences one may provoke a drastic change of parameters, as one can observe in the parameter inlets of Video 5.2. It is interesting to see how the system reorganizes in a few seconds or minutes into a new active, seemingly organized mode of behavior.

The behavior of the BARREL, in particular the sweeping mode, will be discussed in more detail in Sect. 7.3.3. Beside rolling, the BARREL will be seen in different

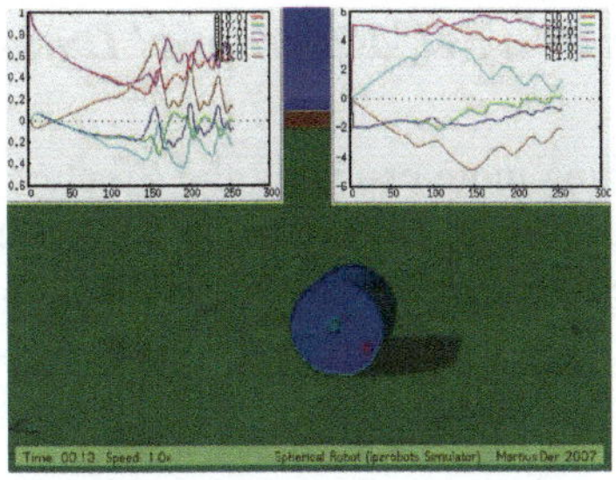

Video 5.2: Destabilization and Emergence. Behavior of the BARREL when starting in a situation where the center of gravity of the BARREL is very low so that the situation is physically stable. The homeokinetic learning is seen to destabilize the system quite rapidly (a few seconds real time), getting the BARREL into moving. The parameter dynamics of both the forward model and the controller can be followed in the panels at the left and right upper corners, respectively. The panels depict the course of the parameters in a time window of 250 steps corresponding to about 10 sec. After some time stable rolling patterns are emerging. Later on (at time 01 : 05) the BARREL was stopped by applying a physical force to it (note the red dot that pulls the robot). After releasing the BARREL, the system recovers again in a very short time and resumes the rolling patterns in a slightly modified form. The video can be watched at `http://playfulmachines.com`.

experiments to produce many other interesting modes. Rolling is the most natural behavior for a barrel when lying on a surface so that obtaining these behaviors is quite "cheap." As we will see in Sect. 8.5 homeokinetic learning is always good for a surprise. In fact we will see there quite unexpected behaviors that disclose a certain amount of creativity of the system in impasse situations. This result is not exceptional. In fact in many of the experiments provided in the following chapters, the emergence of surprising, though body related and seemingly meaningful behaviors can be observed.

5.3 Key Features of Homeokinesis

The interplay between state and parameter dynamics introduced by homeokinetic learning, see Eq. (5.16), is determined by the concrete realization, predominantly by the **structure** and the specific parametrization, of the brain-body system. As it will be seen by the numerous experiments to be presented in the later chapters of this book, there is an overwhelming variety of dynamical patterns emerging from this interplay. Nevertheless, one can see the basic features already at a rather formal level. We are going now to outline the most salient characteristics of the new learning paradigm. The discussions given in the following partly anticipate results of specific examples treated later, in particular in Chap. 6 and Chap. 7, so that they can be skimmed over and reread later.

5.3.1 Homeokinesis as a Self-Referential Dynamical System

Let us start with qualifying more explicitly the structure of the combined system of state and parameter dynamics. Complementing the state dynamics as given by Eq. (5.1) with the gradient rule yields quite generally the complete dynamics of our brain-body system as

$$x_{t+1} = \psi(x_t, p_t) + \xi_{t+1} \tag{5.28}$$

$$p_{t+1} = \Gamma(x_t, p_t) \tag{5.29}$$

where Γ is defined in terms of the gradient on the error landscape as

$$\Gamma(x,p) = p - \varepsilon_c \frac{\partial E(x,p)}{\partial p} . \tag{5.30}$$

This expression emphasizes the direct connection between state and parameter dynamics. We may relate it to the general definition of a controller given by Eqs. (3.4, 3.5) in Sect. 3.1. Noting that $\psi(x) = M(x,y) = M(x,K(x))$ and comparing Eq. (5.29) with Eq. (3.4), we may interpret the parameter vector p (that is implicit in K) as the internal state of the controller and the gradient descent on the TLE as the generating rule for the internal state dynamics.

Whatever the interpretation of the parameter dynamics is, Eqs. (5.28, 5.29) represent an adaptive system, the adaptation being driven by a gradient descent on the objective function E. In the conventional supervised learning setting, the objective is given from outside, E measuring the distance to some predefined learning goal, so that the adaptation means lowering the distance between current and goal behavior. In our case, the objective function E is entirely internal to the agent. In fact, E is a function of the state vector (and the prediction error) since its structure is defined by the Jacobian matrix of the dynamical map ψ. The latter is kept up to date by a simultaneous learning process.

Considering Eq. (5.28) alone, the state dynamics is seen as a dynamical system with parameters p. Dynamical systems theory, in particular the theory of bifurcations, studies the properties of the dynamical system under changes of the parameters. We have given (and will give) examples of bifurcation scenarios in Chap. 3, Chap. 6, and Chap. 7. The new feature introduced by Eq. (5.29) is a dynamical rule that changes the parameters as a function of the state itself. In a sense, the state dynamics starts a self-investigation of its behavior driven by a self-induced variation of its parameters. Once the structure of the objective function E is given, the state dynamics changes its parameters completely autonomously. We call systems of the above kind (Eqs. (5.28–5.30)) that are equipped with a self-driven parameter dynamics *self-referential dynamical systems* since they do everything only in reference to themselves.

5.3.2 Sensitization — The Driving Force of Homeokinesis

A key feature of homeokinetic learning can be already obtained at this level of our approach by considering the dependence of the TLE E in Eq. (5.9) on the Jacobian matrix L, ignoring the concrete dependence of the latter on the parameters of the controller for a while. The specific expression for the TLE is analyzed effectively by means of matrix polar decomposition, putting $L = WU$ where $U^\top = U^{-1}$ and W is symmetric. Then, $LL^\top = W^2$ is symmetric with positive eigenvalues, assuming L is regular. Writing the eigenvalues as $e^{2\lambda_i}$, we may identify the λ_i as the local Lyapunov exponents that quantify the stability of the system in the direction of the respective eigenvector over one time step. In fact (assuming all eigenvalues different for simplicity), over one time step, the separation between two neighboring trajectories expanding if $\lambda_i > 0$ and contracting if $\lambda_i < 0$. Using $\left(LL^\top\right)^{-1} = W^{-2}$ we get

$$E = \sum e^{-2\lambda_i} \hat{\xi}_i^2$$

where $\hat{\xi}_i$ is the projection of ξ onto the eigenvector of λ_i. With nonzero noise in all channels, E is minimal if the local Lyapunov exponents $\lambda_i \to \infty$ for all i.

Gradient descending the error w. r. t. L would lead to the total destabilization of the system so that we must not consider the system without the confining effects caused by the potential nonlinearities in the dependence of L on its parameters. In order to get an explicit result, we can mimic this effect by adding some penalty term to the error E, for instance we may use the Frobenius norm of the matrix L so that

$$\begin{aligned} E &= \xi^\top \frac{1}{LL^\top} \xi + \gamma Tr\left(L^\top L\right) \\ &= \sum \left(e^{-2\lambda_i} \hat{\xi}_i^2 + \gamma e^{2\lambda_i}\right) . \end{aligned} \tag{5.31}$$

All terms are positive and independent so that the sum is minimal if each individual term is minimal, which is at

$$2\lambda_i = \ln \sqrt{\hat{\xi}_i^2 / \gamma} \tag{5.32}$$

for all i. In the frame of reference spanned by the eigenvectors, each eigenvalue defines the feedback strength in the sensorimotor loop in the respective dimension and thus is a measure for the sensitivity (the degree of noise amplification) in that dimension. Hence, with i. i. d. noise this result demonstrates that the minimization of the TLE leads to the equilibration of all modes of the system at the maximum level of sensitivity.

Equation (5.32) also shows that, if the noise strengths differ across channels, the destabilization by noise amplification is more effective in those dimensions where the prediction error ξ is high. Noise amplification is, in this setting, responsible for the explorative nature of the system so that exploration is being more enhanced the noisier the channel. The actual flavor of the method, however, develops only if we include the nonlinearities directly. This complicates the whole analysis so that we will start with the elementary sensorimotor loop of the one-dimensional case in Chap. 6.

5.3.3 State-Parameter Interplay

The crucial point of homeokinetic learning is the interplay between the learning dynamics (the gradient descent on the TLE) and the state dynamics. As seen in Sect. 5.3.2, the learning dynamics tries to destabilize the state dynamics as much as possible, increasing thereby the sensitivity of the system against external perturbations. This tendency would drive the system into chaos. This unrestricted growth of the sensitivity is counteracted by the nonlinearities of the system, introduced for instance by the saturation of the neurons in the simple cases considered explicitly in Chap. 3 and Sect. 5.2.2.

The two tendencies can be read off directly from the homeokinetic objective, as given by the TLE in the form of Eq. (5.19), the factors of the scalar product representing the two opposing tendencies: while the first factor, the primary gradient, is aiming at increasing the sensitivity, the second factor containing the derivatives of L is causing the confinement effect by the nonlinearities. In a similar way, the two tendencies can also be recognized in the learning rules, see Eq. (5.22), corresponding to the concrete parametrization given by Eq. (5.11). Here, the update is given by the sum of two terms, the first one driving the sensitivity whereas the confinement is realized by the second term.

It is important to note that the confinement effect depends on the concrete parametrization of the system in a crucial and intricate way—with a different parametrization the self-referential system (state-parameter dynamics) will develop vastly different behaviors. Despite this variety, there are two fundamental scenarios for the interplay between state and parameter dynamics that can easily be identified. The first scenario may be called the fast-slow regime discussed in the following section and the second one is the entanglement regime introduced in the subsequent section.

5.3.3.1 The Fast-Slow Regime

The role of the parametrization can be read off from the bifurcation diagram for the system. The first case we consider corresponds to a system with fixed point attractors and a bifurcation scenario characterized by bifurcation points with maximum instability of the state dynamics. A generic example is the pitchfork bifurcation treated in Sect. 3.3.2 (p. 36). Right at the bifurcation point of the pitchfork bifurcation the stability is minimal so that the gradient descent will drive the parameters towards that point as illustrated by Fig. 5.5.

The speed of the parameter dynamics is regulated by the learning rate ε_c. If that one is not taken too large;[5] the state dynamics may still be faster than the parameter one so that there is a separation of time scales **emerging** in this specific situation. On the slow time scale the parameters change but before there is a substantial change,

[5] Remember that an essential feature of homeokinesis is that the learning rate is being kept at a fixed value, usually quite large.

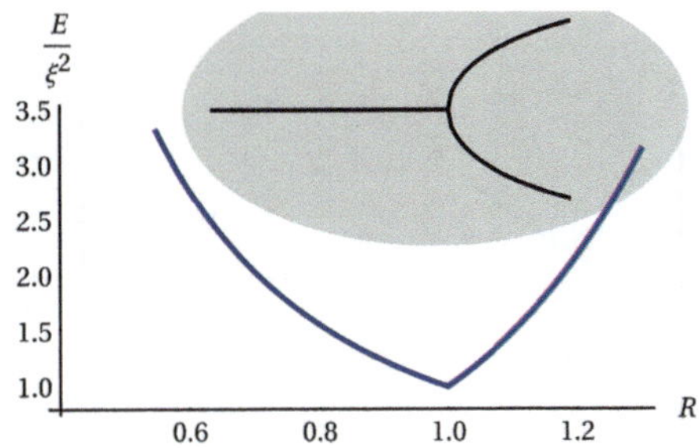

Fig. 5.5: Key features of homeokinesis — the fast-slow regime. If the original system (without learning) has a point of maximum sensitivity like the bifurcation point of a pitchfork bifurcation (inset), the gradient descent on the time-loop error is driving the system to that point. Case of infinitesimal noise corresponding to the limit of very large values for the α parameter in the learning rule.

the state is already more or less converged to the current fixed point. In this so called fast-slow regime the combined dynamics given by Eqs. (5.28, 5.29) is obviously characterized by a dominance of the parameter dynamics over the state dynamics.

These considerations are valid as long as the noise (the prediction error) is infinitesimally small. In the realistic situation with finite noise, as shown in Sect. 6.2.2, the system is driven towards the effective bifurcation point – the point of maximum sensitivity in a noisy, bistable system. In the concrete learning rules, see Eqs. (5.22, 5.23), this is achieved by choosing the `sense` parameter α appropriately.

Systems with different time scales play an essential role in the theory of synergetics [60, 61, 63] where the elimination of fast variables and the so called slaving principle are key features. In that terminology one might say that the parameter dynamics is slaving the system dynamics in this specific regime.

From the more mathematical point of view there is a body of literature on the identification of the so called slow and fast manifolds. This is the topic of geometric singular perturbation theory, which has been put forward mainly by Neil Fenichel [51]. The present dynamical regime could be formulated in terms of that theory in more mathematical terms but this must be left to future work.

5.3.3.2 Entanglement of State and Parameter Dynamics

While the fast-slow regime is exemplified by the pitchfork bifurcation system, the regime to be considered now is typical for the hysteresis system considered in Sect. 3.4 (p. 45) with catastrophic bifurcation points. In the hysteresis regime (with fixed, positive feedback strength, i. e. $R > 1$), the bifurcation diagram is characterized by a saddle-node or catastrophic bifurcation. Let us consider here the case that the hysteresis parameter h is the only parameter driven by the learning rule. The corresponding bifurcation diagram is depicted in Fig. 5.6, which suggests the following scenario. When on or close to one of the stable branches, the destabilization

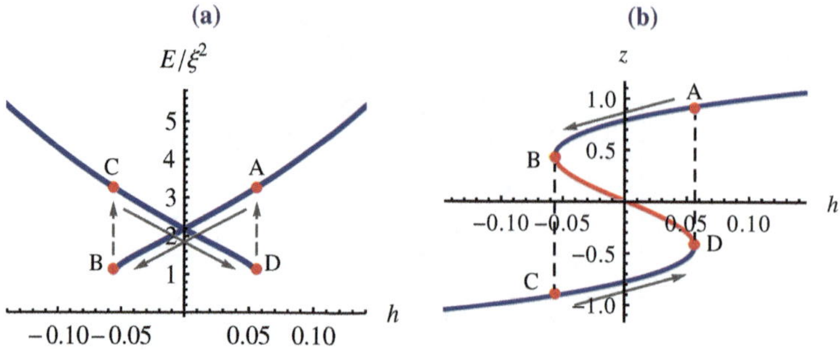

Fig. 5.6: Key features of homeokinesis — entanglement. The interplay between parameter and state dynamics in a hysteresis system with $R = 1.2$ (finite noise case). In (a) the normalized time-loop error E/ξ^2 is depicted as a function of the hysteresis parameter h. Starting in a situation A, the gradient descent drives h towards regions of less stability until in situation B the fixed point disappears altogether, see also Figure (b). In B, the parameter dynamics looses control over the state dynamics and the state jumps back to a fixed point of higher stability. There, the scenario restarts. The essential difference to the fast-slow regime, see Fig. 5.5, consists in the fact that there is no convergence, the state and parameter dynamics entertaining instead a persistent oscillatory pattern. See the text for more details and references to the relevant content of later sections.

effect drives the parameter towards the catastrophic bifurcation point (essentially in accordance with the fast-slow regime discussed above). There the system dynamics loses stability and jumps to the other branch, where the scenario restarts.[6]

In this way, there is no convergence to any definite fixed point and/or an emerging separation of time scales. Instead, this entanglement regime corresponds to a mixing of time scales and is characterized by the competition between an ongoing destabilization (by the parameter dynamics) with increasing resistance of the state dynamics and its eventual breaking out by jumping to another regime that is more stable. This may happen if (i) either the destabilization destroys an attractor the system was in before, see Fig. 5.6 and Fig. 5.7 for a more detailed picture of the error evolution; or (ii) if the parameter dynamics looses its influence on the system dynamics because the latter is getting more and more complex due to the destabilization process. This jumping effect leads to oscillatory patterns with frequencies depending on the learning rate ε in a definite way.

This regime is characterized by the fact that the TLE never converges but keeps oscillating with large jumps if the system "breaks out" from getting more and more instable. The entanglement scenario is controlled by the learning rate ε and is breaking down completely if the parameter dynamics is switched off ($\varepsilon = 0$). We will give in the following chapters numerous examples demonstrating how the involvement of the parameter dynamics is constitutive for the behavior.

---

[6] The situation with both R and h as parameters is analyzed in detail in Sect. 6.3 (p. 112) and is seen to reproduce the one parameter case once R has self-regulated into the hysteresis regime.

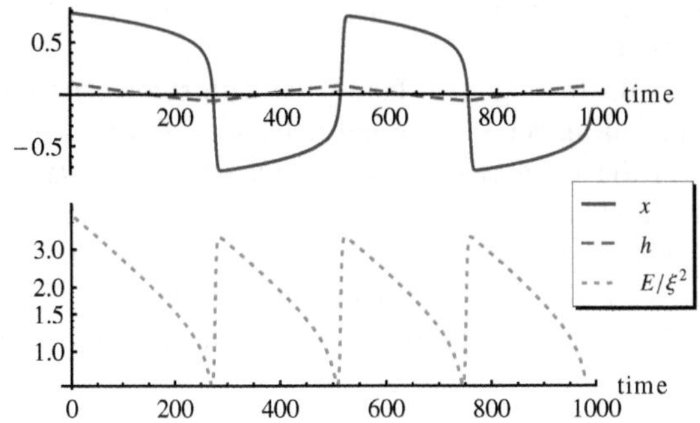

Fig. 5.7: Time-loop error and hysteresis oscillation. In a hysteresis oscillation, see Fig. 5.6, the time-loop error is not converging. When in the limit cycle, the error decays as long as the dynamics is essentially a fixed point flow so that the combined dynamics is dominated by the gradient descent in the parameter space. The jumps to higher error values occur if the current fixed point is destroyed and the state jumps to the other fixed point. These jumps against the error minimization tendencies are a consequence of the hysteresis properties of the sensorimotor loop and may generally occur in systems with catastrophic bifurcations. The figure depicts the behavior of the time-loop error E, the state x and the hysteresis parameter h. The value of the learning rate is $\varepsilon = 0.001$, the fixed value of $C = 1.2$ and the initial conditions are $x_0 = 0.8$, $h_0 = 0.1$. Details may be found in Sect. 6.3 (p. 112).

This specific feature is dominating the behavior for the high dimensional systems considered in Chap. 10 where the combined dynamics is a dynamical system of several hundred dimensions.

5.3.4 Temperature Effect

The TLE depends in a complicated way on the state x of the system, the parameters p of the controller, and the prediction error (noise) ξ. Seen as a landscape over the parameters of the controller alone, the error landscape is fluctuating due to the dynamics of the noise and the state itself. The fluctuations due to the state dynamics are very different in nature and strength. It can be shown analytically that a dynamics which is invariant against special symmetry operations will leave the landscape more or less invariant, see next section and Fig. 5.8. These specific dynamical patterns correspond to large plateaus of the landscape, mostly minima, and are therefore preferred in the development of the system. We will encounter many examples for that scenario.

The fluctuations due to the noise are of a different nature; they are driven most often directly by a large prediction error ξ due to a collision or other unpredictable perturbations, see for instance the example of a robot colliding with a wall as in

Fig. 3.6. The fluctuations of the error may cover one or two orders of magnitude in the different situations.

The picture of a fluctuating error landscape with some calm regions of favorite behaviors is helpful in understanding the behaviors observed in the experiments. Stipulating that in the large error regions with strong fluctuations gradients are rather large and of high directional variance, one may view the gradient flow on the TLE landscape more like a diffusion on a surface with fluctuations corresponding to widely varying temperatures, hot regions being left more rapidly in favor of the more calm or cooler ones where the system may settle down in a kind of dynamical equilibrium.

In our high-dimensional systems with several hundred parameters, we have made the astonishing observation that the picture is relevant and can be used to understand the emergence of low dimensional modes corresponding to active behavior with high sensorimotor coordination. If so, there must be simple reasons behind that scenario. One reason we see is in the very foundation of the principle itself. We have defined an objective function that is domain invariant but manages, due to the sensitization effect, to introduce a general drive for activity. Being completely unbiased and self-referential with everything emerging out of the physical dynamics itself, the combined system may be assumed to settle into the most simple solutions, corresponding to a kind of dynamical synergy between the large number of degrees of freedom in parameter space with those of the physical space. This scenario is additionally supported by the dominating influence of the symmetries inherent in the physical system.

*5.3.5 Symmetry Groups as Attractors**

One of the most prominent observations made with real systems under homeokinetic control is the emergence of low-dimensional modes in high-dimensional systems, manifest in particular by the predominance of oscillatory motion patterns. We argue that the ultimate reason for this phenomenon is rooted in the symmetry properties of the TLE, which structure the error landscape in a characteristic way, fostering thereby the phenomenon of spontaneous symmetry breaking. Let us consider the TLE defined in Eq. (5.9) for a state dynamics in n dimensions. The TLE

$$E = v^\top v = \xi^\top \frac{1}{LL^\top} \xi$$

is defined by the Euclidean norm so that it is invariant against rotations in the n-dimensional space. Let us consider any transformation O of the orthogonal group and replace L with $O^\top L O$ so that

$$E = \xi^\top O^\top \frac{1}{LL^\top} O \xi$$

by virtue of $O^\top = O^{-1}$. Assuming moreover that the noise is invariant under O so that any realization ξ is as probable as $O\xi$ (this is justified for instance if the noise is i. i. d.) we find that, in the stochastic sense, the TLE is invariant under the replacement of L with its transformed $O^\top LO$. As a consequence, the TLE has, in the stochastic sense, the same value for all L fulfilling the symmetry condition

$$L = O^\top LO \tag{5.33}$$

for any O of the symmetry group or a subgroup thereof. As a consequence, the symmetry group defines large plateaus in the landscape of the TLE that belong to equivalent matrices L with related system behavior.

By way of example, we consider the two-dimensional case and assume that O is a matrix

$$O(\phi) = \begin{pmatrix} \cos\phi & -\sin\phi \\ \sin\phi & \cos\phi \end{pmatrix}$$

of the special orthogonal group[7] $SO(2)$. The symmetry condition Eq. (5.33) reads now

$$LO(\phi) = O(\phi)L \tag{5.34}$$

which is fulfilled if L is a scaled rotation matrix since all the orthogonal matrices commute in $2D$. This means in the linear case (where $\psi(x) = Lx$) that the plateau in the error landscape is hosting all oscillatory motion patterns with a fixed but arbitrary frequency. In more dimensions the scenario is a little different since not all orthogonal matrices commute if the dimension is larger than two. However, in the $3D$ case for instance we can describe each $SO(3)$ matrix as a rotation about a fixed axis. Then, the above arguments repeat with respect to the subgroup of all rotations about that axis. In general, one may always be able to find subgroups which leave the error E invariant corresponding to large plateaus of equivalent behavior.

Another example may demonstrate that the symmetry groups do not only define just plateaus but real valleys in the error landscape. Consider the matrix C of the controller given by

$$C = \begin{pmatrix} \cos\alpha & -\sin\beta \\ \sin\beta & \cos\alpha \end{pmatrix} . \tag{5.35}$$

Figure 5.8 clearly demonstrates that, if $\alpha \overset{\mathrm{mod}\,\pi}{=} \beta$, there is a deep valley in the error landscape showing that a gradient descent in the α, β space would drive the system into a pure oscillatory dynamics indeed.

Noteworthy, this is another example where the parameter dynamics converges to a stable attractor (actually a metastable one, see Sect. 7.2.2 (p. 134)) for details, similar to the fast-slow case considered above with the difference that the attractor now is a limit cycle instead of a fixed point. A further difference is seen in the fact that there is a continuum of states of the same minimal error value instead of just one, the pitchfork bifurcation point in the former setting.

---

[7] The $SO(2)$ is the set of all orthogonal matrices with determinant equal to $+1$. It is only these matrices that describe pure rotations.

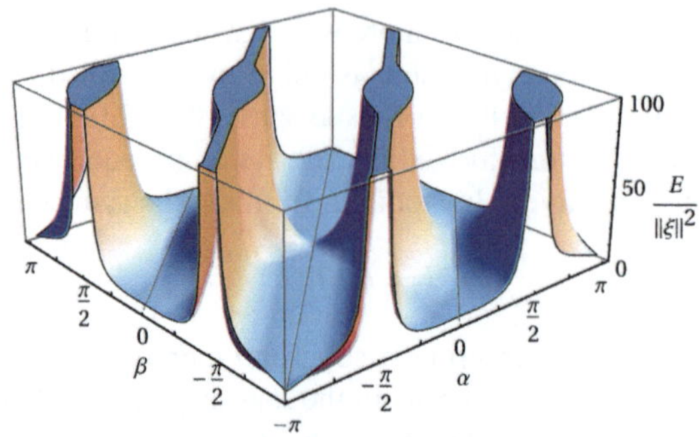

Fig. 5.8: Key features of homeokinesis — the role of symmetries. The normalized error landscape of a two-dimensional system with the matrix of the controller C being parameterized by two parameters α and β, see Eq. (5.35). For $\alpha \overset{\mathrm{mod}\,\pi}{=} \beta$ the controller matrix is an $SO(2)$ matrix (shown with gray lines). The message is that rotation matrices supporting oscillatory behavior correspond to deep valleys of the error landscape.

It is important to note that the symmetry condition poses a direct requirement only to L, see Eq. (5.34). In the general case, L may depend in a complicated way on the parameters so that the findings on L must be translated into the "language" of the parameter space before being able to draw conclusions on the influence of the symmetry condition on the dynamics of the system.

This has a direct bearing on nonlinear systems where the translation depends on the current state x_t of the system since the Jacobian matrix L is a function of x_t that changes in time. However, let us consider a special subgroup $\mathbf{G}$ that leaves the TLE invariant in the sense of Eq. (5.33). Let us consider moreover a sequence of states $x_t, \bar{x}_{t+1}, \bar{x}_{t+2}, \ldots$ chosen such that you can always find an element G of the group so that, for any transition $\bar{x}_{t'} \to \bar{x}_{t'+1}$ between two subsequent states, the symmetry condition

$$L(\bar{x}_{t'+1}) = G^\top L(\bar{x}_{t'})\, G$$

is fulfilled for a specific set of parameters p of the Jacobian matrix. This means that (i) the TLE is invariant under this dynamics and that (ii) this specific dynamics belongs to the plateau of the TLE landscape defined by the symmetry group $\mathbf{G}$. In other words, there is an increased probability that this dynamics is discovered and realized by the homeokinetic learning. A special example of this scenario is the emergence and high stability of period-4 cycles in two-dimensional systems, see Sect. 7.2.1 (p. 132), and the corresponding higher order cycles in higher dimensional systems.

Quite generally, combined with the temperature effect introduced in Sect. 5.3.4 (p. 95), these plateaus play an essential role in understanding the amazing fact of the emergence of low-dimensional modes in high dimensional systems and the preference of periodic motion patterns in all systems realized so far in the applications, see Chap. 7 for properties of idealized systems, Chap. 8 for low-dimensional, and Chap. 10 for the high-dimensional robotic examples.

5.3.6 Spontaneous Symmetry Breaking

The invariance properties reveal a most interesting similarity between the behavior of homeokinetic learning systems and the general scenario known from self-organizing systems in nature. As discussed in more detail in Chap. 2, in nature we generically have on the one hand a force that tries to bring the system state to maximal symmetry. On the other hand, the symmetry conserving tendencies are counteracted by self-amplification mechanisms that blow up local microscopic fluctuations to a macroscopic level.

The above considerations have shown that the landscape of the homeokinetic objective function (the TLE) is characterized by vast plateaus reflecting the invariance of the TLE under groups of symmetry operations. Prominent example is the special orthogonal group $SO(n)$ (or subgroups thereof) which reflects the isotropy of the state space, i. e. its invariance under rotations. Partly in combination with the temperature effect, these plateaus may attract the system towards dynamical patterns of high symmetry. This corresponds to the symmetry conserving drives observed in nature. On the other hand, we know that the homeokinetic learning also increases the instability in the system. In fact, it is designed to minimize the TLE, which is done best if the controller enables the system to most sensitively react to its sensor values, see Sect. 5.3.2. In the sensorimotor loop, this corresponds to the self-amplification mechanism.

In this way we register a close correspondence to systems in nature with something like 10^{23} dimensions that are known to generate low dimensional patterns in space and/or time of high regularity. The high regularity of these patterns, observed for instance in the hexagonal structures in so many natural phenomena, may be attributed to a kind of economy in symmetry breaking, i. e. symmetries are broken preferentially in the least possible way, conserving as many as possible of the pertinent invariance properties. An impression of this economy principle can be gained by the applications treated in later chapters, see for instance the example of the SPHERICAL in a circular basin in Sect. 8.2.3 (p. 165).

Appendix 5.A Formalizations

The time-loop error (TLE) E defined by Eq. 5.9 (p. 79) can be rewritten in many different ways that will prove helpful in the upcoming investigations. Often it is helpful to use the trace operation to simplify derivations, writing for instance $v^\top v = Tr\left(vv^\top\right) = \sum_{i=1}^{n} v_i^2$ for any vector $v \in \mathbb{R}^n$. Using the cyclic invariance property of the trace, we may write the TLE as

$$E = Tr\left(vv^\top\right) = Tr\left(\frac{1}{L}\xi\xi^\top\frac{1}{L^\top}\right) = Tr\left(\xi\xi^\top\frac{1}{LL^\top}\right) . \tag{5.36}$$

This is helpful when taking the average over the noise. Assuming the average[8] $\langle \xi \rangle = 0$ (this is reasonable since any systematic dependence on ξ can be absorbed into ψ), we write the covariance matrix of the noise as

$$D = \left\langle \xi \xi^\top \right\rangle \tag{5.37}$$

so that the average of E becomes

$$\langle E \rangle = Tr\left(\frac{1}{L} D \frac{1}{L^\top}\right) . \tag{5.38}$$

We introduce a scalar product between two $n \times n$ matrices A and B as[9]

$$\langle A, B \rangle = Tr\left(A^\top B\right) = \sum_{ij} A_{ij} B_{ij} . \tag{5.39}$$

We may generalize this by introducing a metric by means of any positive symmetric matrix M, writing

$$\langle A, B \rangle_M = Tr\left(A^\top M B\right) \tag{5.40}$$

so that

$$\langle E \rangle = \left\langle \frac{1}{L}, D \frac{1}{L} \right\rangle = \left\langle \frac{1}{L}, \frac{1}{L} \right\rangle_D = \left\| \frac{1}{L} \right\|_D^2 \tag{5.41}$$

with the matrix norm $\|.\|_D^2$ defined by the scalar product $\langle .,. \rangle_D$.

In the case of i. i. d. noise, the covariance matrix is defined as $D = \sigma^2 \mathbb{I}$ so that

$$\langle E \rangle = \sigma^2 \left\| \frac{1}{L} \right\|^2 \tag{5.42}$$

with $\|.\|^2 = \langle .,. \rangle$, correspondingly. This case is of general importance since the noise ξ can always be brought to this standard form by a convenient coordinate transformation in state space. Note however that in practice the covariance matrix may rapidly change in time so that the coordinate transformation is dynamical.

Appendix 5.B Effective States

In this section we will discuss how the approximation leading to Eq. (5.8) can be improved or even made exact. For that let us introduce the mean value theorem. It

[8] Note the difference between the average of a quantity η written as $\langle \eta \rangle$ and the scalar product between two matrices A and B written as $\langle A, B \rangle$.

[9] The idea is to interpret the elements M_{ij} of a matrix M as the elements m_α of a column vector m so that the scalar product between matrices is reduced to the normal scalar product between vectors.

states that under very mild assumptions the expansion of a function $f:\mathbb{R}^1 \to \mathbb{R}^1$ can be written as

$$f(s+a) = f(s) + f'(\bar{s})\,a \tag{5.43}$$

with $\bar{s}$ somewhere between s and $s+a$. This can immediately be translated to the multi-dimensional case.

Transferred to our case, in Eq. (5.8) we have to choose an *effective state* $\bar{x}$ in the Jacobian matrix so that the input shift v becomes

$$v = \frac{1}{L(\bar{x})}\xi_{t+1}\,.$$

If chosen correctly v may even become the exact solution of Eq. (5.2). However, it is of course not trivial to find the appropriate effective state. In our elementary setting one might use the explicitly known inverse of (tanh) and thus calculate the exact effective state, as done in [102], but this is extremely complicated in more general cases. Moreover, in the applications it appeared not to be so important, such that we do not employ it here, see also the discussion below Eq. 5.7 (p. 78).

Interestingly, the effective state scheme can be adapted consistently to many different situations by using the effective state as a way to deal with more global issues instead of enhancing the approximation in each individual time step. An important example is given by the effective state under strong noise in the one-dimensional sensorimotor loop discussed in Sect. 3.3.4 (p. 38). As we will demonstrate in Sect. 6.2.2 (p. 110), the use of an effective state, obtained by averaging in a sliding time window, shifts the minimum of the TLE from the classical towards the effective bifurcation point. This is an important feature given the very favorable properties of the sensorimotor dynamics when at the effective bifurcation point.

Quite generally, we prefer the definition of the TLE as given by Eq. (5.8) over the rigorous but often rather useless exact mathematical one. Among many other advantages, the definition is singled out by utmost simplicity, which is a very important point for the application to the high dimensional systems studied in Chap. 10. In fact, we have created in this way a pseudolinear structure that can be made more exact if desired by just using an effective state x in Eq. (5.8).

Appendix 5.C Average Gradient

In many cases, the learning rate is small enough so that there is a certain self-averaging of the noise in our on-line gradient descent rules. We can make this explicit by directly taking the average over the noise ξ in the gradient expressions. By explicitly averaging the gradient, we get more direct information on the nature of homeokinetic learning. This point will be exploited in Chaps. 6 and 7 in order to derive basic effects in homeokinetic learning in idealized worlds.

The noise ξ enters only the primary gradient introduced in Eq. (5.20) with average

$$\left\langle \frac{\partial E}{\partial L} \right\rangle = -2\left\langle \chi \nu^\top \right\rangle = -2\frac{1}{LL^\top} D \frac{1}{L^\top} . \tag{5.44}$$

the covariance matrix D being defined in Eq. (5.37). Introducing this expression into the general rule Eq. (5.19) immediately yields the average gradient. In particular, the explicit learning rules Eqs. (5.25, 5.26) become now[10,11]

$$\Delta C = \varepsilon_c G' A^\top \frac{1}{LL^\top} D \frac{1}{L^\top} - \epsilon y x^\top \tag{5.45}$$

$$\Delta h = -\epsilon y \tag{5.46}$$

where ϵ is a diagonal matrix defined by its matrix elements as

$$\epsilon_{ij} = 2\varepsilon_c \alpha \delta_{ij} \left(A^\top \frac{1}{LL^\top} D \frac{1}{A^\top} \right)_{ij} . \tag{5.47}$$

In the special but important case that the noise is i. i. d. so that $D = \sigma^2 \mathbb{I}$, we obtain the learning rule as

$$\Delta C = \varepsilon G' A^\top \frac{1}{L^\top L L^\top} - \epsilon y x^\top \tag{5.48}$$

$$\Delta h = -\epsilon y \tag{5.49}$$

with

$$\epsilon_{ij} = 2\varepsilon \alpha \delta_{ij} \left(A^\top \frac{1}{LL^\top} \frac{1}{A^\top} \right)_{ij} \tag{5.50}$$

and

$$\varepsilon = \varepsilon_c \sigma^2 , \tag{5.51}$$

ε may be called a normalized learning rate. In particular, we can infer the limit of vanishing noise by adapting ε_c as

$$\varepsilon_c = \frac{\varepsilon}{\sigma^2} \tag{5.52}$$

with ε kept fixed. In this way, we can study the most typical behaviors with the learning dynamics at a finite speed whereas the noise in the state dynamics is equal to zero so that the **combined dynamics** (comprising state and parameters) **is deterministic**. This procedure can be readily generalized to arbitrary noise by considering σ^2 as a common factor for all noise strengths in the different channels.

---

[10] We can directly derive the learning rule from Eqs. (5.25–5.27) by writing the terms $\langle \mu \nu^\top \rangle = G' A^\top \langle \chi \nu^\top \rangle = -G' A^\top \left\langle \frac{\partial E}{\partial L} \right\rangle$ and $\langle \mu_i \zeta_i \rangle = (G' A^\top \langle \chi \zeta^\top \rangle)_{ii} = \left(G' A^\top \frac{1}{LL^\top} \langle \xi \xi^\top \rangle \frac{1}{G' A^\top} \right)_{ii}$ by means of Eq. (5.24). The common factor of 2 is absorbed into the learning rate ε_c.

[11] Use pseudoinverses if necessary.

Appendix 5.D Remarks on Reconstruction vs. Postdiction

Let us discuss the different definitions of the TLE given above on a more general level. Actually, we are free to define the structure of the dynamics model rather arbitrarily. Let us discuss this point in terms of the so called generative models for a (stationary) stochastic process $\{X_t\}_{t=0}^{\infty}$. Any realization $\{x_t\}_{t=0}^{\infty}$ can be represented by a so called generative model [127] as

$$x_{t+1} = \Phi(x_t, \rho_t)$$

with ρ_t being the realization of another stochastic process ρ_t that can be considered as the noise. Actually, the structure of the function Φ can be chosen more or less arbitrarily, its choice being reflected in the properties of ρ. If the modeling function Φ is chosen reasonably, the noise will be small and as "noisy" as possible, meaning that both its size and the dependence on the state vector x_t is minimized. This leads us to our second remark concerning the relation to what is usually understood by postdiction. Quite generally, postdiction is the task of giving an estimate of previous instances of a time series in terms of the current one. The standard approach is given by learning a function μ (that might contain internal states) so that

$$x_t = \mu(x_{t+1}) + \rho_t \tag{5.53}$$

with ρ "small and noisy." In terms of μ one can model the behavior of the system backward in time. However, if the process X_t is stationary the estimated value $\hat{x}_t = \mu(x_{t+1})$ is qualitatively different from the reconstructed value $\psi^{-1}(x_{t+1})$. The reason is that, while ψ represents the systematic part extracted by following the dynamics forward in time, the function μ represents the systematic part backward in time. However, assuming detailed balance, see [55], the process looks the same forward and backward in time. Technically, this means that μ is **not** equal to ψ^{-1} so that postdiction in the sense of Eq. (5.53) is different from our procedure of reconstruction as inverted prediction.

Appendix 5.E Derivation of Eq. (5.16)

We need as a first step the derivative of the inverse matrix L^{-1}. Considering that $LL^{-1} = \mathbb{I}$ we find

$$0 = \frac{\partial}{\partial p}\left(L\frac{1}{L}\right) = \left(\frac{\partial L}{\partial p}\right)\frac{1}{L} + L\frac{\partial}{\partial p}\frac{1}{L}$$

so that

$$\frac{\partial}{\partial p}\frac{1}{L} = -\frac{1}{L}\left(\frac{\partial L}{\partial p}\right)\frac{1}{L}.$$

As the next step we consider for any two vectors a and b

$$a^\top \left(\frac{\partial}{\partial p} \frac{1}{LL^\top} \right) b = 2a^\top \frac{1}{L^\top} \left(\frac{\partial}{\partial p} \frac{1}{L} \right) b = -2a^\top \frac{1}{LL^\top} \left(\frac{\partial L}{\partial p} \right) \frac{1}{L} b$$

where we used that $a^\top Q^\top b = b^\top Q a$ for dealing with the inverse of $L^\top$. Using the gradient in Eq. (5.13) and absorbing the factor 2 into the learning rate yields Eq. (5.16).

Appendix 5.F Derivation of Eqs. (5.22, 5.23)

In order to derive Eq. (5.22) we start from Eq. (5.16), use L from Eq. (5.12), and consider

$$\begin{aligned} \chi^\top \frac{\partial L}{\partial C_{ij}} v &= \chi^\top A G' \frac{\partial C}{\partial C_{ij}} v + \chi^\top A \frac{\partial G'}{\partial C_{ij}} C v \\ &= T_1 + T_2 \,. \end{aligned}$$

The first term is treated easily, writing it with $\chi^\top A G' = \mu^\top$ as

$$T_1 = \mu^\top \frac{\partial C}{\partial C_{ij}} v = \sum_{kl} \mu_k \frac{\partial C_{kl}}{\partial C_{ij}} v_l = \mu_i v_j$$

In order to evaluate the second term we have to consider that G' is a diagonal matrix, $G' = \operatorname{diag}[g'(z_1), \ldots, g'(z_m)] = \operatorname{diag}[g'_1, \ldots, g'_m]$ so that with $z = Cx + h$ and

$$\frac{\partial z_k}{\partial C_{ij}} = \delta_{ik} x_j \;\text{ and }\; \frac{\partial g'_k}{\partial C_{ij}} = \delta_{ik} g''_i x_j$$

we get

$$T_2 = \chi^\top A \frac{\partial G'}{\partial C_{ij}} C v = \left(\chi^\top A \right)_i g''_i (Cv)_i x_j \,.$$

In the case of the tanh activation function we have $g''_i = -2 g_i g'_i = -2 y_i g'_i$ so that

$$T_2 = -2 \left(\chi^\top A G' \right)_i (Cv)_i y_i x_j = -2 \mu_i \zeta_i y_i x_j \tag{5.54}$$

with $\zeta = Cv$.

The derivative w. r. t. h is obtained as

$$\chi^\top \frac{\partial L}{\partial h_i} v = \chi^\top A \frac{\partial G'}{\partial h_i} C v = \left(\chi^\top A \right)_i g''_i (Cv)_i \,.$$

In the tanh case this becomes

$$\chi^\top \frac{\partial L}{\partial h_i} v = -2 \mu_i \zeta_i y_i \,.$$

Actually, the derivation is not yet complete since the state x featuring in L is in general also depending on the parameters of the controller. This is obvious for a fixed point system with the fixed points being given by $x = \psi(x)$ or, in the specific setting,

$$x = Ag(Cx + h)$$

ignoring the noise (prediction error). We have analyzed this system for the special case of the pitchfork bifurcation in Sect. 3.3.2 and found that already in this elementary case the dependence of the state on the parameters is very complicated so that a general rule for these effects is prohibitively complicated in the general case.

Therefor we need an approximate rule for considering these effects. In Sect. 6.2 we analyze the situation and give arguments for lumping the effects of the state dependence into an empirical factor modifying the strength of the term T_2, i. e. we rewrite Eq. (5.54) as

$$T_2 = -\varepsilon_i x_i y_j$$

with the channel dependent learning rates ε_i defined now as (in the case of the tanh activation function)

$$\varepsilon_i = 2\alpha\mu_i\zeta_i$$

where α is the empirical factor (default value 1) weighting the T_2 term.

Appendix 5.G Moore-Penrose Pseudoinverse

The pseudoinverse yields solutions of ill posed linear inverse problems. Given two vectors $u \in \mathbb{R}^m$ and $v \in \mathbb{R}^n$, the inverse problem of finding v as the solution of the equation $u = Qv$ where $Q \in \mathbb{R}^{m\times n}$, is ill posed if $m \neq n$. This linear system of equations is overdetermined if $m > n$ (more equations than unknowns) so that there is no solution in general. However, we can ask for some best solution, usually this is the one fulfilling the least squares criterion that $\|u - Qv\|$ is minimal,

$$\|u - Qv\| \stackrel{!}{=} min$$

More generally, we can impose the additional condition that the solution is of minimal norm, corresponding to the objective function

$$\|u - Qv\|^2 + \lambda \|v\|^2 . \tag{5.55}$$

The minimum fulfills

$$Q^\top (u - Qv) - \lambda v = 0$$

with solution

$$v_\lambda = \frac{1}{Q^\top Q + \lambda \mathbb{I}} Q^\top u .$$

With finite λ we have a regularized inverse, which is always existing. In the limit, this solution is written as $v = Q^{+}u$ with the pseudoinverse Q^{+} defined by

$$Q^{+} = \lim_{\lambda \to 0} \frac{1}{Q^{\top}Q + \lambda \mathbb{I}} Q^{\top} . \tag{5.56}$$

Note that Q^{+} is a so called left inverse, i. e. $Q^{+}Q = \mathbb{I}$. If the problem is underdetermined, i. e. $m < n$ (more unknowns than equations), there are, in general, infinitely many solutions. The one we choose is again derived from the objective Eq. (5.55) and, in the limit $\lambda \to 0$, is obtained as

$$Q^{+} = \lim_{\lambda \to 0} Q^{\top} \frac{1}{QQ^{\top} + \lambda \mathbb{I}} \tag{5.57}$$

so that the matrix to be inverted is again a square matrix of the minimal dimension. Equation (5.57) defines the right inverse, i. e. $QQ^{+} = \mathbb{I}$. If $QQ^{\top}$ is not singular we immediately see that Eq. (5.57) is a solution of $u = Qv$ since

$$u = Qv = QQ^{\top} \frac{1}{QQ^{\top}} u = u$$

for any u that is not the null-vector. The advantage of these two expressions is that one can always find the pseudoinverse by inverting a matrix of dimension $\min(m, n)$.

The concept of pseudoinverses is generalized in Sect. 15.4 (p. 277).

Chapter 6
From Fixed-Point Flows to Hysteresis Oscillators

Abstract: Homeokinesis realizes the self-organization of artificial brain-body systems by gradient descending the time-loop error, a quantity that is truly internal to the robot since it is defined exclusively in terms of its sensorimotor dynamics. Homeokinesis can therefore be considered as a self-supervised learning procedure with the special effect of making the brain-body system self-referential. We will study this phenomenon here in an idealized one-dimensional world in order to identify key features of our self-referential dynamical systems independently of any specific embodiment effects. In particular, we will gain some insight into the entanglement of state and parameter dynamics and investigate the way how the latter induces behavioral variability.

One of the key features of homeokinesis, as introduced in the preceding chapter, is self-referentiality. A discussion was given already in Sect. 5.3.1 at a general level. This chapter studies concrete features of our one-dimensional self-referential dynamical systems. We consider for that purpose idealized worlds stripped from any embodiment to get the most basic effects. As it turns out, the parameter dynamics modifies the original system dynamics in a systematic way. The most prominent effects are to make fixed points flow towards bifurcation points and to turn hysteresis systems into hysteresis oscillators (by a self-induced sweeping of the hysteresis parameter). We will develop these fundamental properties of our approach in an essentially self-contained way.

Many of the specifics of homeokinetic systems can be seen already in the case of the one-dimensional sensorimotor loop that was introduced in Sect. 3.3. By way of example we may again consider the single-wheel controller of a mobile robot. Let us reconsider our general sensorimotor dynamics given by

$$x_{t+1} = \psi(x_t) + \xi_{t+1} \tag{6.1}$$

with

$$\psi(x) = AK(x) = Ag(Cx+h) \ , \tag{6.2}$$

$x \in \mathbb{R}^1$, and the controller outputs

$$y_t = K(x_t) = g(Cx_t + h) \ , \tag{6.3}$$

R. Der and G. Martius: The Playful Machine, COSMOS 15, pp. 107–125.
springerlink.com

where $g:\mathbb{R}^1 \to \mathbb{R}^1$ is the activation function $g(z) = \tanh(z)$. The forward model is nothing but the hardware constant A relating motor to sensor values.

6.1 Time-Loop Error in One-Dimensional Systems

In the time-loop error as introduced in Eq. 5.9 (p. 79) the Jacobian is given now in terms of the 1×1 matrix

$$L(x) = AK'(x) = ACg'(z)$$

with $z = Cx + h$ so that the time-loop error (dropping time indices) is

$$E(x) = \frac{\xi^2}{L^2(x)} \tag{6.4}$$

treating all 1×1 matrices as scalars. Learning consists in gradient descending the time-loop error so that we can get a first impression of the learning dynamics from the error landscape, considering first E as a function of $R = AC$ with $h = 0$. The error landscape is characterized by the two singularities, given by the zeros of L at either $C = 0$ or $|z| \to \infty$ (so that $g'(z) = 0$), i. e. at the maximum saturation of the neuron. Both singularities correspond to the maximum insensibility of the neuron against changes of the input. At all other values of C, the error is obtained by defining the fixed point x and evaluating the value of $L(x)$. This must be done numerically, since the fixed point equation is transcendental. The resulting curve was already presented in Fig. 5.5 (p. 93) and is repeated here in Fig. 6.1. Obviously, when following the gradient the system is driven to the point of maximum sensitivity as indicated by the fact that $L = 1$ at the bifurcation point ($R = 1$) [1]

6.2 Explicit Learning Rules and Fixed Point Flows

The learning rules for the adaptation of the parameters C and h are given by the gradient descent on E. We could of course use the generally valid rules given by Eqs. (5.22, 5.23) but it is instructive to repeat the derivation here in more elementary terms.

[1] Note that the local Lyapunov exponent λ is given by $\lambda = \ln L$ in the one-dimensional case so that $\lambda = 0$ right at the bifurcation point.

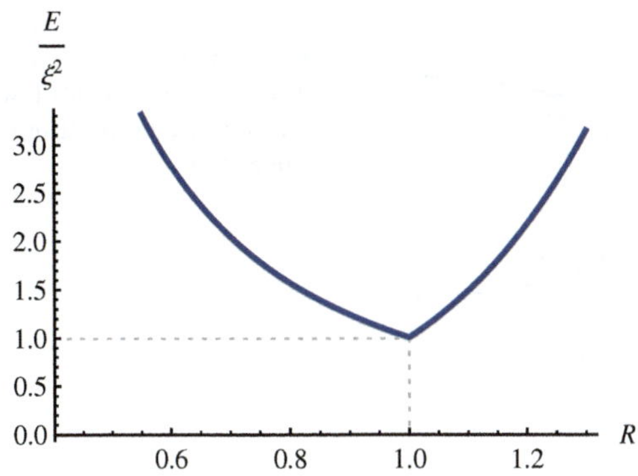

Fig. 6.1: The time-loop error over the feedback strength R in a one-dimensional loop. Gradient descent drives the feedback strength towards $R = 1$, the point of minimal stability of the fixed point. The error is normalized w. r. t. the noise strength.

6.2.1 Learning Rule

Starting from the general gradient descent formula for updating C, i. e.

$$\Delta C = -\varepsilon \frac{\partial E}{\partial C} \tag{6.5}$$

we obtain by using Eq. (6.4) and putting $h = 0$ for the moment

$$\frac{1}{2}\frac{\partial}{\partial C}E = -\frac{E}{L}\frac{\partial L}{\partial C} = -\frac{E}{L}\left(Ag' - 2gg'AC\frac{\partial z}{\partial C}\right) = -E\left(\frac{1}{C} - 2g\frac{\partial z}{\partial C}\right)$$

where $g'' = -2gg'$ is the second derivative of the tanh activation function. In order to get an explicit learning rule we have to evaluate the derivative $\partial z/\partial C$.

Let us assume that the state is fluctuating around a fixed point of the deterministic dynamics and the noise is sufficiently low so that we can use the fixed point equation $z = CAg(z)$ to obtain $\partial z/\partial C = x + L\partial z/\partial C$ using $x = Ag(z)$. Rearranging yields $\partial z/\partial C = x/(1-L)$ so that by introducing the (state dependent) relative learning rate

$$\alpha_x = \frac{1}{1 - L(x)} \tag{6.6}$$

the learning rule can be written as

$$\frac{1}{\varepsilon E}\Delta C = \frac{1}{C} - 2\alpha_x yx\,. \tag{6.7}$$

The detailed analysis of this learning rule is delegated to the appendix, Sect. 6.A, since the message is the one already given by Fig. 6.1 and its essence is depicted in Fig. 6.2. Noteworthy is the discontinuity of the gradient at $R = 1$ so that the

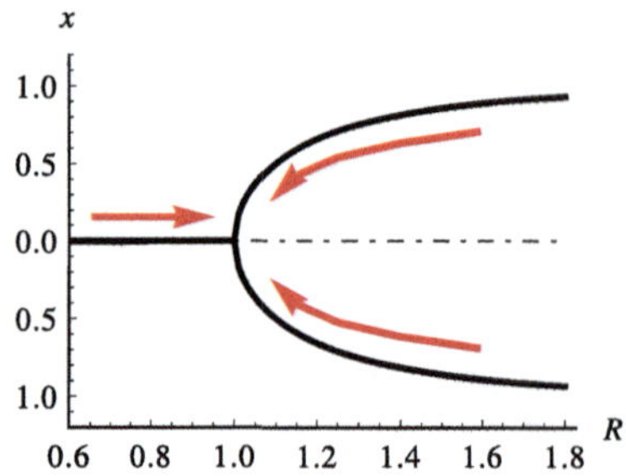

Fig. 6.2: Fixed point flow with homeokinetic learning. The learning dynamics drives the fixed points towards the pitchfork bifurcation point in the case of vanishing noise.

feedback strength R is driven with always finite speed towards the bifurcation point. The discontinuity is a direct consequence of the kink in the error curve seen in Fig. 6.1. We note again that the learning process converges towards the point of maximum sensitivity of the sensorimotor loop against external perturbations.

The above results, in particular the discontinuity of the gradient at the bifurcation point, is only true as long as the noise is vanishingly small. In fact, the derivation of α_x rests on the fact that the state variable x is fluctuating around one of the two fixed points of the bistable system. Rigorous results can only be achieved in the limit of $\xi^2 \to 0$ and $\varepsilon \to \infty$ such that the product $\varepsilon\xi^2$ remains finite. Therefore, the result is more of academic interest and we have to modify it for the finite noise situation.

6.2.2 *Learning under Real World Conditions*

In any real system the noise (prediction error) is finite, such that the clear fixed points of the deterministic system may be destroyed and the effective bifurcation point becomes important. Detailed considerations are given in the appendix, Sect. 6.B, but the message is again very simple: We can introduce a constant relative learning rate α in Eq. (6.7),

$$\frac{1}{\varepsilon E}\Delta C = \frac{1}{C} - 2\alpha xy \tag{6.8}$$

to define the point of stationarity for the feedback strength R such that the parameter dynamics converges towards the desired working regime. In the specific case the point of convergence is characterized by[2]

$$z \propto \sqrt{\frac{1}{\alpha}}\,. \tag{6.9}$$

for small values of α. As a consequence, the case of infinitesimal noise is covered by the limit of $\alpha \to \infty$.

[2] Note that the motor value $y = g(z)$ decrease with increasing α. This might have to be corrected for in practical applications.

By this mechanism the system can be driven to the "interesting" region, discussed as the effective bifurcation point in Sect. 3.3.4, as we will see below. For an intuitive picture check Fig. 3.10 (p. 40). The considerations remain valid also in the multi-dimensional case to a certain extent, see for instance Sect. 7.2.4. We therefore keep in the general algorithm α as a sensibility parameter for open choice. In the `sox` algorithm (Sect. 15.1) the parameter is called `sense`. The name is chosen because, with too large values of α, the working point is shifted in the direction of a subcritical behavior, where the system reacts very sensitive to the external perturbations, whereas within the supercritical region (too small α) the system sticks more resolutely to the selection of one of the modes (fixed points or oscillations) the system is capable of.

Moreover, there is a severe practical reason for using an empirically fixed value of α. In most applications we cannot assume that our state is at a fixed point or, if so, that we cannot know enough on its dependence on the parameters of the controller. When ignoring this dependence altogether, we get $\alpha = 1$, which is what we used in most of our practical applications with high-dimensional and compliant robotic systems. Nevertheless it is often helpful to enhance or damp the influence of the second term in the learning rule so that the α parameter is important for the general algorithm.

*6.2.3 Relation to the Effective Bifurcation Point**

The relation of the α parameter to the effective bifurcation point can be established by using an appropriate effective state in L, see Sect. 5.B (p. 100). A working definition is given by a moving average of the state x_t in a certain time window. This is suggested by the results of Sect. 3.3.4 demonstrating that the averaging defines the effective bifurcation point in a rather obvious way. The effect is immediately clear if we use the averaged state values $\bar{x}$ instead of x_t in the definition of E (6.4). The result is visualized in Fig. 6.3 displaying the normalized error E/ξ^2 as a function of $R = CA$ for the case of both vanishing and finite noise. The TLE with the effective state concept has the minimum at the effective bifurcation point (with a specific χ, see below) and this is where the rule Eq. (6.8) is driving the system.

The theory presented in the appendix, Sect. 6.B, can even give a rough estimate of the appropriate choice as

$$\alpha \propto \frac{1}{\chi\sqrt{D}} \tag{6.10}$$

where $D = \langle \xi^2 \rangle$ is the variance of the noise. Remember the parameter χ determines the concrete working point depending on the preferred timescales. Eq. (6.10) gives us a feeling for the correct choice of α, which, however, is appropriate only if the noise is quite small.

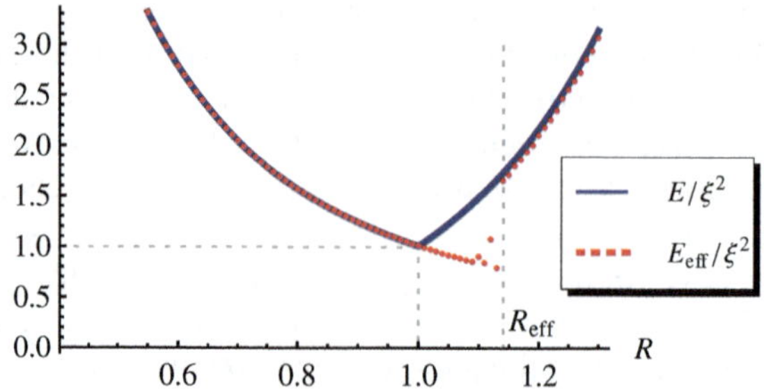

Fig. 6.3: Time-loop error with converging parameter dynamics. The normalized time-loop error for the one-dimensional loop as a function of the feedback strength $R = CA$ in the case of vanishing noise (line) and uniform noise in $[-0.1, 0.1]$ (dots). The minimum is at the bare and effective bifurcation point, respectively. In the finite noise case, the Jacobian matrix was taken at the effective state obtained by averaging in a sliding time window of 20000 steps. The particular effective bifurcation point at $R_{\text{eff}} = 1.14$ was obtained from Eq. (3.29) with $\chi = 3.5$ ($D = 0.00\overline{3}$), see Fig. 3.11.

6.3 Extending the Parameter Space — The Hysteresis Oscillator

The learning dynamics so far is quite close to the conventional one: we have an objective function E and change the parameters of the controller in order to reach convergence at the minimum of E, reaching thereby a specific working regime defining the behavior of the robot. The principle of Homeokinesis means more than that. A generic example is obtained if we extend the parameter space of the controller. We consider now the controller, see Eq. (6.3), including h, the bias of the neuron. In the learning scenario, h is just another parameter to be driven by the gradient descent. If included we obtain the following state-parameter dynamics (putting $\alpha = 1$ for simplicity as discussed in Sect. 6.2.2)

$$x_{t+1} = Ay + \xi_{t+1} \tag{6.11}$$

$$\frac{1}{\varepsilon E}\Delta C = \frac{1}{C} - 2xy \tag{6.12}$$

$$\frac{1}{\varepsilon E}\Delta h = -2y \tag{6.13}$$

with $y = g(Cx + h)$ and all quantities without time index taken at time t.

Figure 6.4 shows the emergence of a new effect. Starting at a subcritical value of the feedback strength $R = AC$ of the loop, R is driven rapidly towards larger values just in the same way as without bias. However, as soon as the loop becomes supercritical, we observe the onset of regular oscillations driven by h that reach a finite amplitude after some time. Both shape and frequency of the oscillations depend on the learning rate ε in a definite way. This oscillatory regime has already been considered in Sect. 5.3.3.2 (p. 93) but with the difference that there the feedback strength R was kept fixed so that there was only the single parameter h driven by the learning

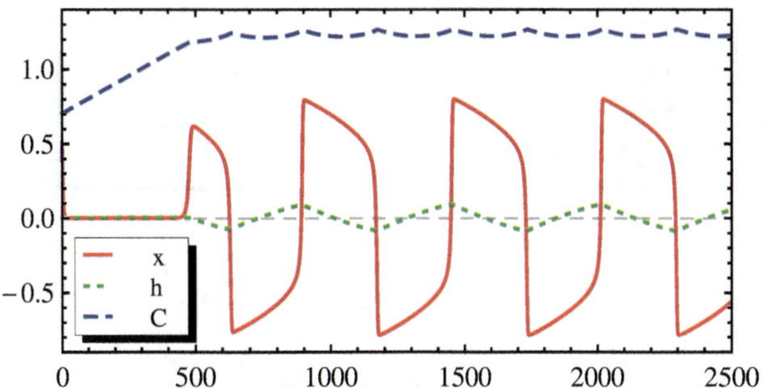

Fig. 6.4: State and parameter dynamics in the one-dimensional case. C increases until it reaches its fixed point around 1.2. The bias h oscillates around zero and causes the state x to jump between the positive and negative fixed points at $x^* \approx \pm 0.65$ (for $h = 0$). Setting: Average gradient, $A = 1$, $\varepsilon = 0.001$, $\alpha = 1$, infinitesimal noise.

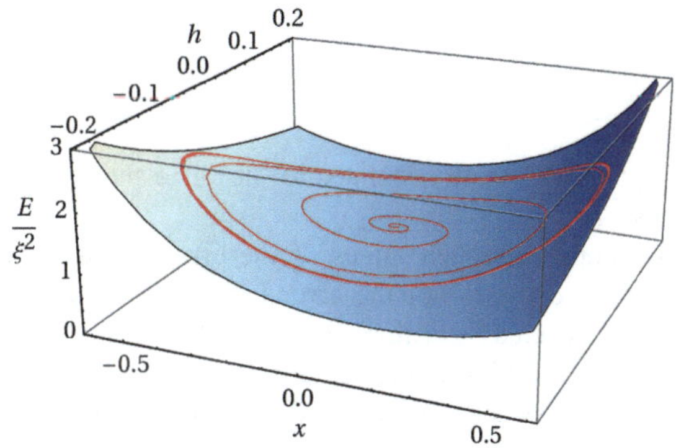

Fig. 6.5: Time-loop error with limit cycle oscillations. Error landscape of the normalized time-loop error of the system Eqs. (6.11, 6.13) with $C = 1.05$ and a trajectory starting at $x = .001$ and $h = 0.0$. As in Fig. 6.6, in the limit cycle we observe an oscillatory behavior of the error.

rule. The difference now is that obviously the feedback strength self-regulates into this critical value where the oscillations is stably realized.

One can show that the frequency ω of the oscillations is

$$\omega = \sqrt{2\varepsilon} + O\left(\varepsilon^{\frac{3}{2}}\right) ,$$

see Sect. 6.C for the conditions and the derivation.

The behavior of the error can be understood by studying Fig. 6.5 that depicts the error landscape over the variables x and h with R fixed at the slightly supercritical value $R = 1.05$. Starting from some small values for both state and parameter, the system very soon reaches the limit cycle. Following the trajectory, the varying values for the error can be read off. Note that the velocity of the trajectory on the surface is

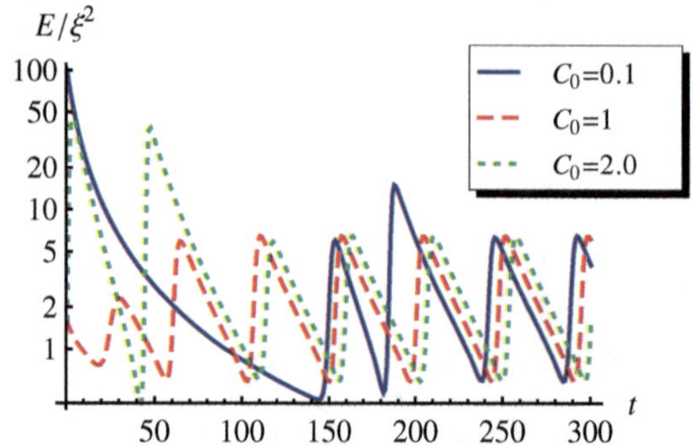

Fig. 6.6: Time-loop error with extended parameter space. The figure depicts the behavior of the normalized time-loop error if both h and C are included into the combined dynamics. Please note the logarithmic scale, showing that the error dynamics, before reaching the limit cycle, is extending over more than one order of magnitude. The curves belong to different initial values of C, see legend, and starting values $x_0 = 0.5$, $h_0 = 0$ in all cases. Setting: Average gradient, $A = 1$, $\alpha = 1$, $\varepsilon = 0.05$.

depending on its position, which generates for instance sharp rises in the evolution of the error values.

These sudden jumps to higher values of the error are seen very distinctly in Fig. 6.6, depicting the evolution of the error for different values of the feedback strength R. We have chosen a logarithmic scale for the error in order to underline its vast variations of the error. Figure 6.7, depicting the foliation of the error landscape in a hysteresis system, may serve as yet another clarification of the situation.

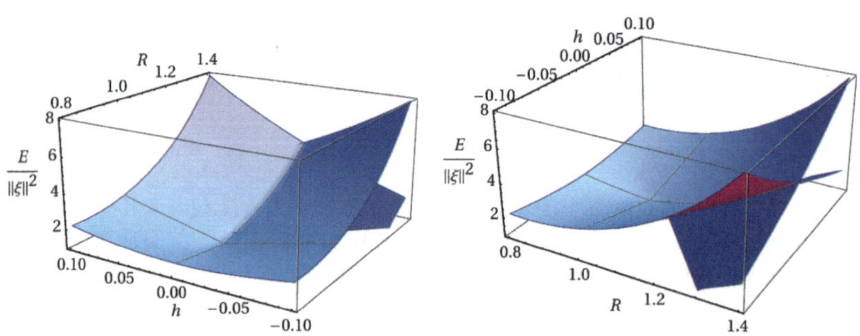

Fig. 6.7: Time-loop error over bias h (left) and feedback strength R. If the bias is included, the sensorimotor loop becomes a hysteresis system with either a single or two stable fixed points. The error landscape, seen from two different angles, clearly reflects this property by foliating in the latter region. The landscape has a minimum at $R = h = 0$ in the case of vanishing noise considered here. With finite noise and effective states this minimum is shifted like in Fig. 6.3. Then, the gradient descent becomes a hysteresis cycle due to jumps between the folios.

We have given in Sect. 5.3.3.2 (p. 93) already an explanation for the effect in a more general context that can now be supplemented by a detailed mechanism for the entanglement between state and parameter dynamics. As seen from Eq. (6.13), the h dynamics is easy to understand since its direction of change is always opposite to the sign of y. So, if y is in the saturation region ($|y| = 1 - \delta$ with $delta \ll 1$), the h dynamics will drive it out of that region. This is in correspondence to the general rule that homeokinetic learning is trying to make the neuron more sensitive. However, due to the hysteresis property (Fig. 5.7), there is an overshooting. After h has crossed the zero line, the state does not immediately switch to the other fixed point, but requires a certain value of h (with opposite sign of x). This effect leads to an oscillatory dynamics of the combined system.

Altogether, the h dynamics generates a periodic running through the hysteresis cycle introduced in Sect. 3.4. The difference to the usual hysteresis scenario is that the phenomenon is self-generated by the h dynamics inherent in the learning rule instead of being controlled by hand as in Fig. 3.17(c) and Fig. 3.18. There is a close similarity of the above dynamics with so called relaxation oscillators like the van der Pols oscillator or the FitzHugh-Nagumo-model of a single neuron, see Sect. 6.5. In our case, the phenomenon rests on the hysteresis properties of the system so that we may call it a hysteresis oscillator.

6.4 Embodiment and Situatedness — Robotic Experiments

Sensitization and self-induced hysteresis oscillations have been established above as phenomena generated by the homeokinetic controller acting in an idealized world. We are now going to apply the controller to some more realistic settings with real robots, in simulation at least. These investigations will demonstrate that the balancing of the controlled system between instability and regularity leads to interesting effects, like spontaneous cooperation in multi-dimensional systems, that are directly related to the specific embodiment and the situatedness of the robot. Actually, so far we have investigated only the one-dimensional sensorimotor loop but we can apply those findings to much more complicated systems by adhering to the paradigm of decentralized control, meaning that each degree of freedom (DoF) has its own independent homeokinetic controller. The robots we will consider are wheeled robots, a chain of those, and the so called ARMBAND robot, which is a complicated physical object with 18 degrees of freedom.

6.4.1 Wheeled Robots

Our first topic is to exemplify the role of the embodiment with different robots under homeokinetic control. In Experiment 6.1 we consider three mobile robots with different embodiment, the TWOWHEELED, the LONGVEHICLE, and the FOUR-

WHEELED, see Video 6.1. Methodologically, we study the influence of the embodiment by using identical controllers always with the same initial condition and learning rate so that the only difference is the embodiment. Nevertheless, as you may find out by doing the experiments or by watching Video 6.1, the behavior of the robots is very much different depending on both the embodiment and the situatedness (maze or empty arena).

We consider the case of decentralized control, namely the motors are driven each separately by the above introduced controller with parameter dynamics given by Eqs. (6.12, 6.13). Each controller receives only the readings of the corresponding wheel velocity sensor. The TWOWHEELED robot shows a pattern of on-site rotations followed by occasional forward or backward movements. Quite differently, the FOURWHEELED robot clearly prefers the straight motions since the on-site rotations are more difficult to realize due to a kind of resistance against turning around caused by the specific morphology. In the same way, the motion pattern of the LONGVEHICLE robot are largely depending on the length of its body since the longer the vehicle, the more "resoluteness" is required to change the direction of the motion. As a first conclusion, the homeokinetic learning rule generates a self-regulated search of the behavior space that is dominated by the morphology and the physical properties of the robotic system in its interaction with the environment.

Experiment 6.1: Emerging search with mobile robots

For this experiment you have two choices: (a) Different wheeled robots (learning) in a square arena without obstacles and (b) five LONGVEHICLEs in a maze (learning), see Video 6.1. During the simulation you may change the parameters `epsC` to speed up or slow down the learning speed. Try `>epsC=.005` or `>epsC=.05`
Use the GUILOGGER to watch sensor values and motor values and parameter evolution. Consider the dependence of the motion patterns on both the morphology of the robot, the obstacle situation in the playground and the parameters.

6.4.2 *Spontaneous Cooperation in a Chain of Wheeled Robots*

In the single robot experiments we have observed a sensitive reaction to physical forces caused either by collisions or by the inertial effects due to the mass of the body itself. These forces hinder the function of the motors so that the wheels may get totally or partly blocked, thereby influencing the sensor values of the wheel counters. What happens if we connect many of the TWOWHEELED robots by a passive joint? We have already seen in Sect. 3.3.5.2 that, with a fixed closed loop controller, the individual elements can spontaneously synchronize. We observed a collective behavior enabling the chain to explore a maze, provided the feedback strength was chosen around the critical point. At the critical point, the sensorimotor dynamics is very sensitive to external perturbations so that each individual control loop is easily switched by physical forces, exerted by the other robots in the chain so that the wheels may synchronize their actions.

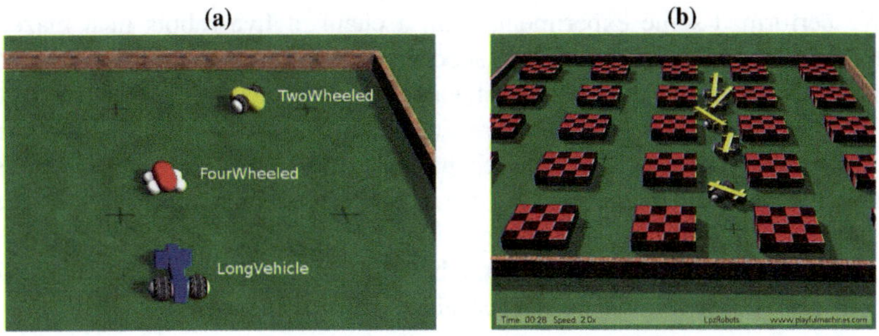

Video 6.1: The role of embodiment and situatedness with wheeled robots. (**a**) Three different wheeled robot with each wheel controlled by its own homeokinetic controller. The TwoWheeled mostly rotates, the FourWheeled drives mostly straight, and the LongVehicle does both equally likely; (**b**) LongVehicles of different length in a maze. The robots may get stuck, but manage to find out again. Upon being blocked, actions are very gentle because the sensor values are small and thus the activity in the sensorimotor loop goes down rapidly. Due to the sensitization effect in the learning dynamics, the feedback strength is increasing rapidly so that stronger and stronger actions are performed. Settings: $\varepsilon_c = 0.02, \varepsilon_A = 0.01$. The videos can be watched at `http://playfulmachines.com`.

Will the synchronization be further enhanced if we include learning or will the entanglement with the parameter dynamics rather serve as an irritation of this subtle synchronization phenomenon? Interestingly, with homeokinetic learning there is even an additional tendency to synchronize. With enabled bias dynamics, each learning controller is able to initiate the switching from one fixed point to the other with its bias dynamics. The switching frequency is proportional to the size of the modeling error, which is low if all wheels cooperate, but is high if the wheels are working against each other, because then the wheel velocity does not follow the motor actions. If out of synchrony, the controller therefore will be more ready for switching the fixed point and thereby getting in synchrony with the other loops. We can understand this also as an example of the general temperature effect as introduced in Sect. 5.3.4.

Video 6.2: Chain of wheeled robots exploring a maze. Five TwoWheeled robots connected to a chain by passive joints. Control is decentralized with each wheel being controlled individually by a homeokinetic controller. The position of the *red* robot is used for the analysis in Fig. 6.8. Settings: $\varepsilon_c = \varepsilon_A = 0.01$. The video can be watched at `http://playfulmachines.com`.

We performed some experiments with a chain of five robots in a maze, see Video 6.2. After initialization in the subcritical region, the destabilization drive of homeokinetic learning slowly but systematically increases the feedback strength in the loop so that the chain starts to move after some time. Upon hitting an obstacle the chain typically inverts its velocity, such that the maze is well explored. In Fig. 6.8 the quantitative data is displayed and we see that with homeokinetic learning the arena is explored very well for a large range of learning rates, the learning system visits nearly twice as much of the area as the best static controller (that was determined empirically by scanning the parameter range), i. e. the one used in Sect. 3.3.5).

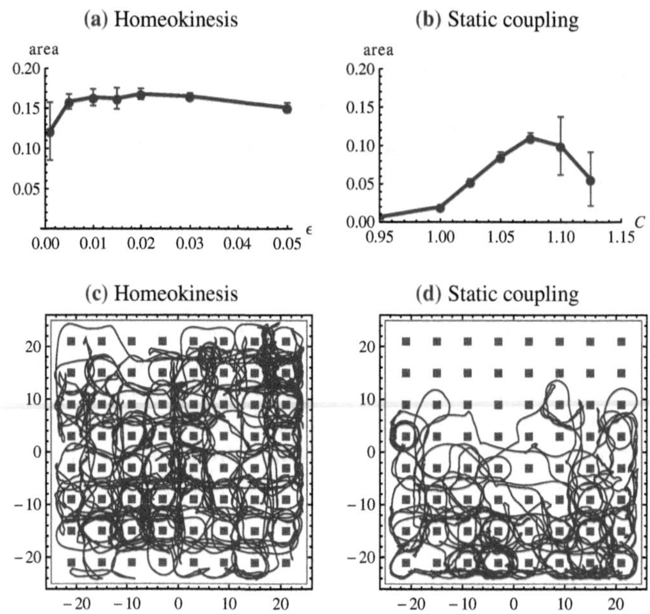

Fig. 6.8: The chain of robots explores a maze much more effectively with homeokinetic learning. **(a)** Area coverage (box counting method, normalized to full area including obstacles) for the homeokinetic control in dependence on the learning rate (smallest value $\varepsilon = 0.001$). Mean and standard deviation of 10 runs of 30 min length is plotted; **(b)** The same as in (a), but for the fixed controller in dependence on coupling strength C; **(c)** Sample trajectory with learning and $\varepsilon = 0.02$ with area coverage of 0.18; **(d)** Sample trajectory for static controller with $C = 1.075$ with area coverage of 0.1. In all figures the position of the second robot in the chain (drawn in *red* in Video 6.2) was used.

Experiment 6.2: Spontaneous cooperation in a chain of robots

Start the simulation Spontaneous cooperation in a chain of robots. The setting is as shown in Video 6.2. Try different parameters for
`>epsC`.
Also check the difference to the case without learning by putting
`>epsC=0` and `>coupling=1.1`
The latter sets the coupling strength C to the given value and resets the bias to $h = 0$.

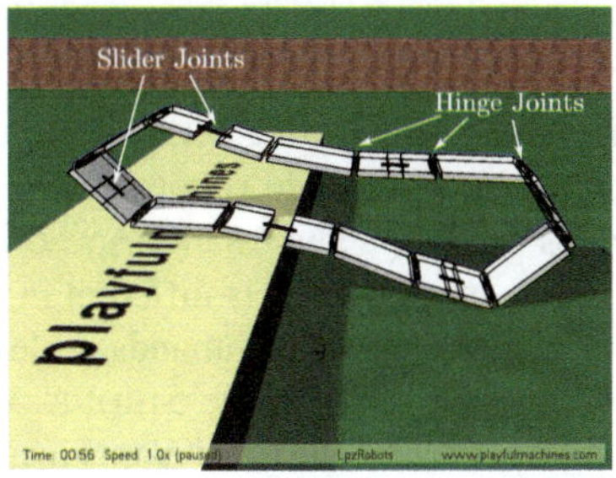

Fig. 6.9: SLIDER ARMBAND moving over an obstacle. The robot has 12 hinge joints and 6 slider joints, each actuated by a servo motor and equipped with a proprioceptive sensor. The name comes from the German word *Armband* meaning a wristband or bracelet of a watch.

6.4.3 Emergent Locomotion of the SLIDER ARMBAND

The cooperation and the emerging search for behavioral modes can be also observed in more complex objects. In Experiment 6.3 we realize the decentralized control of a bracelet-like robot called SLIDER ARMBAND as depicted in Fig. 6.9. In this experiment, each joint is controlled by an individual controller as described above. The controller receives the measured joint angle or slider position, the output of the controller defining the target joint angle or target slider position to be realized by the motor. The motors are realized as servomotors, see Sect. 16.4.4 for details, with a limited force so that the target angle may differ substantially from the true joint angle due to the load on the joint caused by the interaction with obstacles or the interaction between the system's different degrees of freedom. It is only by these deviations that the current physical state of the entire body is influencing both the control and the learning process.

Video 6.3: Locomotion of SLIDER ARMBAND with decentralized control. Each of the 18 joints is controlled individually by a one-dimensional controller with the corresponding joint position as the only input information. Nevertheless, after some time a spontaneous cooperation of the segments occurs that causes the robot to locomote. It can even surpass obstacles. At the walls the robot manages to invert its velocity. The video can be watched at `http://playfulmachines.com`.

Experiment 6.3: Spontaneous cooperation in high-dimensional systems.

The simulation Spontaneous cooperation in high-dimensional systems starts with a the SLIDER ARMBAND placed in a square arena. You may change the parameters `epsC`, `epsA`, and `noise`. Try

```
>epsC=0
```

to switch off learning of the controller. Try also other values between 0.01 and 10. Use the GUILOGGER for watching sensor values, motor values and the parameters for the controller of the first joint.(The parameters of the other controllers are not visible). Study the dependence of the emerging motion patterns on the learning rate and the noise strength.

The mechanism for getting into a collective locomotion mode is different from that of the chain of wheeled robots. In the mobile robot case, the fixed points of the dynamics determine the rotation speed of the wheels. In this way a fixed point causes a steady motion. This is different in the SLIDER ARMBAND where the sensor state x_t and the motor commands y_t define a static configuration of the robot. Thus, a locomotion mode of the SLIDER ARMBAND can only be achieved by actively changing between the fixed points in an organized manner. The closer analysis shows that it is the bias dynamics that plays the central role in this process. In our underactuated setting, those segments of the chain that are currently lying on the ground do not have enough force to move in any direction. So, it is the physical situation that keeps the sensor values constant. However, the bias dynamics is running anyway and as we know from Sect. 6.3 the bias dynamics aims at destabilizing the current fixed point. As long as the joint is blocked it cannot switch even though the motor commands are already shifted by the bias onto the opposite side; but as soon as the joint is getting free the still outstanding switching can eventually take place.

With a hinge joint this means that it actively bends once it is getting physically free. In this way the individual elements (hysteresis oscillators) get in synchrony with any potential movement of the robot and thus the entire SLIDER ARMBAND develops a robust locomotion behavior. Video 6.3 gives an example. In the experiment, we start with a subcritical initialization without learning in order to demonstrate that the robot will come to rest very quickly. Note that with very strong motor power, the SLIDER ARMBAND could stabilize a wheel-like shape, but this is not possible in our underactuated, weak-force setting. After enabling the learning, the robot starts to move since the learning drives the system towards instability. The movement is seen to become more and more regular. The locomotion is smoothly adapted when surpassing an obstacle. Later on we switch off the learning and the robot comes to rest, even though the control parameters were tuned appropriately by the learning. Thus the ongoing learning and the entanglement between state and parameter dynamics is essential for this behavior. You can convince yourself by performing Experiment 6.3.

6.5 Relation to Other Systems

The homeokinetic neuron has a close relation to well known physical systems and neuron models. Let us consider the neuron in its feedback loop with the combined dynamics given by Eqs. (6.11–6.13). As shown above, the coupling strength C reaches a slightly supercritical value after a short time and stays there forever with a low amplitude oscillation. We therefore assume C to be fixed and consider only Eqs. (6.11, 6.13) describing the interplay between the bias and the state dynamics. The relation to the physical systems is seen best if using the z dynamics obtained by multiplying Eq. (6.11) with C and adding h on both sides.

Dropping the noise, using $R = CA$, and absorbing all positive factors into ε we can write

$$z_{t+1} = Rg(z_t) + h_t \tag{6.14}$$

$$h_{t+1} = h_t - \varepsilon g(z_t) \ . \tag{6.15}$$

Rewriting this in terms of the updates $\Delta z_t = z_{t+1} - z_t$ and $\Delta h_t = h_{t+1} - h_t$, we may consider Eqs. (6.14, 6.15) as the Euler approximation of the system of differential equations

$$\dot{z} = Rg(z) - z + h \tag{6.16}$$

$$\dot{h} = -\varepsilon g(z) \ . \tag{6.17}$$

With $g(z) = \tanh(z)$, Eqs. (6.16, 6.17) realize the dynamics of a so called relaxation oscillator. In particular, with $R = 1 + \delta$ and $0 < \delta \ll 1$, the amplitude of the system can be kept small so that we may approximate $g(z)$ by $z - z^3/3$. With a convenient rescaling of both time and variables, Eqs. (6.16, 6.17) can be given the standard form of the famous Van der Pol oscillator [84]

$$\dot{z} = z - \frac{z^3}{3} + h \tag{6.18}$$

$$\dot{h} = -\varepsilon' z \tag{6.19}$$

using only the leading term of $g(z)$ in the equation for $\dot{h}$ since the amplitude of the h oscillations is inferior to that of z. This is a well studied self-exciting non-linear oscillator, used originally to describe stable limit cycles in electrical circuits employing vacuum tubes.

We can also establish a relationship to the FitzHugh-Nagumo-model, a simplified version of the Hodgkin-Huxley neuronal model that describes the neuronal dynamics in terms of the membrane potential z and a recovery variable that can be identified with our h. However, we are not going into details here. The point of interest we want to emphasize by this little excursion is to demonstrate the consequences of making a system self-referential. In the example, a fixed point system as given by Eq. (6.14) with parameter h is converted into a relaxation oscillator when adding a parameter dynamics for h by gradient descending the TLE.

Appendix 6.A Fixed-Point Flow

We use the learning rule (6.7) in the form

$$\frac{1}{\xi^2 \varepsilon} \Delta C = -L^{-2} \left(-\frac{1}{C} + 2\alpha_x xy \right)$$

In case that $0 < R < 1$ with $R = CA$, we are below the bifurcation point where $x = y = 0$ implying $g' = 1$ and $\Delta C = -\varepsilon \xi^2 L^{-2} C^{-1} = -\xi^2 A^{-2} C^{-3}$. This means that C is driven away from $C = 0$. Note the singularity at $C = 0$ where the velocity of the parameter dynamics is diverging. This has to be considered for the practical application. Slightly above the bifurcation point we put $R = 1 + \delta$, $0 < \delta \ll 1$ and use the approximated fixed point Eq. (3.23)

$$z^2 = 3\delta + O\left(\delta^2\right)$$

which is appropriate since $|z|$ is small. Hence,[3] $L = Rg'(z) = R\left(1 - z^2\right) + O\left(\delta^2\right)$ and

$$L = 1 - 2\delta + O\left(\delta^2\right)$$

As a consequence, using Eq. (6.6) we get

$$\alpha_x = \frac{1}{1 - L(x)} = (2\delta)^{-1} + O(1) \tag{6.20}$$

and

$$\begin{aligned} L^{-2}\left(-\frac{1}{C} + 2\varepsilon_x xy\right) &= \left(-\frac{A}{1+\delta} + \frac{3A\delta}{\delta}\right) + O(\delta) \\ &= 2A + O(\delta) \end{aligned} \tag{6.21}$$

where $xy = Ay^2 = Az^2 + O\left(\delta^2\right)$ was used.

Moreover, with $R \gg 1$ we may approximate the fixed point as $z \approx \pm R$ and[4] $\tanh' z \approx 4e^{-2R}$, so that $L \approx 4R\exp(-2R)$ is infinitesimally small, $\alpha_x xy \approx A$ so that finally we obtain (assuming the noise is infinitesimally small)

---

[3] Again, using a Taylor expansion in lowest order $g(z)$ is substituted by z.

[4] The approximation is obtained by writing, in the $z > 0$ case,

$$\tanh' z = 4\frac{u}{(1+u)^2},$$

where $u = \exp(-2z)$, and expanding for small u to obtain

$$\tanh' z = 4u + O\left(u^2\right).$$

A similar consideration is valid for $z < 0$ so that, altogether, in the limit of very large $|z|$ we obtain $\tanh' z \approx 4e^{-|2z|}$ and $\tanh' z \approx 4e^{-2R}$ for $R \gg 1$.

$$L^{-2}\left(-\frac{1}{C}+2\varepsilon_x xy\right) \approx \frac{1}{16R^2\exp(-4R)}\left(-\frac{A}{R}+2A\right) \tag{6.22}$$

$$= \frac{A}{16R^3\exp(-4R)}(-1+2R)\,. \tag{6.23}$$

Altogether we obtain the following approximate expressions for the update of C

$$\frac{1}{\varepsilon\xi^2}\Delta C \approx \begin{cases} A^{-2}C^{-3} \text{ for } 0<R<1 \\ -2A \text{ for } R \gtrsim 1 \\ -\frac{A(2R-1)}{16R^3}e^{4R} \text{ for } R \gg 1 \end{cases} \tag{6.24}$$

with $R = AC$. The gradient diverges for both $R \to 0$ and $R \to \infty$ so that the parameter C is driven away from these regions towards the bifurcation point at $R = 1$. There, ΔC displays a discontinuity, of width $\varepsilon\xi^2\left(A^{-2}C^{-3} - 2A\right)$ when crossing the bifurcation point from below. Formally the discontinuity is a result of α_x being diverging at $R = 1$, whereas the state $x \to 0$ when approaching the bifurcation point from above. The product $\alpha_x xy$ is finite but does not match the case when approaching the bifurcation point from below.

Appendix 6.B Towards the Effective Bifurcation Point

The main reason why we had to assume vanishing noise in deriving the learning rule was the validity of $\partial z/\partial C = x + C\partial x/\partial C$ which is fulfilled exactly only at the fixed points of the **deterministic** system. With any finite noise, there is always a region where the assumption is not valid any longer, see Fig. 3.10 so that a special consideration is necessary for that region. The idea for the correction is drawn from Fig. 3.10, which demonstrates that the time averaged state variables behave rather similar to the fixed points of the deterministic system if only we replace the bifurcation point with its effective counterpart. In fact, the time average $\langle x_t \rangle$ is roughly zero up to the effective bifurcation point R_{eff} and after that we have two stable fixed points, which are behaviorally similar to the deterministic ones. The main difference is that there is not any longer a classical pitchfork bifurcation but we have a jump at the effective bifurcation point. Based on this behavior, we put

$$\alpha_x = \begin{cases} 1 & \text{for } R \le R_{\text{eff}} \\ (1 - L(\langle x \rangle))^{-1} & \text{for } R > R_{\text{eff}} \end{cases} \tag{6.25}$$

so that α_x will always be finite. If using this in the learning rule we may simplify the latter further by replacing α_x with its maximal value, i. e. its value at or slightly above the effective bifurcation point, since this influences only the speed of the learning but not the position of the equilibrium point (where α_x is maximal). So, we can replace the value of α_x with a constant value denoted by α chosen such that the stationary state of the learning dynamics

$$\frac{1}{\varepsilon E}\Delta C = \frac{1}{C} - 2\alpha xy \tag{6.26}$$

is reached at the effective bifurcation point.

This is made more transparent by considering the equilibrium point, which is defined by $\Delta C = 0$, explicitly so that, using $Cx = Ry = Rg(z)$,

$$1 = 2\alpha R g^2(z) \; .$$

We consider this at the fixed point (state dynamics is converged), i. e. with the additional information $z = Rg(z)$ and obtain numerically with $g(z) = \tanh z$ the solution $z \approx 0.77$ and $R \approx 1.19$ in the special case $\alpha = 1$, see Sect. 3.3.4 (p. 38). We get an analytical solution by using the expansion for the fixed point $z = \sqrt{3\delta} + O(\delta)$ (see above), putting again $R = 1 + \delta$. Using that in

$$1 = 2\alpha R g^2(z) \approx 6\alpha\delta$$

we obtain

$$R_\alpha \approx 1 + \frac{1}{6\alpha}, \qquad z \approx \sqrt{\frac{1}{2\alpha}}$$

where $z \approx \sqrt{3\delta}$ was used for the final step. This means that we can shift the stationary point of the learning dynamics by choosing α appropriately. In Sect. 3.3.4 we derived the position of the effective bifurcation point as a function of the noise strength. In particular, using Eq. (3.49) we obtain Eq. (6.10).

Appendix 6.C Frequency of Hysteresis Oscillations

Let us assume that $A = 1$ and C in Eq. (6.12) has already reached its limit cycle regime so that we may give it a fixed value of $C = 1 + \delta$ with $0 < \delta \ll 1$. With δ small, we may assume that the amplitude of the oscillations of both x and h is small, so that we may linearize Eqs. (6.11, 6.13) to get

$$\Delta x_{t+1} = \delta x_t + h_t \tag{6.27}$$

$$\Delta h_{t+1} = -2\varepsilon\left((1+\delta)x_t + h_t\right) \tag{6.28}$$

with Jacobian matrix

$$J = \begin{pmatrix} \delta & 1 \\ -2(1+\delta)\varepsilon & -2\varepsilon \end{pmatrix} .$$

The eigenvalues of J are

$$\mu_{1,2} = -\varepsilon + \frac{1}{2}\delta \pm \frac{1}{2}\sqrt{4\varepsilon^2 - 4\delta\varepsilon + \delta^2 - 8\varepsilon}$$

In the limiting case of very small δ ($\delta > 2\varepsilon$ though, otherwise damped oscillation) we obtain

$$\mu_{1,2} = -\varepsilon \pm \sqrt{\varepsilon^2 - 2\varepsilon} + O(\delta)$$

so that the frequency ω of the oscillations is in the limit of small ε

$$\omega = \sqrt{2\varepsilon} + O\left(\varepsilon^{\frac{3}{2}}\right) . \tag{6.29}$$

Chapter 7
Symmetries, Resonances, and Second Order Hysteresis

Abstract: This chapter is a continuation of the preceding chapter to two-dimensional systems. We will identify new features of our self-referential dynamical system again independently of any specific embodiment effects. The additional dimension opens the possibility for state oscillations, where the entanglement of state and parameter dynamics will lead to interesting phenomena and induces behavioral variability. The most prominent effects are driving oscillatory systems into a second order hysteresis (by a self-organized frequency sweeping effect) and getting into resonance with latent oscillatory modes of the controlled system.

In continuation to the previous chapter we study concrete features of our self-referential dynamical system in two-dimensional idealized worlds. The one-dimensional sensorimotor dynamics has revealed already some basic features of homeokinetic control. However, these examples have shown merely the situation with a fixed point as the central dynamical pattern. Additional dimensions in the state dynamics bring new patterns into play, in particular oscillatory ones. Generic structures emerge already in the 2D case, which we are going to discuss in some detail now. The parameter dynamics will be seen to modify the original system dynamics in a complex but systematic way.

7.1 Properties of the Two-Dimensional Sensorimotor Loop

7.1.1 Two-Dimensional Loop

Let us repeat first the generic dynamical system Eq. (3.7), i. e.

$$x_{t+1} = \psi(x_t) + \xi_{t+1} \tag{7.1}$$

now with $x_t \in \mathbb{R}^2$, the function $\psi : R^2 \to R^2$ being given by

$$\psi(x) = Ag(Cx + h) \tag{7.2}$$

R. Der and G. Martius: The Playful Machine, COSMOS 15, pp. 127–151.
springerlink.com

so that the two-dimensional system with $g(s) = \tanh(s)$ and $A = \mathbb{I}$ reads in detail (we raise the time index when writing the vectors componentwise)

$$x_1^{t+1} = \tanh\left(C_{11}x_1^t + C_{12}x_2^t + h_1\right) + \xi_1^{t+1} \tag{7.3}$$
$$x_2^{t+1} = \tanh\left(C_{21}x_1^t + C_{22}x_2^t + h_2\right) + \xi_2^{t+1} \tag{7.4}$$

This system, with fixed parameters without noise, was studied in the context of recurrent neural networks [123, 124] where the dynamics for the complete parameter space and the resulting bifurcations are presented. Its ability to produce various waveforms makes it very useful as central pattern generator in the context of evolving gates for bipedal [70], four-legged [101] and six-legged [99] walking machines. The latter two used a reduced parametric form of Eqs. (7.3, 7.4), which has only two parameters $a = C_{11} = C_{22}$ and $b = C_{21} = C_{12}$ to produce sinusoidal waveforms. In [86] a key-framing technique was introduced to easily tune the parameters of the $SO(2)$-network to produce a desired output wave. The hysteresis properties were exploited in [74] and in [177] reflex oscillators were created for single leg controllers in a stick insect model.

The examples where chosen for illustration before making the system self-referential by introducing the homeokinetic learning (see next section), as they show the wealth of the dynamics of this system, as well as applicability to technical realizations.

7.1.2 Bifurcation Scenario

The first step in the analysis consists in finding the eigenvalues of the Jacobian matrix

$$L = AG'(z)C \tag{7.5}$$

with $z = Cx + h$ and the diagonal matrix $G'(z) = \text{diag}[g'_1, g'_2]$ where $g'_i = g'(z_i)$ as before. It proves convenient to write the eigenvalues in terms of the trace (Tr) and the determinant (Det) of the matrix as

$$\lambda_{1,2} = \frac{1}{2}\left(\text{Tr} \pm \sqrt{(\text{Tr}^2 - 4\text{Det}}\right) . \tag{7.6}$$

The dynamics is asymptotically stable if the two eigenvalues lie inside the unit circle in the complex plane, however it bifurcates if an eigenvalue λ—during parameter variation—leaves the unit circle. Depending on the exact location where $\lambda(\text{Tr}, \text{Det})$ crosses the unit circle, the following three bifurcation types can be distinguished:

- Saddle node (fold) bifurcation $\lambda = 1$: It creates or destroys a pair of fixed points.
- Period doubling(flip)bifurcation $\lambda = -1$: It creates or destroys a period-2 cycle from a fixed point.

- Neimark-Sacker bifurcation[1] $\lambda = \exp(\pm i\omega)$ where $\exp(\pm i\omega) \neq 1$ for both signs: It creates an asymptotically stable invariant curve with periodic or quasi-periodic dynamics on it, depending on whether the rotation number is rational or irrational, see [92, 123] and the discussion below.

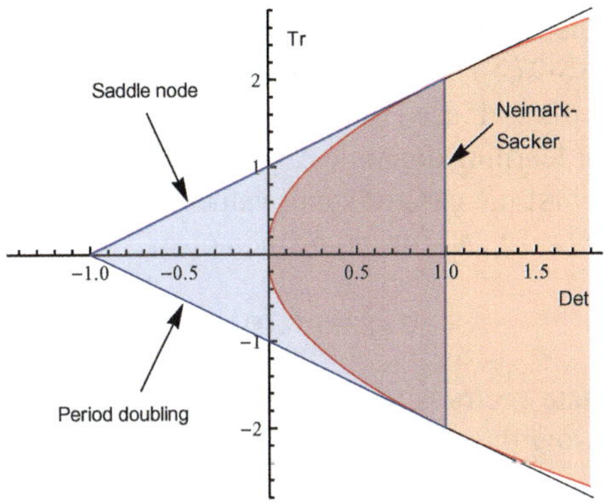

Fig. 7.1: Bifurcations in the 2-D system. Displayed is the trace (Tr) and the determinant (Det) of the Jacobian L. The domain of stability of the fixed point x^* is given by the *blue triangle* bounded by the three lines $\mathrm{Tr} - \mathrm{Det} = 1$, $\mathrm{Tr} + \mathrm{Det} = -1$, and $\mathrm{Det} = 1$. The system is able to perform oscillations in the *red domain* (region of nonzero imaginary part of the eigenvalues).

According to the above definitions, the region of stability is defined in the (Tr, Det) plane by a triangle bounded by the three straight lines $\mathrm{Tr} - \mathrm{Det} = 1$, $\mathrm{Tr} + \mathrm{Det} = -1$, $\mathrm{Det} = 1$, see Fig. 7.1. For instance, along the line $\mathrm{Tr} + \mathrm{Det} = -1$ (on which the eigenvalues are real and one of them is -1) there will be a period-doubling bifurcation. Along the line $\mathrm{Det} = 1$, $|\mathrm{Tr}| < 1$ (on which the two eigenvalues are complex and of modulus 1), there will be Neimark-Sacker bifurcations from a fixed point attractor to a periodic or quasi-periodic attractor, see [161].

7.1.3 Properties of $SO(2)$ Dynamics

For the rest of the chapter we stick to our toy world given by Eqs. (7.3, 7.4) that we call the SHORT CIRCUIT in the LPZROBOTS simulator. A specific form of this two-dimensional dynamics is obtained if C is given by

$$C = uO(\phi) \tag{7.7}$$

[1] Neimark-Sacker bifurcation is the discrete counterpart of the Hopf bifurcation.

where O is an orthogonal matrix and $u > 0$ a scaling factor. In particular, we choose O as a special orthogonal matrix

$$O(\phi) = \begin{pmatrix} \cos\phi & -\sin\phi \\ \sin\phi & \cos\phi \end{pmatrix} \tag{7.8}$$

which rotates a vector by an angle ϕ. These matrices are closed under multiplication with each other and the inverse operation[2] so that they form a group called the special orthogonal group $SO(2)$. With this specific choice of C, the dynamics becomes a nonlinear oscillator if $u > 1$, see [124]. Interestingly, we will find below that under various conditions our learning dynamics drives the parameters to just this structure.

Let us consider at first the case of small values of x so that the system Eq. (7.2) can be linearized ($g(z) \approx z$) and (dropping the noise)

$$x_{t+1} = uO(\phi)x_t \,. \tag{7.9}$$

This means that the state vector is rotated in each step by the angle ϕ and multiplied by the factor u. Choosing the starting vector as

$$x_0 = \begin{pmatrix} \cos\theta \\ \sin\theta \end{pmatrix}$$

in this linear setting the state vector after t iterations will be

$$x_t = u^t \begin{pmatrix} \cos(\theta + t\phi) \\ \sin(\theta + t\phi) \end{pmatrix} \tag{7.10}$$

meaning that the vector is rotated by the angle $t\phi$ and scaled by u^t.

If there is an integer k with $2\pi = k$, the state vector returns to its initial direction after k steps so that we have a periodic behavior with frequency $\omega = 1/k$ or

$$\omega = \frac{\phi}{2\pi} \,. \tag{7.11}$$

If there is no such k we have a quasi-periodic behavior since the angle of the state vector never returns exactly to its initial direction.

Equation (7.10) also directly shows that the length of the state vector is decreasing (damped oscillation) if $u < 1$ and increasing if $u > 1$. The case of $u = 1$ is of special interest since the amplitude is conserved under the dynamics, i.e. the vector rotates with constant length and we have a periodic or quasi-periodic oscillation with constant amplitude.[3]

---

[2] The multiplication of two matrices $O(\phi_1)O(\phi_2)$ is again a of the same form, namely $O(\phi_1 + \phi_2)$. For the inverse holds the same: $O(\phi)^{-1} = O(-\phi)$.

[3] This is a good point to explain why we choose the special orthogonal matrix, which form a subgroup, defined by $\mathrm{Det}O = +1$, of the group of all orthogonal matrices. Actually, oscillatory behavior can also be achieved when using the orthogonal matrices with determinant -1. The reason is that the latter not only rotate the state vector by some angle but also invert it. In a robotic

The present considerations are valid for the linearized dynamics. In the nonlinear system with $u > 1$ the length of x increases so that, very soon, the linearization is not longer valid. However, the expansion of x is confined by the nonlinearity introduced by the function $g(Cx+h)$ in Eq. (7.2) so that the frequency is still roughly given by Eq. (7.11). More details will be given in numerical simulations below.

7.1.4 The Effect of the Bias

Before we turn to the learning dynamics, let us consider the effect of a fixed bias term h on the oscillatory dynamics. Remember from the one-dimensional system Sect. 3.4, that the bias changes the stability of the fixed points. In the two-dimensional system we observe a change in the shape and frequency of the nonlinear oscillations in the first place, but the most important effect is the break down of the oscillations if h exceeds a critical value. For illustration we use $C = 1.2O(\pi/30)$, see Eq. (7.7), ($C_{11} = C_{22} \approx 1.2$ and $C_{12} = -C_{21} \approx -0.125$) corresponding to $\omega = 1/60$. In numeric simulations, see Fig. 7.2, we observe without bias ($h_{1,2} = 0$) an oscillation with the frequency $\omega \approx 1/80$. Due to the nonlinearities there is a deviation from the expected $1/60$. Already when using $h_{1,2} = 0.02$ the oscillation frequency decreases to $\omega \approx 1/100$. For $h_{1,2} = 0.04$ the oscillations are destroyed and the state settles into a fixed point. This most sensitive dependence of the oscillations on the value of the h vector was already observed in [64], see also [65] and was shown to be useful in robotics for triggering oscillatory behavior depending on context. Interestingly, the authors found that the oscillatory region predicted by the theory is much larger than that obtained by computer simulations.

Our numerical studies corroborate these findings. We will see further below that including the learning dynamics of h into the combined system of state and parameter dynamics has a tremendous effect onto the behavior of the system.

7.2 Homeokinetic Learning

Let us now study the homeokinetic learning rules in two dimensional systems, especially in the short circuit setup. We will elaborate on different effects and provide analytical and numerical results. We will use the learning rules based on the time-loop error as derived in Sect. 5.2.2, see Eqs. (5.22, 5.23)

$$\frac{1}{\varepsilon}\Delta C_{ij} = \mu_i v_j - \varepsilon_i y_i x_j \tag{7.12}$$

$$\frac{1}{\varepsilon}\Delta h_{ij} = -\varepsilon_i y_i x_j \tag{7.13}$$

---

application this would mean that the state variable and with it the controller signals change sign in every time step, which might lead to the destruction of the robot.

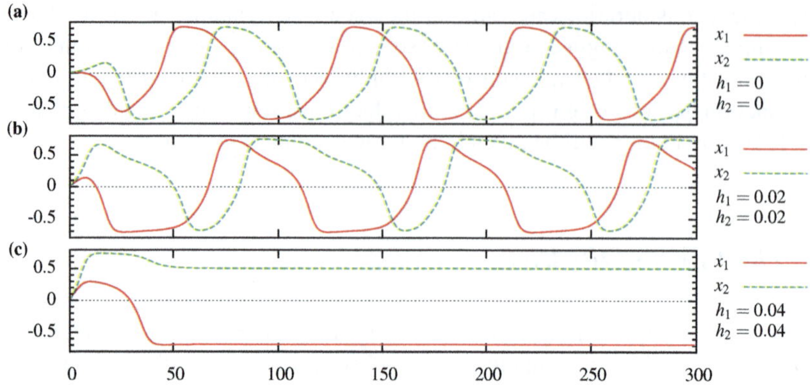

Fig. 7.2: The influence of the bias h on the stability of a limit cycle of an $SO(2)$ controller. $C = 1.2O(\pi/30)$ is a supercritical rotation matrix, the initial condition is $x_1 = x_2 = 0.01$. **(a)** shows the non-linear oscillatory dynamics without bias ($h_1 = h_2 = 0$); **(b)** bias $h_1 = h_2 = 0.02$ changes the shape of the oscillation; **(c)** bias $h_1 = h_2 = 0.04$ suppresses the oscillation.

with $v = L^{-1}\xi$, $\varepsilon_i = 2\mu_i (Cv)_i$, $\chi = \left(LL^\top\right)^{-1}\xi$, and $\mu = G'\, A^\top \chi$. In most of the applications we use the average gradient as introduced in Sect. 5.C with infinitesimal noise in the sense explained there.

In the nonlinear system there are many different dynamical regimes already in the two-dimensional case We will show in the following that there are different kinds of limit cycles, corresponding to both regular and irregular oscillations, and moreover that there are also metastable limit cycles. As in the one-dimensional case, the inclusion of additional parameters, the bias, deeply changes the nature of the combined dynamics and introduces completely new phenomena based on higher order hysteresis effects.

7.2.1 State-Parameter Dynamics

At first we will confine ourselves to the case that the bias of the neurons is equal to zero so that the full parameter dynamics is given by Eq. (7.12). We will be working with the average gradient, see Sect. 5.C (p. 101), in the example so that the combined system is a deterministic self-referential dynamical system in 6 dimensions. Such a system can show the full scale of behaviors known from deterministic dynamical systems ranging from fixed points (in 6-d space) to fully developed chaos. Nevertheless, the range of behaviors realized by this system is restricted somewhere in between these alternatives. The heuristics developed in Sect. 5.3 proves very helpful in understanding the emerging behaviors. In particular we will retrieve the attractive properties of the $SO(2)$ symmetry group reflected by the plateaus in the error landscape and the entanglement between state and parameter dynamics in weak and strong forms.

We have seen above that the case of $SO(2)$ networks is of special interest since it produces oscillations. Let us therefore ask how this specific structure interacts with the learning dynamics.

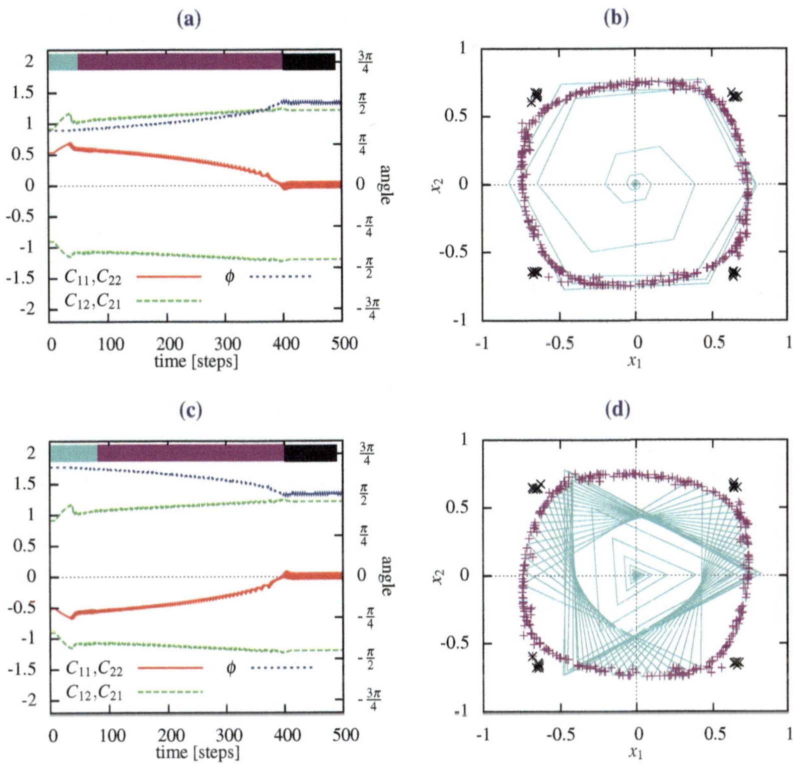

Fig. 7.3: Frequency drift to period-4 cycle. The panels **(a,c)** show the evolution of the controller matrix C and the rotation angle ϕ, **(b,d)** display the corresponding state vector x initialized with $\phi = \pi/3$ (60°) **(a,b)** and initialized with $\phi = 2\pi/3$ (120°) **(c,d)**. In both cases the system very rapidly self-regulates towards a Neimark-Sacker bifurcation with an ensuing slower drift towards a stable period-4 cycle. The color code in **(b,d)** corresponds to the time intervals marked with respective colored bars in **(a,c)**. Settings: $\varepsilon = 0.02$, average gradient, infinitesimal noise.

7.2.2 Oscillatory Behavior

Most of the observed dynamical patterns are oscillatory in nature. This can be attributed to either the fact that the controller matrix C is proportional to a rotation matrix or to a strong entanglement like in the case of the hysteresis oscillations in the one-dimensional system considered above. Let the controller matrix C be given by a scaled rotation matrix, i. e. $C = uO(\phi)$ with O given by Eq. (7.8). A first observation is based on the explicit expression of the error

$$E = \xi^\top \frac{1}{LL^\top} \xi = \zeta^\top \frac{1}{CC^\top} \zeta \tag{7.14}$$

where $\zeta = (AG')^{-1}\xi$. With the above C we find $CC^\top = u^2\mathbb{I}$ (unit matrix). The rotation angle only appears in the nonlinearities (G'), such that in a linear approximation the error is invariant to the rotation angle. The influence of the nonlinearities is kept weak by homeokinetic learning (since the neurons are not sensitive in the saturation region) so that the error landscape is in fact showing a plateau corresponding to the $SO(2)$ symmetry group that is heated up slightly by the residual effects of the nonlinearities in the sense described in Sect. 5.3.

As a consequence, when starting with C in a specific but very large region of the space of all C matrices, we observe a rapid convergence towards an $SO(2)$ structure with ensuing slow drift in the $SO(2)$ subspace due to entanglement provided by the nonlinearities. The direction of this drift is depending on the starting condition as can be seen in the following findings.

Experiment 7.1: Oscillatory behavior in 2D

When you start the simulation Oscillatory behavior in 2D short-circuit you won't see any graphics in this simulation. Instead the GUILOGGER opens and displays the evolution of the state and the parameters. You can interrupt by pressing <Ctrl>+C and change parameters. With
`>rotation=`ϕ
you can re-initialize the controller matrix with the specified angle ϕ (in degree). Try `rotation=` 30, 60, 90, and 120. Try also to randomize the controller matrix with `random=1` several times.

In computer simulations we observe that there is a critical rotation angle $\phi_c = \pi/4$ acting as a repeller in this metastability scenario. Starting a simulation with $\phi > \phi_c$ drives the system towards the period-4 cycle where $\phi = \pi/2$. You can try it yourself in Experiment 7.1. In concrete runs, we start with the matrices $C = uO(\pi/3)$ and $C = uO(2\pi/3)$. The evolution of the state and parameters dynamics is depicted in Fig. 7.3. As mentioned already above, the approach towards the period-4 cycle is also observed when starting from a large set of matrices that are not in the $SO(2)$ group. Instead, quite generally in many experiments one observes a very rapid convergence towards an $SO(2)$ structure with an ensuing slow increase of ϕ towards the period-4 cycle so that the latter is found to have a very wide basin of attraction.

We can understand this theoretically, at least the fact that the learning is stationary at this point. Qualitatively this can be directly read from the expression Eq. (7.14) of the error. As seen from Fig. 7.3, in the converged situation, the state vector is jumping between the corners of the square, meaning that the factors g'_i occurring in the vectors ζ of the error are always the same since $g'(z) = g'(-z)$ so that learning and state dynamics decouple (no entanglement) and stationarity is reached, see Fig. 7.4a.

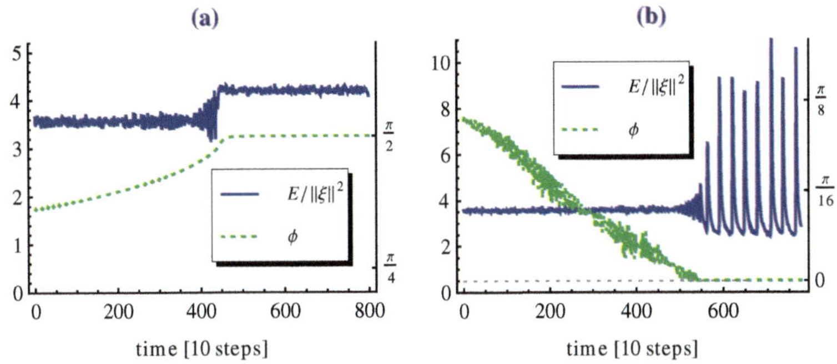

Fig. 7.4: Evolution of the time-loop error E for frequency drifting. **(a)** Drift to the period-4 cycle. A part of the time interval of experiment $46° \to 90°$, see Fig. 7.5, is shown. The normalized error $E/||\xi||^2$ (averaged over 10 steps) is invariant to the rotation angle unless close to the period-4 cycle. There is a slow drift to that situation and the error jumps to a higher plateau in the transition region. The state dynamics gets locked in this limit cycle, so that the parameter dynamics cannot decrease the error further. **(b)** Drift towards low frequencies. Corresponds to experiment $44° \to 0°$, see Fig. 7.5. As the angle approaches zero, the normalized error starts to oscillate and becomes smaller except for the peaks induced by the fixed-point switching of the x dynamics, see Fig. 7.6(a). Settings: average gradient, deterministic dynamics, $\varepsilon = 0.001$.

What happens if we initialize C with a smaller rotation angle, $\phi < \phi_c$? It turns out that the learning dynamics drives ϕ systematically down to lower rotation frequencies. The learning dynamics for different starting conditions is illustrated in Fig. 7.5 in the determinant-trace plane, as introduced in Sect. 7.1.2 and the evolution of the error can be followed in Fig. 7.4(b).

When starting with subcritical values of u (the Frobenius norm of the C matrix) we also observe that the system self-regulates towards the Neimark-Sacker bifurcation and further towards the slightly supercritical value of about 1.2. If the rotation angle is not small, than the state dynamics performs an oscillation and gets largely independent of the gradient descent on the time-loop error (weak entanglement), apart from the slow drift effect, that drives the rotation angle slowly towards either 0 or $\pi/4$ depending on the initial conditions.

However, different from the period-4 cycle that corresponds to a high frequency oscillation, the drift towards lower frequencies is ending in a kind of phase transition to a new regime where so to say the gradient descent takes over and an intensive entanglement of state and parameter dynamics is established, see the following.

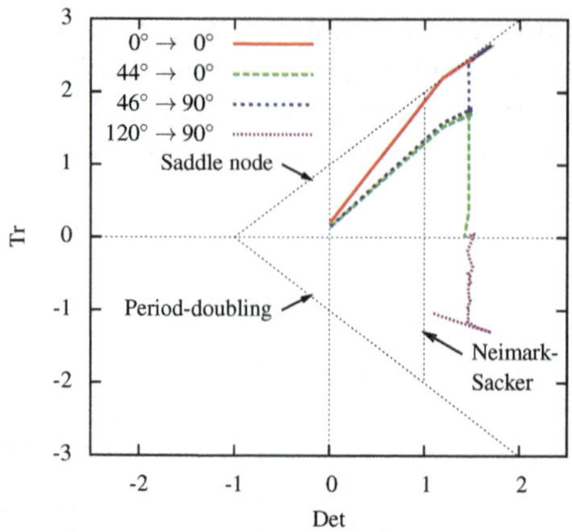

Fig. 7.5: Learning dynamics of the controller matrix in Det − Tr **(determinant-trace) plane.** The evolution of the controller matrix C for different initial conditions is shown. The curves are labeled with the initial rotation angle and the stationary rotation angle ($\phi_{t=0} \to \phi^{\star}$). Settings: average gradient, for the first three lines: $u = 0.1$, $\varepsilon = 0.001$. The last line ($120° \to 90°$) corresponds to the experiment Fig. 7.3(c),(d).

7.2.3 Low Frequency Oscillations

In order to study the new dynamical regime, let us start in the following with $C = \mathbb{I}$ if not otherwise stated. The learning dynamics first increases the diagonal elements of C to a value of around 1.2, as before. However, the off-diagonals do not develop opposite signs as in the $SO(2)$ picture but start oscillating with equal values so that the state-parameter dynamics enters a common periodic oscillation. Interestingly, this can be understood by the situation in 1-d with bias, i. e. each neuron uses the other one as a tonic input corresponding to a bias term that can be regulated by the off-diagonal connection. This can work only if the neurons synchronize their activities and this is just what happens with small values of the learning rate ε, as depicted in Fig. 7.6(a). If the learning rate is increased a different situation occurs, see Fig. 7.6(b), namely that one neuron remains in its fixed point and the other one performs an oscillation. This resembles the one-dimensional case with bias. The break down of the former regime can be attributed to the synchronization necessary for the common oscillation scenario and this is probably only operational if the time scale set by ε is long enough.

This is not the full spectrum yet. Increasing the learning rate even more we observe a switching of the roles of both neurons from time to time, see Fig. 7.7(a). Finally an irregular and frequent switching occurs as shown in Fig. 7.7(b).

7.2.3.1 Remark

We would like to emphasize that this spectrum of different dynamical patterns is achieved in the deterministic state-parameter system, i. e. with the average strength

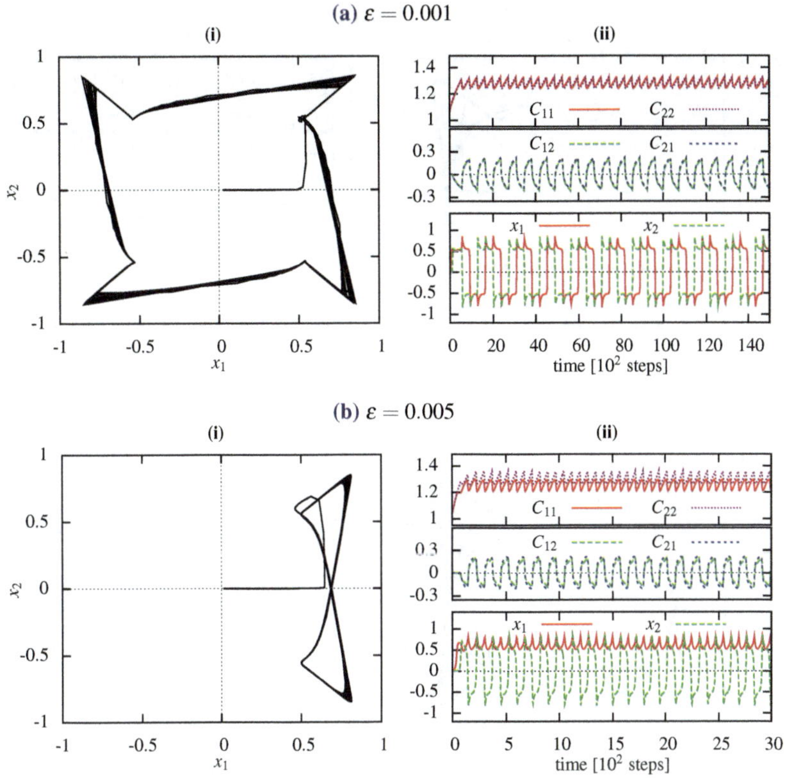

Fig. 7.6: Deterministic state-parameter dynamics in the SHORT CIRCUIT setting. **(a-b)** Dynamics with different learning rates (see Fig. 7.7 for a continuation); **(i)** phase plots of the state vector x; **(ii)** evolution of controller parameters C_{ij} and state x. Settings: average gradient, $x_{t=0} = (0.01, 10^{-5})^{\top}$ and deterministic dynamics.

σ^2 of the noise going to zero and the learning rate to infinity so that the product $\varepsilon = \sigma^2 \varepsilon_c$ stays constant, see Sect. 5.C for details. In the deterministic system, the emergence of the irregular switchings in the case of high ε seems a little astonishing, but they can be considered as an indication for a specific kind of deterministic chaos in the 6 dimensional dynamical system. Interestingly, similar behaviors are also observed if we add some noise to the state dynamics. Figure 7.8 shows that, with ε kept fixed and very small, the noise introduces similar effects as observed in the deterministic system by increasing the learning rate ε. Probably, we may consider ε in the present context as a parameter regulating the distance to the chaotic regime of the state-parameter dynamics but this needs still clarification by considering the Lyapunov exponents of the 6 dimensional system.

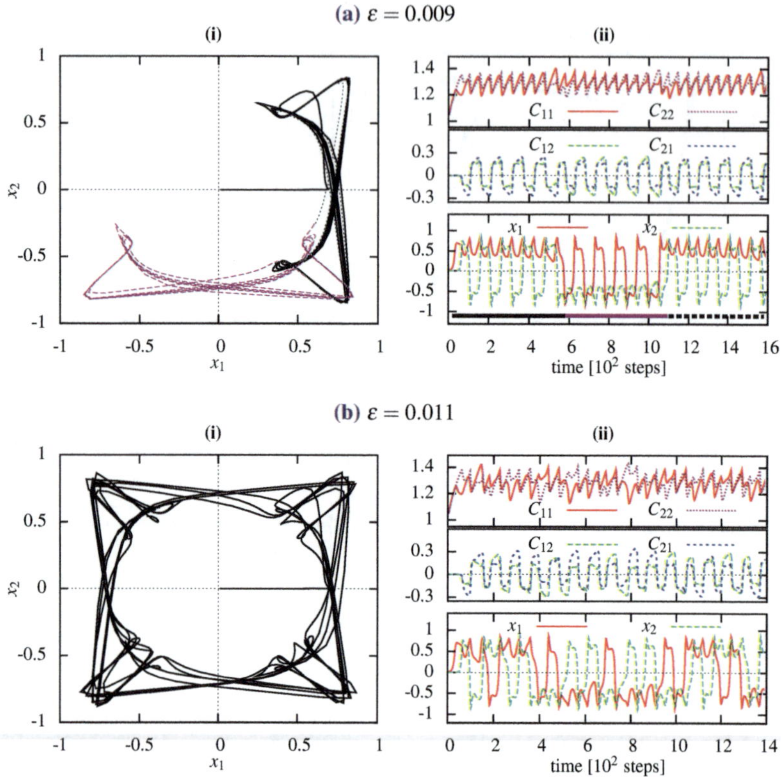

Fig. 7.7: State-parameter dynamics in the SHORT CIRCUIT setting — continuation. See Fig. 7.6 for description. The color code in **(a)(i)** corresponds to the time intervals indicated by the colored bars at the bottom of **(a)(ii)**.

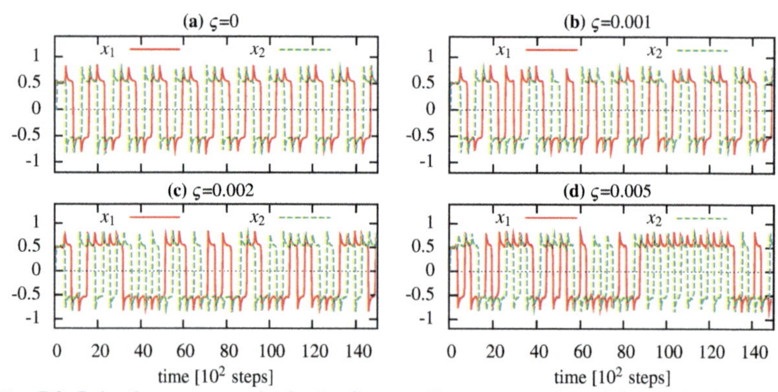

Fig. 7.8: Injecting sensor noise in the SHORT CIRCUIT setting. **(a)** no noise (same as in Fig. 7.6(a)); **(b-d)** increasing noise strength. The noise induces a switching similarly to the one observed when increasing the learning rate, see Fig. 7.7. Settings: average gradient, noisy dynamics, and $\varepsilon = 0.001$.

7.2.4 *Theory of Oscillatory Modes**

The emergence of oscillatory modes is a ubiquitous phenomenon in many of the applications done so far. The reasons for that can be seen from different perspectives: At a general level we identified the drive for oscillations in the invariance of the primary gradient (Sect. 5.2.1) of the TLE to rotational transformations, see Sect. 5.3.5. There we considered the linear system with the Jacobian matrix being a rotation matrix ($L \in SO(2)$). However, in order to prevent the explosion of the system dynamics we had to introduce there an artificial confinement. Let us here substantiate these considerations by showing the self-amplification and the confinement in the nonlinear setting and how oscillatory modes persist.

Let us start with the average gradient dynamics (Sect. 5.C) for the sake of simplicity, and consider moreover the case that the C matrix already is of the orthogonal structure, i. e. it is a rotation matrix $O(\phi)$ multiplied by a factor u, so that $CC^\top = u^2\mathbb{I}$. Note that the Jacobian matrix L is typically not a scaled rotation matrix due to the non-linearities. Introducing C into Eq. (5.48) through Eq. (5.50) with i. i. d. noise, yields the following learning rule[4]

$$\frac{u^2}{\varepsilon}\Delta C = G'^{-2}\left(\frac{1}{C^\top} - 2u^2 y x^\top\right) . \tag{7.15}$$

In Sect. 7.A.1 we show that in the average over a time window (exceeding the length of a period) we obtain

$$\Delta C \propto \left(1 - 2\kappa u^2\right) C \tag{7.16}$$

assuming additionally that the g' factors are all equal. A discussion may be found in Sect. 7.A.2 (p. 150). κ is an empirical constant proportional to $\|x\|$. Using $C = uO(\phi)$ this is an equation driving just the norm u of C towards a stationary value reached at $1 = 2\kappa u^2$.

The first point to note is that the update stays in the space of orthogonal matrices even in the nonlinear case so that the assumption is corroborated and we may derive the following conclusions on the behavior of the matrix elements. Firstly we see from this expression, that the matrix elements increase by absolute value when starting with a C matrix normalized such that $2\kappa u^2 < 1$. Increasing $\|C\|$ increases $\|x\|$ and hence κ so that the diagonal elements become stationary if $2\kappa u^2 = 1$ is reached. As it turns out, this is the same condition as in the $1D$ case revealing a **self-regulation towards a limit cycle regime slightly above the Neimark-Sacker[5] bifurcation point**. Moreover, the exact position of that regime can be defined by the α parameter entering the general learning rules Eqs. (5.22–5.24). This is another justification for using the α parameter in more general situations than the one-dimensional case where it was introduced.

---

[4] Proof: With $A = \mathbb{I}$ and $LL^\top = u^2 G'^2$ we have $\Delta C = \varepsilon u^{-2} G'^{-2} \left(C^\top\right)^{-1} - \epsilon y x^\top$ with $\epsilon = 2\varepsilon u^{-2} G'^{-2}$, which is Eq. (7.15).

[5] The Neimark-Sacker bifurcation is the discrete counterpart of a Hopf bifurcation.

This self-regulation mechanism is well confirmed by the above experiments, e. g. Fig. 7.5. However, the experiments also reveal that, on a (much) longer time scale, the frequency of the oscillations changes. Depending on the initial condition for ϕ we observe a systematic drift either to lower angles or to the period-4 cycle. The explanation of the drift effects is found in the existing correlations between the terms g' (in G') and the state x, which have been omitted in the transition from Eq. (7.15) to Eq. (7.16). The treatment of these terms is more involved so that we leave this to future work. Nevertheless, we can understand why the period-4 cycle is stationary: the correlations between the nonlinearity and the state vanish and thus there is no drift, see Sect. 7.A.2 for details.

7.3 Second Order Hysteresis

In the one-dimensional system we have seen that the inclusion of the bias dynamics introduces an hysteresis effect, which change the fixed point dynamics into a limit cycle. Similar effects are to be expected in the two-dimensional case. In order to get some systematics into the phenomena we at first keep the controller matrix C fixed and study the effect of updating the bias h only. After that we come to the full dynamics, i. e. updating both C and h by Eqs. (7.12, 7.13), to discover the enormous enrichment of the state-parameter dynamics by the inclusion of the adaptive bias.

7.3.1 *Bias Dynamics and* $SO(2)$ *Oscillations*

We have seen in Sect. 7.1.4 that in an $SO(2)$ network, the bias h can be used as a kind of switch for changing between oscillatory and fixed-point behavior. Let us now consider such a network but with h changed by the parameter dynamics of homeokinetic learning as defined by Eq. (7.13). This results in a novel phenomenon: the coexistence of two limit cycle attractors with different frequencies and basins of attraction defined by regions of different size in x-h space. An example is given in Fig. 7.9, depicting the behavior for $C = uO(\phi_C)$ with $O(\phi)$ defined in Eq. (7.8). We choose the rotation angle as $\phi_C = \pi/90$ so that the state oscillates with a period of 180 steps, with some deviation depending on the strength of the nonlinearity, see Sect. 7.1.3 (p. 129). How is it possible that there are oscillations with two different frequencies observed?

For a discussion, we introduce the rotation angle ϕ_x of the state vector x_t in one time step. The size of ϕ_x is obtained empirically by finding the frequency of the nonlinear oscillations (in our case by counting the zero crossings of the state variables) its sign being defined by the phase relation between the components of the state vector x_t. In the bias free dynamics we have $\phi_C \approx \phi_x$ if the nonlinearities are not too dominant so that the state vector indeed follows the rotation dictated by C.

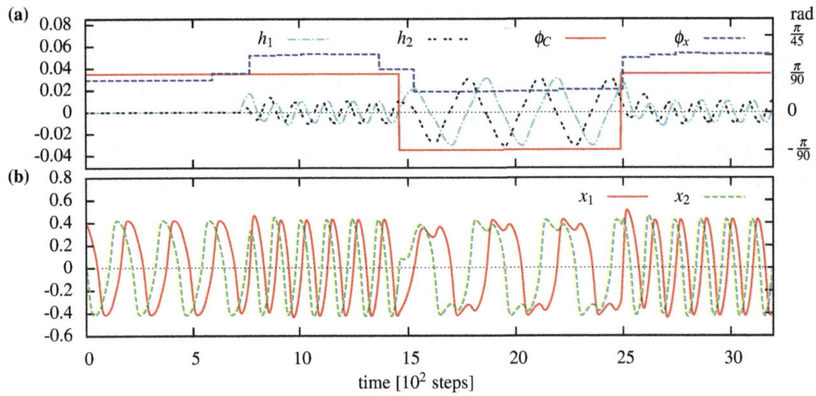

Fig. 7.9: Frequency switching by bias dynamics. **(a)** Bias h, rotation angle ϕ_C of controller matrix C and angle ϕ_x calculated from the state x; **(b)** State x. The controller matrix was fixed to $C = 1.2O(\phi_C)$ with $\phi_C = \pm\pi/90$ (± 2 degree). Up to time 7 (700 steps) the bias dynamics was disabled and an oscillation following the rotation matrix C is observed (nonlinearities cause small deviation: $\phi_x \approx \pi/105$). Then the bias dynamics was switched on ($\varepsilon - 0.0005$) and the frequency increases ($\phi_x \approx \pi/60$). Here we have a weak entanglement between state and parameter dynamics. At about time 15 the rotation angle of the controller matrix is inverted. The bias dynamics continues to rotate the state in the same direction as before but with a lower frequency ($\phi_x \approx \pi/165$), such that we speak from a strong entanglement. After switching back the high frequency regime is entered again. Settings: Average gradient, deterministic dynamics, $\varepsilon = 0.0005$.

The different nature of these two scenarios is understood by looking at the phase relation between ϕ_x and ϕ_C. In the high frequency case we have $\phi_x \gtrsim \phi_C$ so that both vectors rotate into the same direction, whereas in the second case the signs of ϕ_x and ϕ_C are different. In the simulation we have switched between both behaviors by inverting the rotation angle of the C matrix from $\phi_C = \pi/90$ to $\phi_C = -\pi/90$ and back. In both cases the state continues to rotate in positive direction ($\phi_x > 0$) but with a much lower rotation frequency. Obviously, the phase relation influences the interplay between the state dynamics and the h dynamics in an essential way and is therefore responsible for the different frequencies of the limit cycle attractors.

The effect can tentatively be rooted back to the different strength of the entanglement between state and parameter dynamics. It is a recurrent pattern that the latter can get hold of the former if the state dynamics is or can be kept slow. Obviously this is happening if the phase relations are appropriate. The decisive influence of the phase relations on the behavior of the system will be seen to play an essential role also in the more complicated cases considered below. Thus this seems to be an essential ingredient of the present approach. In particular, one may use the phase shift in order to switch between these two limit cycle attractors. In other words this opens the possibility of the robot to react in a definite way to environmental influences. In fact, if the system is in one of its attractors, a simple shift in phase between the sensor values may lead to a switching of the frequencies making the robot to try a different dynamical regime.

These findings can be seen as another piece of a pattern in the interplay between state and parameter dynamics. In the in phase regime, the state dynamics cases the error to be essentially on the $SO(2)$ plateau of its landscape, so that perturbations by the parameter dynamics are weak (weak entanglement). In the special case it leads to a slight frequency increase. The anti-phase regime is mainly dominated by the parameter dynamics, managing to rotate the state vector into the direction opposite to that prescribed by the controller matrix C (strong entanglement). As it can be expected, this regime is taking place only if the state is not changing too fast since otherwise the parameter dynamics can not take over. Interestingly, this taking over has to be initiated from outside (e.g. by perturbations or by a particular situation) by setting the phase relations accordingly.

7.3.2 Frequency Sweeping and Hysteresis in Frequency Space

We are now going to study the combined dynamics of state and parameter variables, i.e. that of x, C, and h, which takes place in the now 8-dimensional space. We initialize as before with $C = \mathbb{I}$, $h_{1,2} = 0$ and $x = (0.01, 10^-5)^\top$.

As the numerical studies reveal, there are different dynamics depending on the learning rate ε. If using a very small learning rate, the state-parameter dynamics is found to converge after some time to a periodic limit cycle with a low frequency. The simulation results are displayed in Fig. 7.10. Interestingly, the C matrix develops into an $SO(2)$ structure with relatively high rotation angle ($\phi_C \approx -\pi/20$). The actual oscillation of the state occurs however with $\phi_C \approx \pi/300$, determined by the bias dynamics as in the previous section. Note the emerging anti-phase relation of h and x w. r. t. C in the regime of strong entanglement between state and parameters dynamics.

The most interesting effect of the combined dynamics was discovered from computer simulations when using larger learning rates where the system is found to wander through its different frequencies in a periodic manner. For this reason we call it the frequency sweeping effect. In order to understand how this effect is produced we consider a typical simulation as given in Fig. 7.11. The first observation is that, most of the time, the matrix C realizes an $SO(2)$ structure. However now the rotation angle ϕ_C is not constant but displays also a periodic behavior on a much larger time scale.

When considering the state dynamics it does not correspondingly change the frequency but instead, similarly to Fig. 7.9, there are again two possible relations between the signs of ϕ_C and ϕ_x. However now, the rotation angles ϕ_x and ϕ_C are varying in size over a very large range and, as the Fig. 7.11 shows, the phase relation may be considered as a kind of hysteresis parameter. Different from the usual hysteresis systems, this is not a system parameter but something that is emerging in the phenomenon itself. Therefore we call this a second order hysteresis. Like with

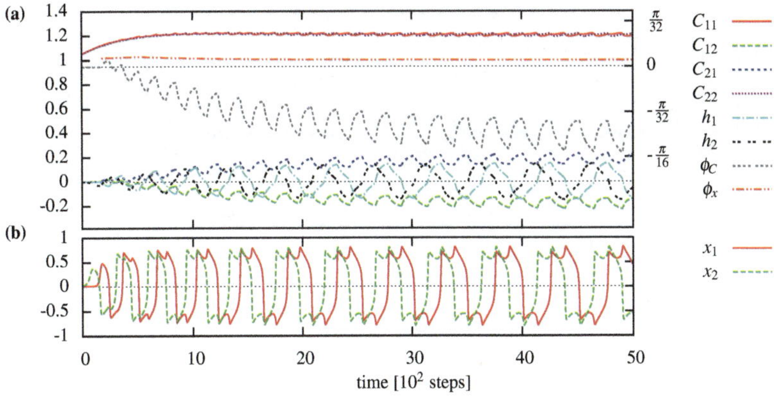

Fig. 7.10: Combined dynamics of C and h for low learning rates. **(a)** Controller parameters C and h, rotation angle ϕ_C (right axis) of controller matrix C and angle ϕ_x (right axis) calculated from the state x; **(b)** State x. The controller matrix C develops into rotation matrix with negative rotation angle (ϕ_C). However, the combined dynamics reaches a periodic orbit with a fixed low frequency with positive rotation angle (see ϕ_C and ϕ_x). In this case the dynamics actually converges (long term behavior checked). Settings: Average gradient, deterministic dynamics, $\varepsilon = 0.0005$.

the hysteresis oscillators introduced in Sect. 6.3 this new hysteresis parameter is swept by the learning dynamics itself and not by hand from outside.

The time-loop error during such a sweeping mode is displayed in Fig. 7.12. As we know from Sect. 7.2.3, the error is essentially the same for all high frequency modes. If, however, the phase of the state does not match the controller matrix (ϕ_x and ϕ_C differ in sign) the error shows a different dynamics. In this regime the state oscillations are strongly influenced by the bias dynamics. Thus the error starts to fluctuate and for very low frequencies the error is decreased. In this period the parameter dynamics has enough time to decrease the error while each component of the state is kept close to its fixed point. Then the state hops again such that the error rises as observed before.

7.3.3 Frequency Sweeping in Real Systems

The sweeping mode discussed above was found in the idealized world system with infinitesimal noise. Most interestingly, the phenomenon is very robust and can be found also in more complicated robotic systems with homeokinetic control. In particular, we may consider the BARREL with its dominating embodiment effects, see Experiment 7.2. As observed in the experiment, where we enable model learning (Eqs. (4.5, 4.6)), under widely differing conditions the BARREL shows the sweeping effect with alterations due to its specific embodiment. In a typical experiment we observe that the velocity (and thus the rotation frequency) is steadily increasing

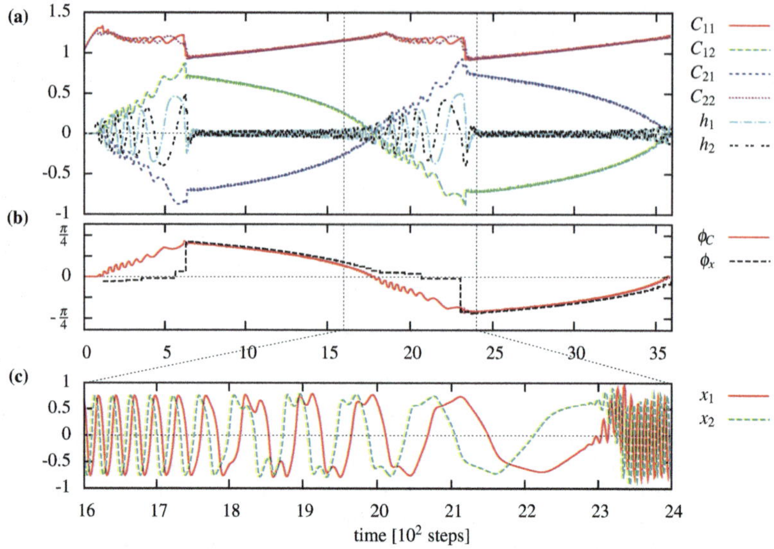

Fig. 7.11: The frequency sweeping effect. Same settings as in Fig. 7.10 with higher learning rate. **(a)** Controller parameters C and h; **(b)** Rotation angle ϕ_C of controller matrix C and rotation angle ϕ_x calculated from the state x; **(c)** State x in a small interval of time. The controller matrix takes the $SO(2)$ structure, running through a wide range of rotation angles **(b)** in a periodic manner. Up to time 7 the bias dynamics slows the oscillations down due to the anti-phase relation as explained in Fig. 7.9. In the slow mode, the destabilization drive of homeokinetic learning is increasing steadily. Eventually a regime of maximum instability is reached with ensuing switching of the phase relation so that the system jumps to the high frequency regime (as dictated by C) at time 7 and 23, see **(c)**. Settings: Average gradient, deterministic dynamics, $\varepsilon = 0.005$.

up to a maximum value, then it decreases, inverts sign and so forth. As a result the velocity of the robot is oscillating periodically at a slow timescale, as displayed in Fig. 7.13. Here we see clearly how the inertia of the robotic system influences the generation of behavior. In contrast to the sharp frequency jumps observed above (Fig. 7.11) we find here a smooth transient as it can be seen in Video 7.1 and/or reproduced by doing Experiment 7.2.

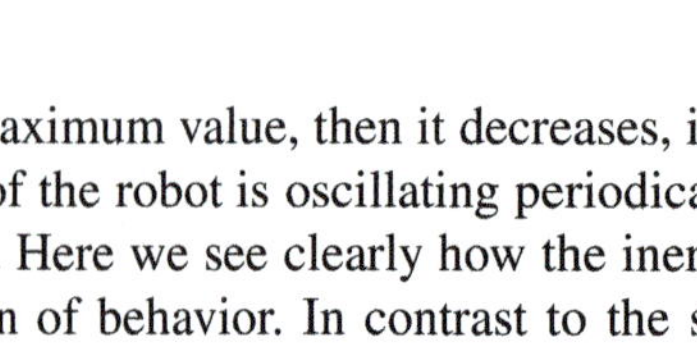

Video 7.1: Sweeping mode. Behavior of the BARREL with homeokinetic control, see also Fig. 7.13 for details. The video can be watched at `http://playfulmachines.com`.

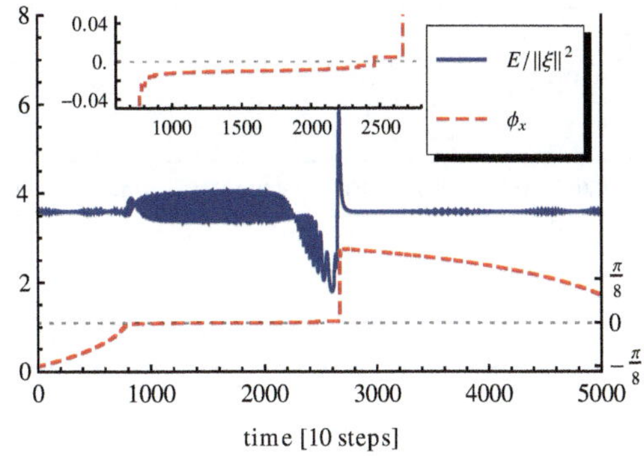

Fig. 7.12: Evolution of the time-loop error E for the frequency sweeping mode. A part of a full sweeping cycle is displayed. In comparison to Fig. 7.11 a much lower learning rate was used to have a better separation between state and parameter dynamics. The rotation angle ϕ_x indicates the current oscillation frequency of the state, note also the inset. The error is nearly constant during the $SO(2)$ scan with matching phase relation at the beginning and the end of the shown time interval. During the time with anti-phase relation (time 900–2850) the error fluctuates and eventually is lowered significantly when the oscillations become very slow. Then the phase relation inverts and thus the error rises and relaxes to the plateau. Settings: average gradient, deterministic dynamics, $\varepsilon = 0.001$.

Experiment 7.2: Homeokinetic control of BARREL

The simulation Homeokinetic control of BARREL starts with the robot at rest. Observe the generated behavior and the parameter dynamics with the GUILOGGER. Remember, that you can change the simulation speed with '+'/'-'. Try different learning rates for instance by entering `>epsC=value`
You may re-initialize both the parameters of the controller and the forward model randomly as described in Experiment 3.2. For many conditions the dynamics will come back to the sweeping behavior, but there are also initial conditions (especially if the $\mathrm{Det}(C) < 0$) where the dynamics will lead to strange and often fixed behaviors.
In order to exert external forces to the barrel use either the <Ctrl>+<left Mouse button> or the <Ctrl>+<right Mouse button>.

7.3.4 Frequency Sweeping in 3D

Even though we are focusing here on 2 dimensional systems, let us have a brief look at a 3-dimensional system, using the SHORT CIRCUIT setting again ($A = \mathbb{I}$). This means $m = n = 3$, such that $x, y, h \in \mathbb{R}^3$ and $C, A \in \mathbb{R}^3 \times \mathbb{R}^3$ and the state-parameter dynamics is realized as a 15 dimensional deterministic dynamical system (we are still working with the average gradient and infinitesimal noise). In Fig. 7.14 the results of a computer simulation are shown where we discover that the frequency sweeping effect also occurs in three dimensions. There are always two channels (dimensions) of the state performing a frequency sweeping behavior analog to the two-dimensional case, the remaining channel being more or less decoupled for that period of time. Interestingly, after such a sweeping mode the pairing changes and an oscillation in a different 2-d subspace occurs. The controller matrix goes through a subclass of $SO(3)$ structures, namely those with an $SO(2)$ form in one of the 2-d subspaces. This corresponds to the rotation around each of the coordinate axes, such

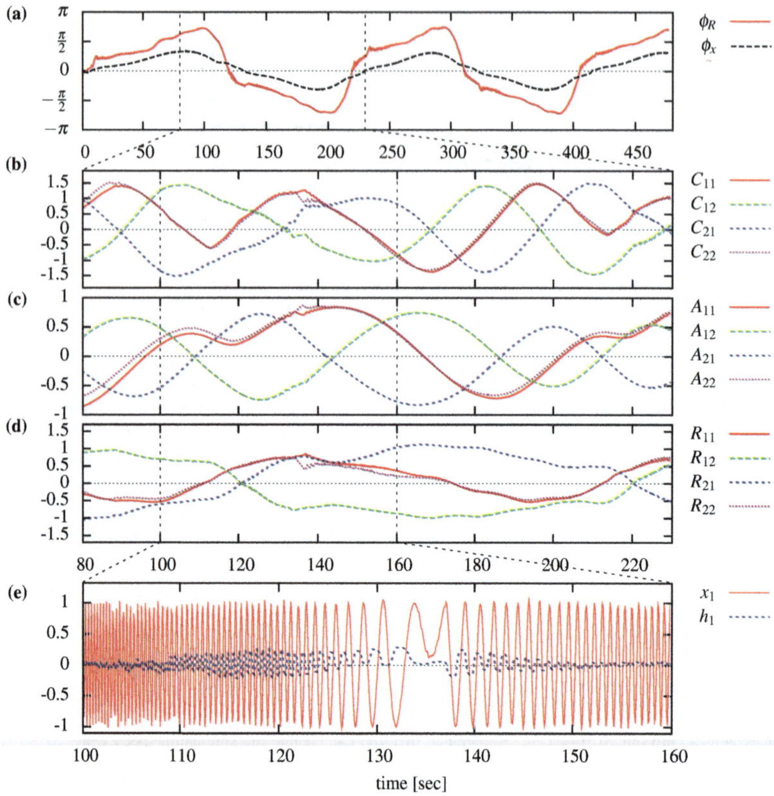

Fig. 7.13: Frequency sweeping of the BARREL. The robot periodically accelerates, then slows down, inverts its velocity and rolls with increasing speed. **(a)** Rotation angle ϕ_R of R and true rotation angle ϕ_x obtained from the sensor values for the entire experiment. **(b), (c)** Parameter evolution: the elements of the controller matrix C and model matrix A, respectively, using a smaller time interval. **(b)** Elements of matrix $R = CA$. **(e)** Sensor value $x_1(t)$ and bias term $h_1(t)$, for clarity shown in an even smaller time interval. Setting: $\varepsilon_c = \varepsilon_A = 0.05$, update rate 25 Hz.

that we have 6 different oscillatory modes, one for each axis running clockwise and counter clockwise. Again we find the regimes with in-phase and anti-phase relation such that the oscillation frequency suddenly rises and then falls slowly.

7.4 Resonances — A Case for Embodiment

Except for the BARREL experiment, we have been working with an extremely simplified world consisting of $A = \mathbb{I}$ meaning that the system's response is a direct copy of the motor signals plus some isotropic noise of infinitesimal strength. Let us now

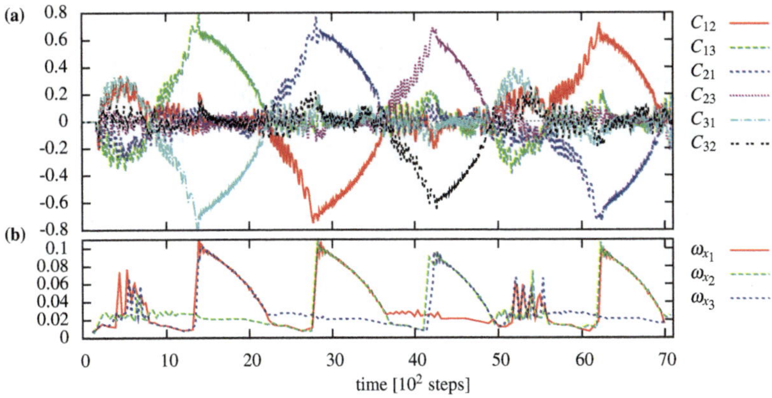

Fig. 7.14: Frequency sweeping effect in 3 dimensions. Same as in Fig. 7.11, but with 3-dimensional state space. **(a)** Off-diagonals of controller parameter matrix C; **(b)** Rotation frequencies ω_{x_i} calculated from the state x (no phase relation extracted here). The diagonal elements of C (not shown) behave analog to the 2-dimensional case. The controller matrix goes through a subclass of $SO(3)$ structures, namely those with an $SO(2)$ form in one of the 2-d subspaces. This corresponds to the rotation around each one of the coordinate axes. The rotation angle is seen to run through a wide range of angles in a periodic manner. Settings: Average gradient, deterministic dynamics, $\varepsilon = 0.005$.

consider homeokinetic control in a world having a dynamics of its own. We consider a system given by

$$x_{t+1} = y_t + Sx_t + \varsigma_t \tag{7.17}$$

where S represents the action independent dynamics of the world. Assume the world is known to the controller and the forward model is thus making a prediction of the next sensor state as

$$x_{t+1} = Ay_t + Sx_t + \xi_{t+1}$$

with $A = \mathbb{I}$. We will study the influence of the world in two different scenarios, starting with the case that the bias $h = 0$ and including the bias dynamics afterwards.

In the following S is manually changed during the experiment. We start with the case that the autonomous ($y = 0$) dynamics of the world[6] (given by S) is a strongly damped harmonic oscillator, i. e. $S = \frac{1}{2}O(\phi)$. Later on we use a magic circle oscillator [58, 70] described by the following matrix

$$S = uO_{\mathrm{MC}}(r) = u \begin{pmatrix} 1 & r \\ -r & 1-r^2 \end{pmatrix}$$

with $0 < r < 2$ realizing an oscillator that is structurally different from the harmonic $SO(2)$ oscillator.

In the $h = 0$ case, the reaction of homeokinetic learning to this new situation and the emerging behavior can be read off from Fig. 7.15. As hoped for, C develops into

[6] The autonomous dynamics of the world is given without actions ($y = 0$): $x_{t+1} = Sx_t + \varsigma_t$.

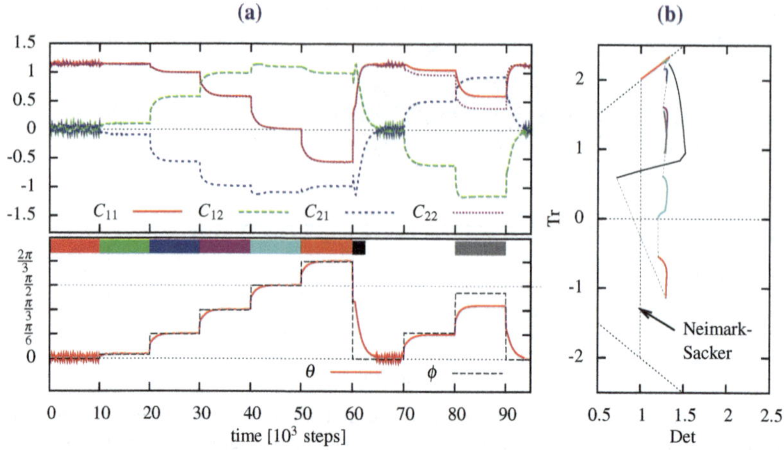

Fig. 7.15: The resonance effect. **(a)** Dynamics of controller matrix C (**upper** panel) and rotation angle ϕ of world (S) and the actual rotation angle θ (of $R = AC + S$) (**lower** panel); Every 10 k steps the world was changed. Up to step 70 k $S = \frac{1}{2}O(\phi)$, after that $S = \frac{1}{2}O_{\text{MC}}(r)$ with $r = 0.5, 1$, and 0 was used. **(b)** Visualization of the dynamics of $R = AC + S$ in Det – Tr plane (see Fig. 7.5). The transitions drawn in *dotted gray* are caused by the change of S. The color code corresponds to the colored bars in **(a)**. Parameters: $\varepsilon_c = 0.001$, $\varsigma = 0$ (average gradient).

a scaled rotation matrix with angle θ that matches the rotation angle of S in the case of the harmonic oscillators and is close to the angle for the magic circle oscillators. In the case that $u = 1/2$ and $r = 1$, C develops into mixture of the magic circle matrix and an $SO(2)$ matrix. Fig. 7.15 demonstrates very clearly that the resonance between internal and external world is established in all cases at a very short time scale despite the fact that the intrinsic oscillations are strongly damped.

When including the bias dynamics, the question is whether the sweeping effect will either overrun or take account of the intrinsic dynamics of the world. In the experiments, we initially set S to zero again so that we observe the sweeping effect as above if we use appropriate initial conditions. Eventually we set S to a strongly damped 20° degree rotation matrix as

$$S = 0.2O(\pi/9)\,.$$

Interestingly, the frequency sweeping continues until the frequency **and** phase relation matches that of the external world. Then, the parameter dynamics becomes stationary, see Fig. 7.16, the controller being locked onto the resonance frequency of the system under control. Interestingly, in comparison to the case without bias dynamics, the intrinsic dynamics of the world was much weaker here (factor 0.2 instead of 0.5). Such a strongly damped dynamics does not lead to a locking without the entanglement with the bias dynamics. The second important novelty here is that

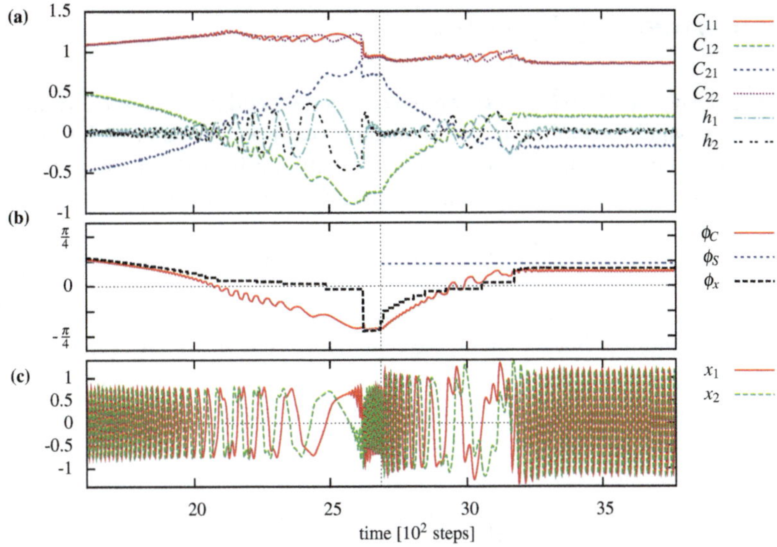

Fig. 7.16: Frequency sweeping detects and locks latent modes. The same experiment as in Fig. 7.11, but with an oscillatory world. **(a)** Controller parameters C and h; **(b)** Rotation angles ϕ_C, ϕ_S of controller matrix C and world matrix S, and angle ϕ_x calculated from the state x; **(c)** State x. Up to time 27 the world matrix S is zero, and after that $S = 0.2O(\pi/9)$. This corresponds to a strongly damped oscillations matrix. The frequency sweeping stops close to the external frequency and locks to it, as seen in **(b)**. Settings: Average gradient, deterministic dynamics, $\varepsilon = 0.005$.

a wide range of frequencies are searched actively by the sweeping effect, so that a broad spectrum of potential resonances can be detected and amplified.

These findings show that homeokinetic learning is able to detect and amplify latent oscillatory modes of the outside physical world. We will encounter in many cases this phenomenon also in the experiments with much more complicated robots both in physical reality and in physically realistic simulations.

Appendix 7.A Properties of Oscillatory Modes

In the following derivations we use $A = \mathbb{I}$ so that the feedback strength $R = C$ for the sake of notational simplicity.

7.A.1 *Derivation of Eq.* (7.16).

In order to derive Eq. 7.16 (p. 139), we have to understand the role of the term $xy^\top$ in Eq. (7.15), which is oscillating since the state vector is rotated in each time step by the angle ϕ. Let us assume that we are sufficiently far from the saturation region of the neurons so that with $g(z) = tanh(z)$ we obtain approximately

$$y_t x_t^\top = g(Cx_t)x_t^\top \approx \gamma C g(x_t) x_t^\top \tag{7.18}$$

where $\gamma > 0$ is a scaling factor taking account of the shrinking effect due to the squashing. Assuming ε is very small, the time dependent quantities in the learning rule can be replaced by their time averages (in an appropriate time window). In order to get the average of $yx^\top$ we neglect possible correlations with the matrix elements of G' and use that with $x_t \propto (\cos\phi t, \sin\phi t)^\top$

$$\lim_{\theta\to\infty} \frac{1}{\theta}\int_0^\theta g(x_t)x_t^\top\, dt = \beta\mathbb{I}$$

since the cross channel terms average out.[7] In particular, the factor $\beta > 0$ is equal to $\|x_t\|^2/2$ if $g(x)$ may be replaced with x (far from saturation). Using this average, we find that the time average of $yx^\top$ is given by

$$\lim_{\theta\to\infty} \frac{1}{\theta}\int_0^\theta y_t x_t^\top\, dt \approx aC$$

with an empirical constant $a > 0$ of order 1. Using moreover $\left(C^\top\right)^{-1} = u^{-2}C$, we obtain Eq. (7.16).

7.A.2 *The Period-*4 *Cycle*

The transition from Eq. (7.15) to Eq. 7.16 (p. 139) is based on the stability properties of the period-4 cycle in the combined state-parameter dynamics In computer simulations the period-4 cycle has been observed to be an attractor with a large basin of attraction, see Fig. 7.5. We can easily understand the stationarity of the period-4 cycle and its local stability in state space. In order to work this out we introduce a vector $s = (1,1)^\top$ and note that the rotation matrix $O(\pi/2)$ acts as

$$Os = \begin{pmatrix}-1\\1\end{pmatrix},\ O^2s = \begin{pmatrix}-1\\-1\end{pmatrix},\ O^3s = \begin{pmatrix}1\\-1\end{pmatrix},\ O^4s = s\,. \tag{7.19}$$

[7] We could also average over one period, i. e. use $\frac{1}{2\pi/\phi}\int_0^{2\pi/\phi} g(x_t)x_t^\top\, dt = \beta\mathbb{I}$, but the self-averaging realized by on-line learning with small learning rate corresponds more to an extended time horizon.

So, in the linear system, once the state has reached one of these vectors, its further fate is given by just changing signs of its components. Now, assuming with the scaling factor ρ that $x_t = \rho s$ at any time t the nonlinear dynamics with $C = uO$ obeys

$$x_{t+1} = g(uOx_t) = Og(ux_t) = Og(u\rho s) = Og(u\rho)s$$

due to the antisymmetry of $g(\cdot)$, the tanh function, and the sign changing property of O, see Eq. (7.19). The same is true with taking any of the vectors in Eq. (7.19) as a starting state. The dynamics is stationary if $x_{t+1} = \rho Os$ meaning

$$g(u\rho) = \rho \, .$$

Let us consider this for u close to one, so that $u = 1 + \delta$ with $0 < \delta \ll 1$ so that the amplitude of the oscillations is kept low (far from saturation). We find the approximate solution by using the third order approximation for the tanh function obtaining

$$\rho = \pm \frac{1}{u^2}\sqrt{3u(u-1)}$$

disclosing an oscillation with finite amplitude ρ for $u > 1$.

The stability of the state dynamics is proven by linearization, putting

$$x_t = x_t^0 + \delta x_t$$

with x_t^0 being the state vector of the stationary oscillation obeying $x_{t+1}^0 = g\left(Cx_t^0\right) = Ox_t^0$ for any t. Using Taylor expansion we get

$$x_{t+1} = g(uOx_t) \approx Ox_t^0 + L\left(x_t^0\right)\delta x_t$$

with $C = uO$, $L(x) = \frac{\partial g(Cx)}{\partial x} = G'C$, and $G' = \mathrm{diag}[g_1', g_2']$. Using that $g_1' = g_2' =: g'$ we obtain

$$\delta x_{t+1} \approx g' u O \delta x_t$$

with $g'u > 0$. Using $g' = 1 - g^2$ and Eq. (7.A.2) we find in the stationary state

$$g'u = \left(1-\rho^2\right)u = u - \frac{3}{u^2}(u-1) = 1 - 2\delta + O\left(\delta^2\right) \, .$$

Convergence is achieved if $g'u < 1$, which is obviously true in the low amplitude case considered here. This proves the stability of the **state** dynamics in the period-4 cycle. The result does not prove the observed attraction of the C matrix towards the period-4 structure by the learning dynamics. However, it proves the convergence towards the equality of the g_i' factors that was essential for the transition from Eq. (7.15) to Eq. (7.16) where the frequency drift vanishes. So, the result at least helps in better understanding the attractive behavior of the combined dynamics in **state-parameter** space.

Chapter 8
Low Dimensional Robotic Systems

Abstract: In this chapter we will demonstrate the performance of the homeokinetic control system when applied to physical robots. We will recognize many of the effects observed in idealized world conditions, shown in the previous chapters, but most dominantly witness new features originating from the interaction of the learning dynamics with the respective embodiment. Among them are non-trivial sensorimotor coordination, excitation of resonance modes, the adaptation to different environments – all emerging from the unspecific homeokinetic learning rules. The entanglement effect is seen to make emerging motion patterns transient so that the behavioral options are explored and a playful behavior is observed. In order to keep things simple enough for analysis we consider here only low-dimensional systems and leave the high-dimensional ones for Chap. 10.

In order not to loose contact to reality this chapter presents some applications of the homeokinetic controller to real and simulated robots. We will restrict ourselves to robots that are driven by two or three motors in order to be able to study various effects in detail.

The common feature of all robots is their compliant nature. As a consequence the state of the robot is not only determined by its actions, but it depends to a large extent on the current situation and the physical interaction of the body in itself and with the environment. New effects are observed that did not occur under the idealized world conditions treated in Chaps. 6 and 7, the most ubiquitous being the establishment of nontrivial sensorimotor coordinations and the astonishing variety of behaviors that emerge from the same principle in robots with different embodiment.

Before starting with the different robots and the experiments let us take a look at the common course of actions we take. We use in all experiments the same one-layer controller using the explicit learning rules given by Eqs. (5.22, 5.23). Specific is only the number n of sensors and m of neurons (given by the number of motors) and the choice of certain algorithmic options that are introduced and explained in more detail in Chap. 15. Except stated otherwise, the forward model is initialized as a unit matrix and the controller as a scaled unit matrix[1] such that the sensorimotor loop (in the robot's internal representation) is subcritical. In this configuration the robot

[1] Depending on the setting the sensors may outnumber the motors, so that C and A are rectangular. We initialize the diagonal elements with a constant and the rest with zero.

R. Der and G. Martius: The Playful Machine, COSMOS 15, pp. 153–182.
springerlink.com

will typically not move and only small actions will occur, caused by the noise in the sensor readings. In this regime neither controller nor forward model will have gained any specific information about the particular robot so that we start with a totally unbiased situation, any further developments being caused alone by the interplay between the physical state and the parameter dynamics. More information about how to initialize and customize the homeokinetic "brain" can be found in Sect. 15.2.

Now we are ready to start! Let us enter the world of playful machines.

8.1 SEMNI

The first robot we want to present here is SEMNI, see Fig. 8.1 and [72], a machine designed and built at the HU Berlin by Manfred Hild and his team. Comparable to the BARREL we have considered earlier the robot also shows strong embodiment effects, given in particular that its sensor readings are related to its physical situation only in a very intricate way, see below for details.

Before we start let us tell a short story on how we came to control this robot. We had a phone call with Manfred about a different matter, when he was eventually saying: "Guys, I have here a robot that you will like!." After looking at the picture there were no questions asked and we took the train for one day to Berlin—very curious about what would happen. First we had to solve the usual technical problems. Thanks to the good organization and the great help of the team in Berlin we managed that until lunch time. The robot was connected with a cable to our laptop and the sensor and motor values were passed back and forth. Then something amazing happened—the robot is getting into motion and already during the first half hour we have seen so many nice but unexpected motion patterns excited by the controller – we could not quite believe it ourselves. The remaining afternoon we experimented with different settings, part of it is presented here.

The main points we want to show with the following experiments are the integration of different sensors into the sensorimotor loop and how the controller picks up resonance modes.

8.1.1 Construction

The robot consists of two parallel frames with an oval shape. Between them a two segmented leg, is mounted with two servo motors see Fig. 8.1. The leg can be moved to either side of the robot and changes such the center of gravity but also the support surface. If moved in the center then it can be completely subsumed into the frame boundary. Above the leg, there is circular "head" in which the controller board is mounted.

The motors are Dynamixel RX-28 [146] servo motors which are, however, controlled by the hardware with the voltage directly, so in fact no servo functionality is

Fig. 8.1: The robot SEMNI by M. Hild. It has strong embodiment effects, is very light and robust, and has an external power supply. See [72] and Fig. 8.2 for more details. © Manfred Hild [71].

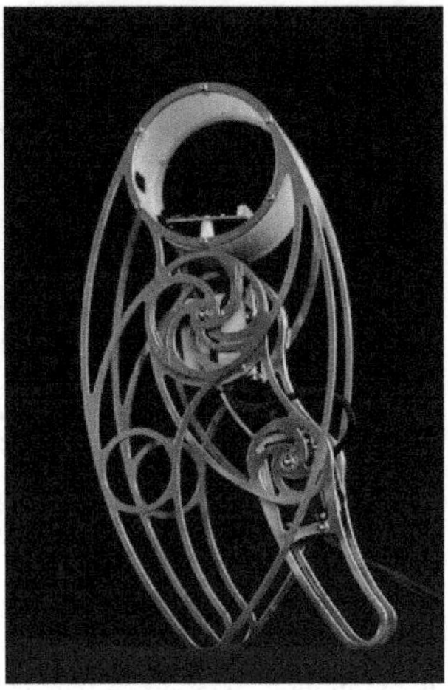

used. In this way the velocity of the motor without any load is controlled, however, the true velocity depends of course on the external forces, the battery voltage and so on. On top of the voltage control we implemented a second control mode setting the target joint angles which are then achieved with a PID controller. So we have two control modes: voltage control and position control. The robot has different sensors as explained in the following.

Joint angle sensors are integrated in the servo motors and measure the actual angle of the joints.

Motor current sensors measure the electric current consumed by each motor individually and are thus a direct measure of their physical performance—the measured current is proportional to the motor torques. The measurement is performed on the embedded board with high precision. In particular, the torques increase abruptly if there is some mechanical resistance against the motion of the leg driven by the motors. When using the current sensors with the homeokinetic controller, one needs a small preprocessing because the current signal is positive[2] independently of the direction of rotation. We therefore multiply the sensor value by the sign of the voltage applied to the motor. The resulting signal has a sign now but we observe a jump when crossing the zero line because the current is never zero due to friction and other mechanical loads.

Acceleration sensors measure the acceleration along the three axes of the robot. In the normal operation (robot not fallen over) the lateral acceleration is always zero so that we only use the two relevant directions a_1, a_2, see Fig. 8.2. The accelerations are caused by the changes of the body position and/or orientation with varying velocity, for instance if the controller drives some periodic motion. Note that the gravitation vector shows up in the sensor values such that the accelerations caused

---

[2] In fact the measured current can be negative if the leg is moved externally against the torque of the motor.

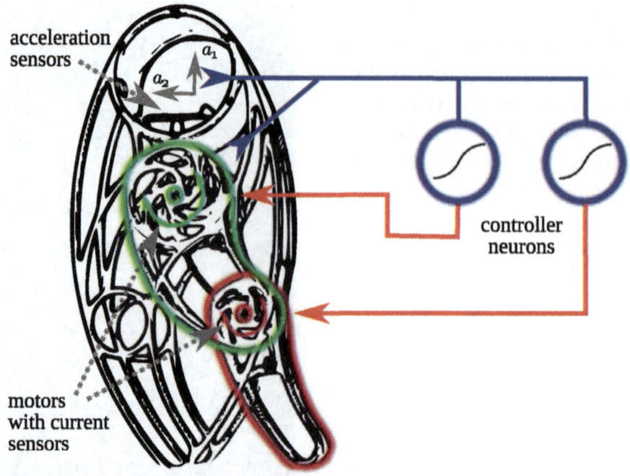

Fig. 8.2: Schematic diagram of the robot SEMNI with controller neurons. The movable leg is highlighted. The *green part* is mounted to the frame and the *red part* is connected to the *green part* both with servo motors. The currents are actually measured by the embedded controller board and not by the servos themselves.

by the movement are overlaid with the static information of the orientation of the body in space.

The sensor values are scaled with a constant factor such that their range is roughly between minus and plus one. Actually this is not necessary in the algorithm since the learning dynamics drives the parameters until the feed back strength in the loop achieves the slightly overcritical value. However, when using the bare sensor values the C_{ij} values increased like $C_{ii} \approx 20$, which means that the size of the gradient decreased by a factor of 400 (due to the $\left(CC^\top\right)^{-1}$ term). This can be compensated by changing ε_c appropriately. In practice, however, it is more convenient to work with some standard learning rates. Moreover, if the sensors have a great variety their relative influence is biased.

8.1.2 Experiments

We have done a number of experiments with the robot with different sensor-motor combinations. The controller network has only two neurons, see Fig. 8.2. The forward model network has the same number of neurons as there are sensors (not shown). We use here the full model ($M(x,y) = Ay + Sx + b$) as introduced in Sect. 9.2.

When connecting the robot to the controller one has to decide which sensor values to use and how to control the motors. The most direct combination is to use the joint angle sensors and send joint angles as target values to the motors. This is best realized by a classical servo motors. In this setting, we observed after a short time the emergence of an $SO(2)$ structure for the controller matrix with ensuing oscillatory behaviors of the leg much in the way of the self-regulation towards the effective bifurcation point of a Neimark-Sacker bifurcation as investigated in Sect. 7.2. In this regime, the robot is typically lying on either of the curved sides so that the movements of the leg excite a rocking behavior. When performed at the natural frequency of the robot the rocking could become stronger. However, this does not happen in this control mode. Thinking about the available information to the controller, we realize that the joint angles in the leg give nearly no feedback from the body about its current pose. Any indirect influences are compensated by the position control of the PID controller so that only little "feeling" for the body can be developed[3] Moreover, the entanglement with the bias dynamics leads to behavioral changes so that a very dynamic rolling of the robot is observed that is realized by consecutive flips, see Video 8.1.

A more suitable configuration to bring the embodiment into play is to use a voltage control. The voltage applied to the motors is directly given by the scaled motor values ($y_{1,2}$). Let us first see which variety of behavior emerges when all available sensors are used, namely the preprocessed current sensors, the joint angle sensors, and the acceleration sensors. Now we observe the balancing of the robot on the lower limp, with the body being horizontal but lifted. The support surface is only given by the lower part of the leg. Oscillatory movements are adapted to the current situation such that the robot's behavior looks like a picking bird, or it slides or steps along the ground. Importantly the robot reacts sensitively to external influences and adapts to new situation very quickly. More details can be found in Video 8.1. Let us look at some effects more closely.

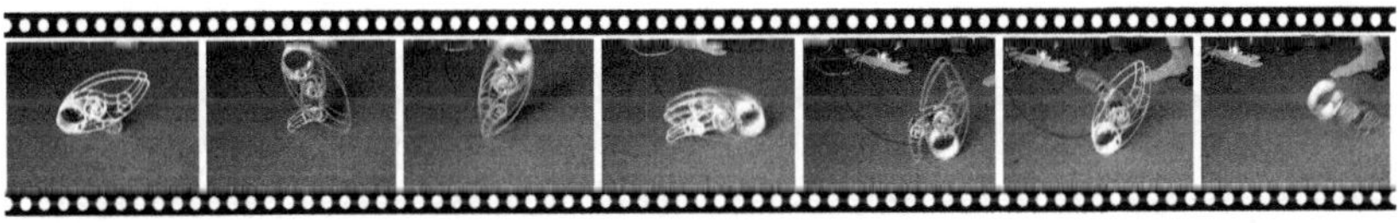

Video 8.1: The SEMNI robot with homeokinetic controller. Until the first cut in the video only joint angle sensors and position control were used. At the beginning a frequency wandering is observed (see Sect. 7.3.2). Eventually the legs hits the ground and the movement is hindered. The system enters a bias dominated dynamics where a slow oscillation occurs. Both joints get synchronized through the mechanical limits and the robot performs a series of flips (as depicted in the screenshots). After the first cut all sensors and the voltage control were used. Note the change in the behavior, which becomes more compliant with the movement of the robot. Different frequencies of oscillations are developing, depending on the current mode of behavior and the pose of the robot. Later in the clip the sensitivity of the controller can be seen. The robot is touched very gently and the behavior changes rather strongly. Also note the reaction when the robot is turned over. Importantly the controller adapts very quickly and engages the body into a different oscillatory mode. The video can be watched at `http://playfulmachines.com`.

[3] The internal error of the PID controller would give some more information on the situation, but they are not accessible to the controller.

8.1.2.1 Sensor Integration

Let us consider the setup with voltage control and all sensors: current sensors (x_1, x_2), acceleration sensors $(x_3 = a_1, x_4 = a_2)$, and joint angle sensors (x_5, x_6). In the initial condition the controller matrix C is such that the motor current sensors are mapped to the motor outputs and the other sensors are ignored. However, after a short time the other sensors start to become integrated as depicted in Fig. 8.3. Thus the robot becomes sensitive to e. g. the accelerations and can quickly pick up arising oscillations or react on external perturbations, see Video 8.1.

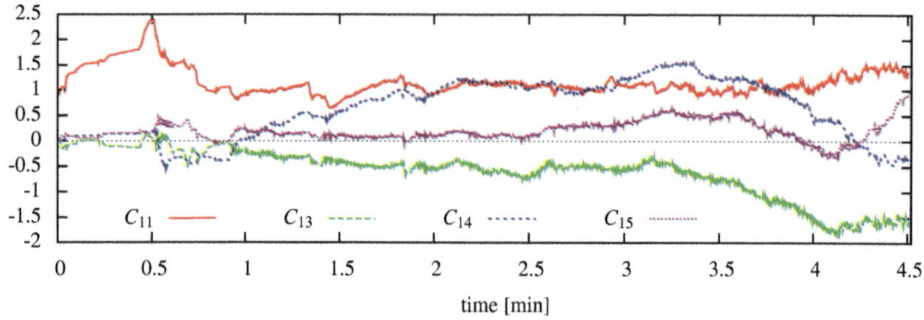

Fig. 8.3: Evolution of the sensor integration by homeokinetic control. Depicted are the coupling strengths to the first motor neuron from: the current sensor of the first motor (C_{11}), the acceleration sensors (C_{13}, C_{14}), and the joint angle sensor of the first joint (C_{15}). Initially only the current sensor is integrated. Later the controller becomes sensitive to all sensors and the integration changes qualitatively with time.

This is a nice effect that helps also in more complex systems to get the robot in dynamical contact with its environment. Some theoretical considerations about the integration of sensors in the loop can be found in Chap. 15, Sect. 15.4.1 (p. 279). It should be noted that we used the so called motor-space realization, introduced in Sect. 15.6 (p. 282), of the homeokinetic learning rule obtained by using the generalized pseudoinverse of Eq. 15.52 (p. 278). This variant of the general algorithm seems most appropriate if the sensors outnumber the motors.

8.1.2.2 Mechanical Limits and Offset Compensation

As seen in Video 8.1 the motors are sometimes driven to the mechanical limits of the joints. This is no surprise given the fact that the controller has no information about the limitations of the control regime. Fortunately, in the case of position control the limits directly correspond to the saturation regions of the neurons. This helps keeping the motors away from the limits since the parameter dynamics is trying to stay away from the saturation regime. In the case of voltage control the situation is different. Especially when the current sensors are used one could expect the motors to drive strongly against the limits, since the current sensor value keeps increasing

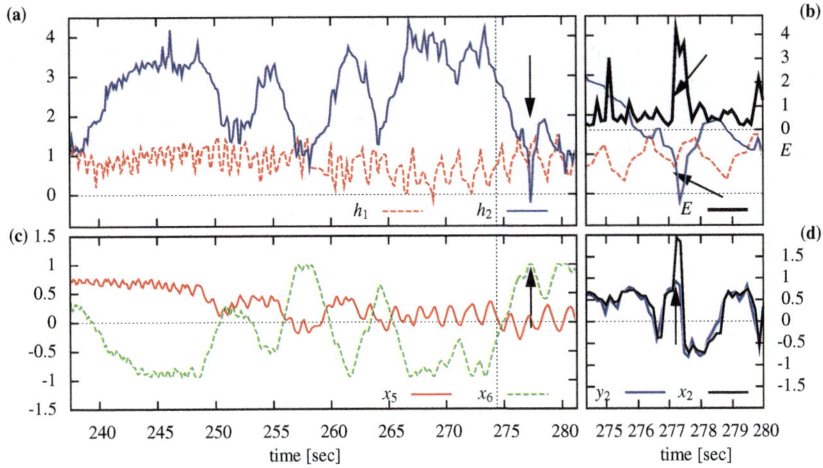

Fig. 8.4: Offset compensation and reversal at mechanical limits with SEMNI. Setting: voltage control and with all sensors. **(a)** Bias dynamics; **(b)** Bias dynamics and error E zoomed in; **(c)** Joint angles; **(d)** Current sensor zoomed in. The plot shows how the bias dynamics (a) compensates the slowly varying mean of the angle sensors (c). At time 277 a mechanical limit was hit at joint 2 (a). The current x_2 rises sharply (d) and thus the error rises as well (b). Consequently the bias h_2 is decreased and the joint moves in the opposite direction (x_2 changes sign).

once the boundary is hit. However, as a rule there is no collision with the boundary observed in the experiments. The point is that the current rises disproportional close to the boundary producing a high modeling error which in turn leads to a fast bias dynamics producing a jump to the other fixed point of the respective control neuron. Also, if the amplitude of the oscillations does not reach both limits equally the mean of the motor value oscillation is automatically shifted away from the physical limits because of the asymmetry in the dynamics. The effect is illustrated in Fig. 8.4. This scenario that also holds for many other sensor qualities can be considered as a specific example of the temperature effect discussed in Sect. 5.3.4 (p. 95).

In the experiments we also encounter sensor values that do not follow the motor signals in a proportional way. Additionally the sensors do not have a zero mean within a certain time window. The bias dynamics compensates the offsets as shown in Fig. 8.4. For example the angle sensors change rather slowly, such that the bias cancels the current offset and enables the system thus to amplify small fluctuations. In this way oscillatory modes can emerge. The same holds for the acceleration sensors as we will see in the next section.

8.1.2.3 Picking up Resonance Frequencies

In the experiments we observe that the robot is often engaged into swinging behavior at the resonance frequency of the body. At this frequency the body would swing by itself if deflected. If now the control manages to stimulate at this frequency the system will come into resonance and the oscillations get a high amplitude. We will now restrict ourselves to the case where only acceleration sensors are used to analyze what happens. Additionally the robot's head was loaded with a weight, which increases the inertia effects lowering the resonance frequency. We will discuss two short sequences where the robot was actively picking up the particular resonance frequency of the body. The data is plotted in Fig. 8.5 and the behavior can be seen in Videos 8.2 and 8.3. In the first sequence the robot was lying on its right (w. r. t. Fig. 8.1) with the leg touching the ground. It first oscillates at a high frequency that fits to the current situation such that the robot is sliding slowly to the left. After manually turning the robot over the head to its left the body starts to swing more slowly. The controller appears to pick up the new frequency immediately and amplifies the rocking behavior by the leg movements.

Video 8.2: SEMNI getting excited at its resonance frequency. Setting: Robot with loaded head, acceleration sensors, and voltage control. The corresponding data is plotted in Fig. 8.5(a-c). At the beginning the leg moves in synchrony with the body at a rather high frequency. The robot slowly slides into one direction. Then the robot is manually turned over and starts to swing at a much lower frequency. The robot's resonance frequency is exited and the amplitude rises such the robot falls over to the previous position. The video can be watched at `http://playfulmachines.com`.

To understand this, let us consider for a moment a person on a swing. In order to keep the swing going (counteract the damping forces) the legs (or the entire body) need to be moved in the right phase relation to the movement of the swing. This is very similar to what happens with the robot. The sensor values correspond to the position of the robot in some way. In order to produce motor commands in the correct phase relation, the C matrix must adapt appropriately. This is obviously what happens in the experiments. As seen in Fig. 8.5, C acquires an $SO(2)$ structure with a rotation angle corresponding to a frequency much higher than the observed physical frequency. Nevertheless, the C matrix produces a suitable phase relation to accelerate the robot so that the controller is exciting the body at its resonance frequency.

Why is the phase relation (the angle of the $SO(2)$ matrix) correct? Despite its $SO(2)$ structure the controller could also be counterproductive (anti-phase) and stop the oscillations! We do not have a good theory about it yet. Nevertheless, we have some hypotheses: first, the system prefers to have an $SO(2)$ structure, as discussed in Sect. 5.3.5, which is a necessary condition to get this behavior. Second, the learning rule optimizes for predictability and sensitivity. If the C matrix damps the oscillations then the effective sensitivity is low. In the other case, when the oscillations are

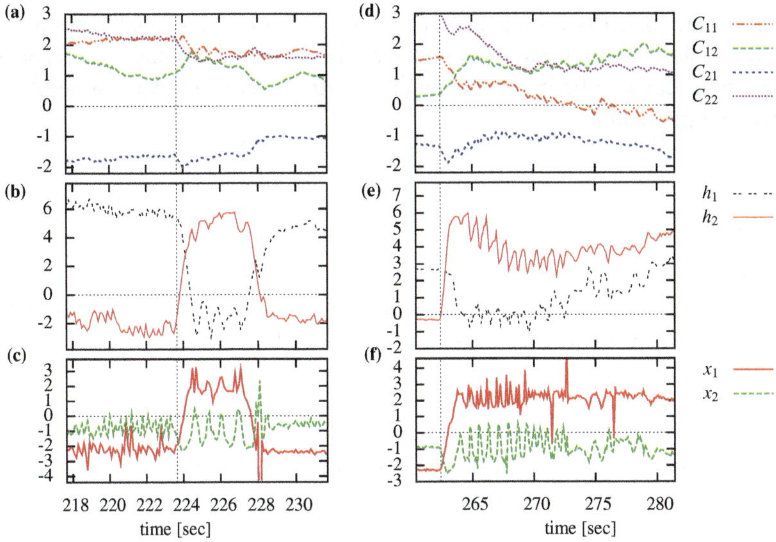

Fig. 8.5: Getting into resonance with the body. The SEMNI robot with loaded head, acceleration sensors, and voltage control. **(a,d)** Controller matrix C; **(b,e)** bias h; **(c,f)** acceleration sensors values ($x_1 = a_1, x_2 = a_2$). Plotted are two time intervals where the robot was manually moved (marked with a vertical line) into a new situation in which the body can strongly swing. Both times the natural frequencies of the robot are taken up and amplified such that a heavy rocking occurs. The corresponding behaviors can be seen in the Video 8.2 (a-c) and Video 8.3 (e-f). In both cases the bias h quickly adapts to the new situation and allows the system to enter an oscillation. In the first interval the robot swings back to the original pose and also the parameters are seen to return. In the second interval the oscillations are interrupted by hand and the robot remains on the other side.

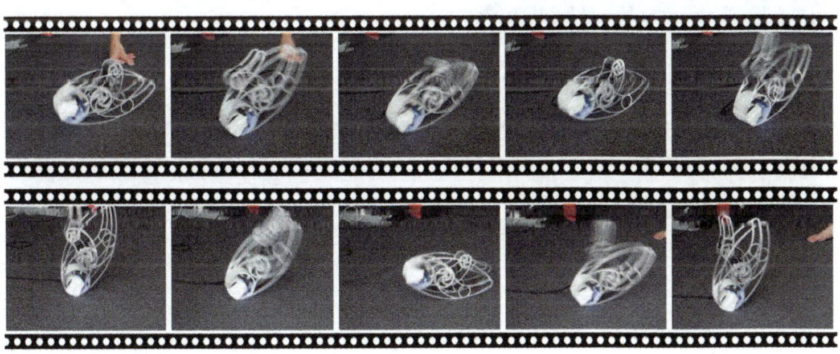

Video 8.3: Oscillatory behavior of SEMNI. Two videos of the SEMNI robot in the same setting as in Video 8.2. **Top:** Here the oscillations are interrupted by hand and the robot remains on the other side. The corresponding data is plotted in Fig. 8.5(e-f). **Bottom:** A longer clip showing the rocking behavior. The videos can be watched at `http://playfulmachines.com`.

amplified, the sensitivity is high and the system is also well predictable because the state vector obeys a strict rotational law. By this argument, the observed resonance phenomenon seems to be in good agreement with the homeokinetic principle—be sensitive but predictable.

8.2 SPHERICAL

Let us now consider a simulated robot with three motors—the SPHERICAL, which is a 3D version of the BARREL (Sect. 3.2). The design is inspired by the artist Julius Popp [141]. We conducted various experiments with the robot in different environments ranging from a free plane to a corridor and complex landscapes with many robots in them. The results are published in [43, 44].

The main points we want to illustrate with this robot are the switching between metastable modes and the role of situatedness.

8.2.1 Construction

The SPHERICAL has a ball shaped body and is equipped with three internal masses whose positions are controlled by motors ($m = 3$), see Fig. 8.6(a). If not stated otherwise each mass has the same weight as the hull alone.

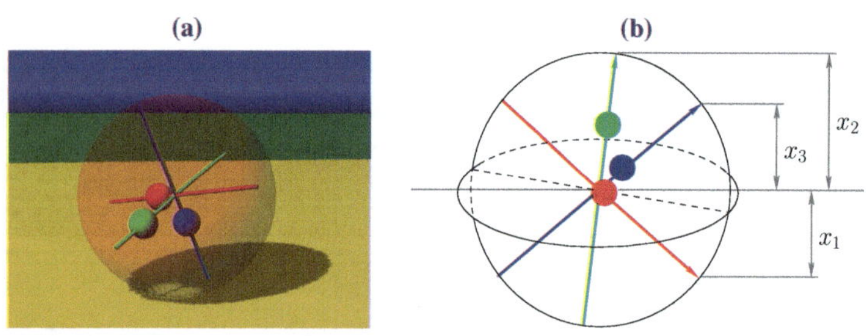

Fig. 8.6: SPHERICAL with axis-orientation sensors. **(a)** Screenshot from a simulation. The *red*, *green*, and *blue* masses are moved by actuators (linear motors) along the axes; **(b)** Schematic view of the robot with axis-orientation sensors (x_i). Note that $x_1 < 0$ here. See also Fig. 3.4.

The position control of the movable masses is done by servo motors. The motor values define the target positions of the masses along the axes. A value of zero stands for a centered position and -1 and 1 correspond to the outer positions. Collisions of these masses especially at the intersection point are ignored in the simulations.

We use this robot in two different sensor configurations. The first one uses axis-orientation sensors ($n = 3$), see Fig. 8.6(b). These are the same sensors used with the BARREL. For each axis the projection of its direction onto the z-component of the world-coordinate system is measured. The second setup uses infrared (IR) sensors and will be explained further below.

In either case, both sensor and motor values are related to the motions of the sphere in a complicated manner. The challenge for homeokinetic control is to develop sensorimotor patterns that are related to the natural motion patterns of the physical system. As we will see, this happens indeed under widely different physical conditions, sensor settings, and initializations.

8.2.2 Self-Exploration of Rolling Modes

In the first experiment we look at the behavior of the SPHERICAL with axis sensors on level ground, see also [44]. The homeokinetic controller has here three motor neurons and receives three sensor inputs. Initially only small fluctuations due to the sensor noise occur. The learning dynamics increases the feedback strength steadily so that the controller is getting more and more sensitive to the sensor values. Once the critical level is exceeded fluctuations get amplified so that the symmetry of the system is spontaneously broken and the body starts to roll into a decided direction. This is the first moment when the sensor values show a defined response to the actions the response depending, however, on the specific embodiment and the physical situation in an intricate way. Nevertheless it is only now that homeokinetic learning can make contact with the nature of the physical system—rolling in particular as the most natural motion pattern of the SPHERICAL.

The most simple of the natural modes is realized by rotating around one of the internal axes with the mass on that axis being used for steering and the other ones for shifting the center of gravity as in the BARREL case. The experiments demonstrate that the controller picks up such a rolling mode and amplifies it very quickly. Due to gyro forces these behaviors are physically stabilized. Nevertheless due to the destabilizing and explorative character of homeokinetic control the steering mass keeps also shifting periodically so that a slightly curvy trajectory is created. Eventually the steering mass starts to oscillate with increasing amplitude such that the current mode is actively destabilized and either a rolling mode around a different axis is found or an alternative rolling mode is realized, controlled by all three masses shifting with equal amplitude. The different modes are waxing and waning as shown in Fig. 8.7 and Video 8.4. Typical trajectories are displayed in Fig. 8.8.

Occasionally we also observe a jumping behavior obtained by a sudden synchronous motion of the masses in the vertical direction, comparable to the "lolloping" mode observed with the BARREL. In addition to the earlier observed sweeping behaviors, which occur here partly while in one of the rolling modes, we find here a rather abrupt switching between different modes. We interpret these findings as an indication that the robot actively explores its behavioral capabilities.

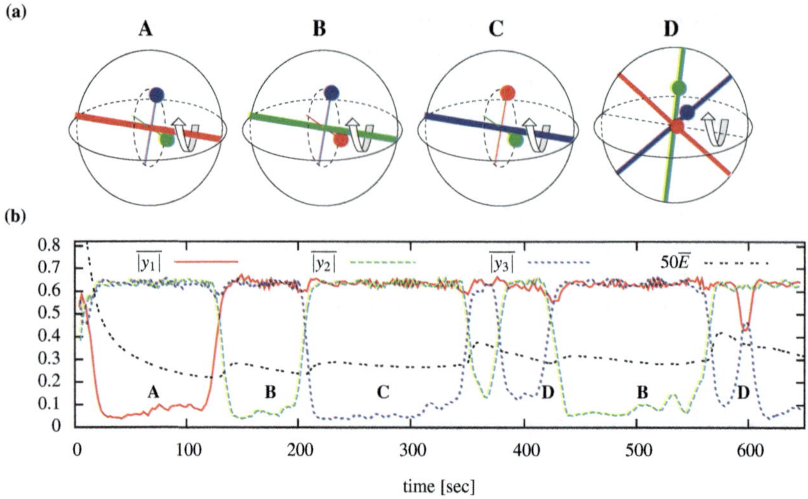

Fig. 8.7: SPHERICAL exploring its behavioral capabilities. (**a**) Sketch of four typical behaviors (**A-D**), namely the rolling mode around the three internal axis (**A-C**) and around another axis (**D**). (**b**) Amplitudes of the motor value oscillations ($y_{1\ldots3}$) and the time-loop error E (scaled for visibility) averaged over 10 and 30 sec, respectively. Corresponding behaviors are indicated with letters A-D. The extended forward model was used, see Sect. 9.2.

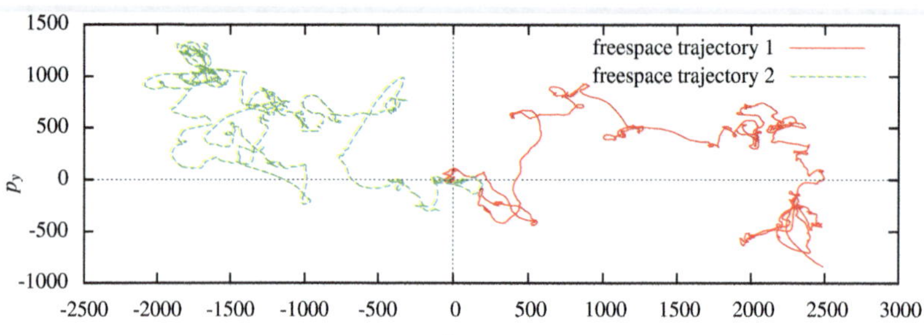

Fig. 8.8: SPHERICAL exploring the space with axis orientation sensors. Two independent trajectories each of a 200 minute run of the SPHERICAL in free space observed from above. Note that the robot has a diameter of 1. There are passages of straight rolling as well as curvy behavior.

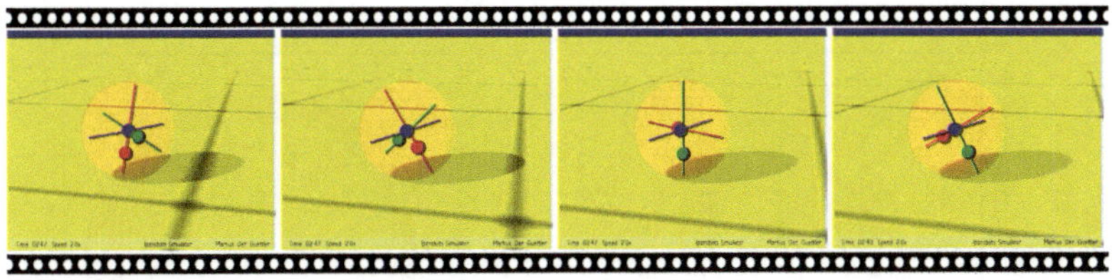

Video 8.4: Different rolling modes of the SPHERICAL. The video can be watched at `http://playfulmachines.com`.

8.2.3 Situatedness — How the Environment Shapes the Behavior

The emergent behaviors are not only dependent on the particular body, but also on the environment. The effect of the environment is well seen in the following two experiments. In the first one the SPHERICAL is put into a basin with a rotational symmetry, see Video 8.5 and [43]. The robot is placed at the bottom of the basin and initialized as usual. The starting phase is the same as on level ground, the robot starts to roll in a random direction, but now the trajectory is influenced by the geometry of the basin. First the robot rolls back and forth and the trajectory crosses the bottom of the basin. However, after some time a circular trajectory is realized, see Fig. 8.9. In this case the robot essentially keeps a constant height.

Video 8.5: SPHERICAL in a circular basin. The robot starts to roll around randomly. Then it enters a circular trajectory as suggested by the environment. The video can be watched at `http://playfulmachines.com`.

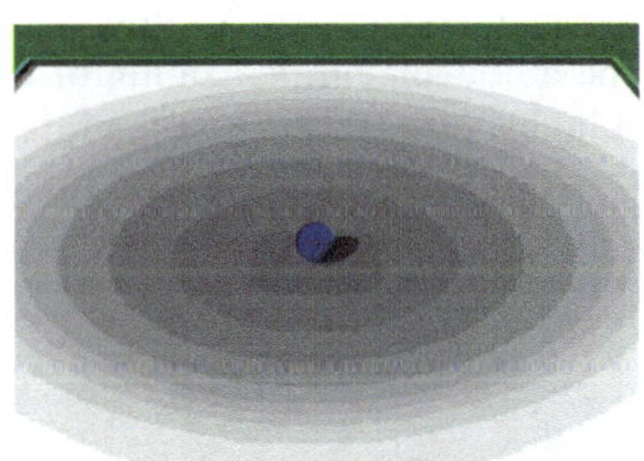

Taking a closer look we find that all three weights are used to control the behavior, see Fig. 8.10. This contrasts the observations on level ground where this mode occurs only rarely, the robot keeping instead one of the weights in the center most of the time. These simple rolling modes are stable on a flat surface, but in the basin the environment disturbs this behavior and thus the robot develops a different rolling mode. You can perform the simulations yourself by doing Experiment 8.1.

At this point one may ask how the controller is able to learn this behavior although it does not have any sensors for directly getting information on the geometry of its environment. We speculate this to be a consequence of the symmetry breaking mechanisms underlying the emergence of the large-scale motions. The point is that the parameter dynamics is generated by gradient descending the time-loop error E, which is defined entirely by the dynamics of the sensorimotor loop. Hence the symmetries present in the sensorimotor dynamics and the environment are conserved

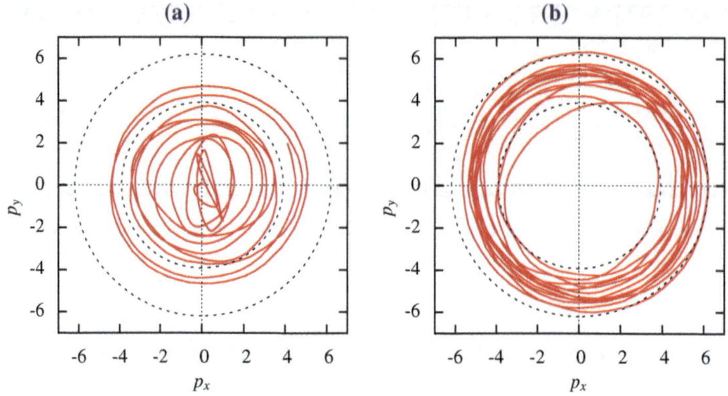

Fig. 8.9: SPHERICAL in a basin. Trajectory viewed from above ($x-y$ coordinates of the position). **(a)** First 120 sec; **(b)** From sec 420–560. The diameter of the robot is 1. The *dotted circles* are the level curves for height 0.5 and 1 of the basin. Later the robot circles at about a constant height along the basin.

also in the full system comprising the parameter dynamics. The sensitization effect, see Sect. 5.3.2, destabilizes the sensorimotor dynamics so that every large scale motion is a result of a spontaneous symmetry breaking. These symmetry breaking effects tends to be economical, i. e. symmetries are broken in the least possible way, see also Sect. 5.3.6 (p. 99). A circular trajectory in a rotational symmetric basin seems to be a convenient realization of this paradigm since we have a large-scale motion while keeping as much of the full symmetry of the system as possible.

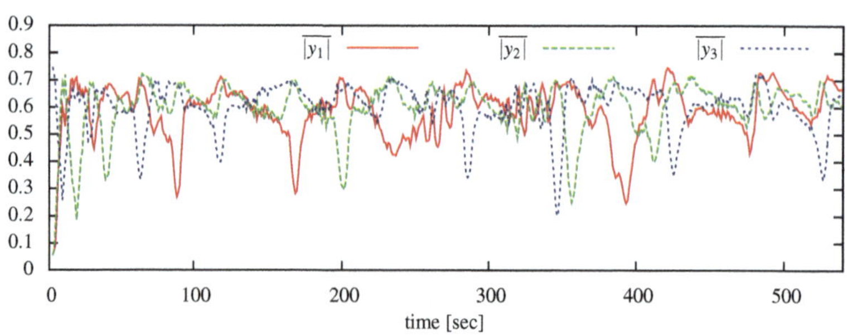

Fig. 8.10: Particular rotation mode when rolling in a basin. The plot shows the amplitudes of the motor value oscillations ($y_{1\ldots3}$) averaged over 5 sec sliding windows. The robot mostly rolls around an inclined axis (not around one of the internal axes), see behavior D in Fig. 8.7. The sensor values and the rotational velocities around the internal axes confirm the finding (not shown).

Experiment 8.1: Self-exploration and situatedness with SPHERICAL.

For this experiment you have two choices: (a) SPHERICAL on flat ground, or (b) SPHERICAL in a basin. In (a) observe how the controller explores the different rolling modes. The learning rate is set very high ($\varepsilon_c = 1$), such that the modes are quickly changed. You may start GUILOGGER or MATRIXVIZ and select C, A and y. Decrease the learning rate to 0.3 or below and see how the modes get more stable. Eventually the robot might have left the virtual world and the colored ground ends. Use <Ctrl>+H to get it back to the initial position. In (b) you can reproduce the results of the robot in the circular basin. You can press * on the Keypad or `>realtimefactor=0` on the console to set to maximum speed. After about 2 hour simulation time (few minutes on your computer) the robot will manage escape from the basin.

8.2.4 *Situatedness — Adapting to the Environment with External Sensors*

In the following experiment we will replace the axis orientation sensors by infrared sensors as depicted in Fig. 8.11(b). In total six sensors are used that extend the internal axes towards the outside of the robot ($n = 6$). Each sensor measures the distance to the ground or to another objects. The range of the sensors is the 5 fold of the hull radius. For distances exceeding the range the sensor values are zero and for close objects they are equal to 1. In between a linear characteristic is used.

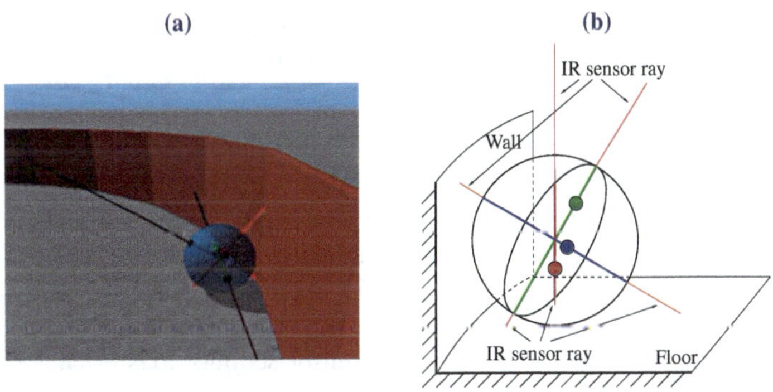

Fig. 8.11: SPHERICAL with infrared sensors. **(a)** Screenshot from a simulation. The IR sensors are drawn in *black* if they measure no obstacle and in *red* if they do. **(b)** Schematic view of the robot with the sensor values x_i.

The relation between the orientation of the sphere and the sensor values is largely different from the angular sensor setup. Nevertheless our controller in a short time learns again to generate nice rolling modes. In free space there is no substantial difference in the behavior between the two sensor cases, only that the sphere is rolling more straight ahead in the IR case. However a qualitative difference is observed if we put the robot into a non-trivial environment. In particular, the robot is placed into a circular corridor with a diameter of 20 and a width of 2 sphere diameters.

The robot now has to change its strategy if it wants to be active and nevertheless predictable. In fact, if the robot is rolling straight ahead it will collide with the walls

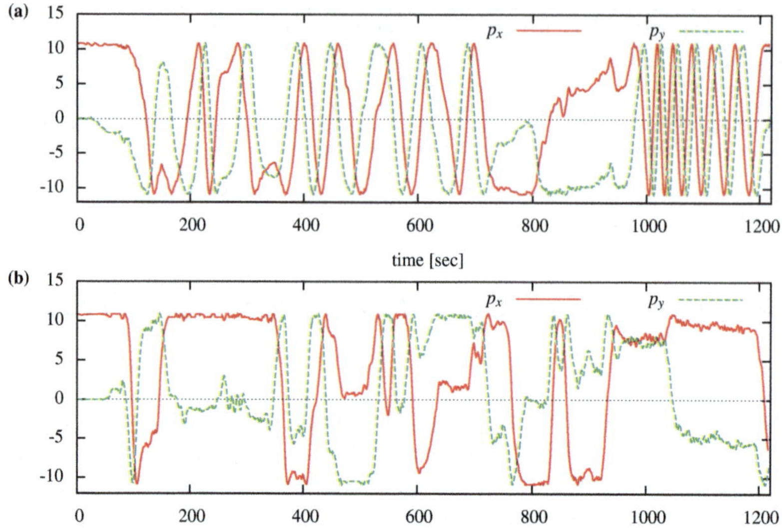

Fig. 8.12: Smooth circulation in the corridor with infrared sensor. Plot of the coordinates of the SPHERICAL in a circular corridor over a total length of 20 min. The robot was equipped with: **(a)** infrared sensors, and **(b)** axis orientation sensors. At the sine-like shape of the upper plot one can clearly see that after some initial time the robot with infrared sensors is passing several rounds in a row in a smooth manner. The robot with z-axis sensor as described above is much less successful.

quite often (the outer wall of the corridor is constructed from 24 straight segments). The collisions affect the rotation axis substantially. In order to go around the environment smoothly the robot needs to steer in a circular manner or find a way to roll smoothly along the wall.

In Fig. 8.12 two trajectories with both sensor settings, axis orientation and infrared sensors, are displayed. The result is distinguishing and shows that the robot with infrared sensors is performing much better in this environment. By observing the robot we have encountered a mode where the sphere is rolling sensibly along the wall realizing a stable rotation around one axis like on level ground, however now with a definite axis inclination, see Video 8.6 (the corridor is a bit wider to allow easier video recording). This balancing is much better possible with infrared sensors than with axis sensors, because the walls can actually be sensed by the former ones.

8.3 SLINGING SNAKE

Here we will consider the SLINGING SNAKE, a simulated robot with two active and many passive degrees of freedom, in order to show how hidden modes can be discovered. The robot is underactuated in the sense that its actuators are not

Video 8.6: SPHERICAL adapts to a circular corridor. Later in the clip the robot repeatedly balances along the wall for some time. The video can be watched at `http://playfulmachines.com`.

powerful enough to set the robot (especially the passive parts of the body) in motion if the physical properties of the body are neglected. Thus a whole-body motion can only be excited by exploiting the embodiment.

As before we connect our controller to the robot using the explicit learning rules, see Eqs. (5.22, 5.23), with the add-on for enhancing dynamical harmony, see Sect. 15.1.1.4. Hence the controller has initially no specific information about the physics of the body. In order to generate active behavioral modes of the whole body it has to become sensitive to the sensor values which show the effects of the body properties if appropriately excited.

In the following section the mechanical configuration of the robot and the setup of the controller will be described before having a closer look at the experimental run. The construction of the robot and the simulations were done by Frank Hesse, see [68, 69] for further details.

8.3.1 Construction

The snake-like artifact, called SLINGING SNAKE, is simulated in our LPZROBOTS simulator (Chap. 16), like the other virtual robots in this book. The SLINGING SNAKE consists of 10 spheres, see Fig. 8.13. The head element (red sphere) is connected to a tail of 9 spheres, which are connected to each other by a link (frictionless ball-and-socket joint) so that each of the spheres can rotate passively like in a string of beads. The red sphere is the actuated head of the SLINGING SNAKE to which a force vector

$$f = \begin{pmatrix} f_x \\ f_y \end{pmatrix} \tag{8.1}$$

is applied. The force vector is parallel to the ground (lying in the plane). All other parts of the body are passive. The spheres have a friction for sliding motion ($\mu = 0.3$) and optionally there is a small rolling friction. The available forces are not strong enough to enable the head element to pull the tail. Hence a whole-body motion can only be generated if certain lateral motions are excited as we will see below.

The robotic device provides feedback from two exteroceptive sensors measuring the components of the velocity vector of the head element

$$v = \begin{pmatrix} v_x \\ v_y \end{pmatrix} \qquad (8.2)$$

given in global coordinates, see Fig. 8.13(b).

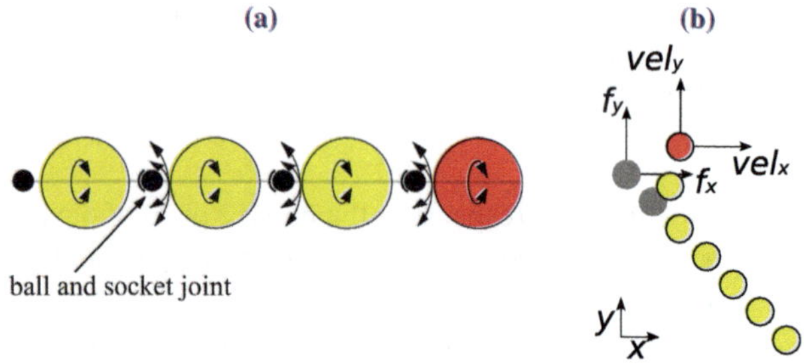

Fig. 8.13: Mechanical construction of the SLINGING SNAKE. **(a)** Ball-and-socket joints between the spheres are drawn. Arrows indicate the possible rotations: in every direction at the joint and around the axis between joints. **(b)** Sketch of the SLINGING SNAKE in the global coordinate system. After execution of a time step in which a force vector f was applied to the head element the velocity vector v is read back as feedback of the system. Both quantities are in global coordinates.

8.3.2 Experiments

During the first experiment three robots are placed on a flat ground within a square arena of a simulated world. After the typical unit-initialization the robots ly on the floor and barely move since the feedback strength in the sensorimotor loop is subcritical. The controller values increase by the learning dynamics and we observe some movements, which are mostly sideways, since the robots cannot simply move straight ahead. Eventually the movements of the head become strong enough to cause the tail to sway and to produce a (delayed) feedback on the head. The controller starts to respond to the swaying of the body and manages to amplify these slinging movements. Thereby a rolling motion of all, active and passive, elements of the body is generated, see Video 8.7(a). This motion is characterized by a rotation of all spheres around their axes, where the first half of the spheres rotate in the opposite direction than the other half, see Fig. 8.14. Hence a collective rotational mode of behavior of the underactuated robotic device with many degree of freedom was excited. While the whole snake is rotating around its center, it becomes stiff like a stick due to centrifugal forces and gyro effects. The frequency of this whole body rotation is altered with time. Later on we observe collisions of fast and slowly rotat-

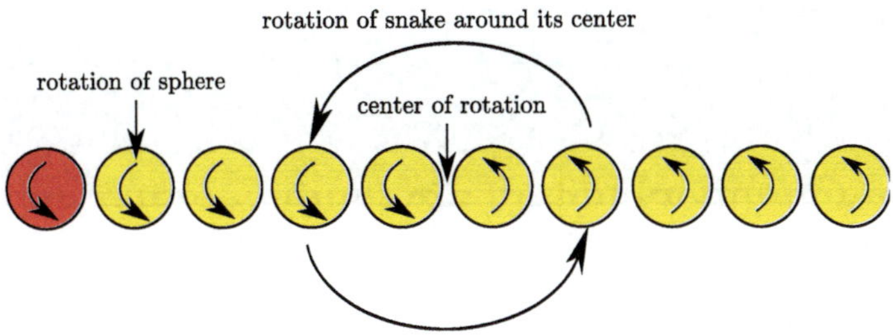

Fig. 8.14: SLINGING SNAKE in the rotational mode. This behavior is characterized by a rotation of all spheres around their axis (connection between the joints), where the first half of the spheres rotate in the opposite direction than the other half. Hence the whole snake rotates around its center and becomes stiff like a stick due to the centrifugal forces.

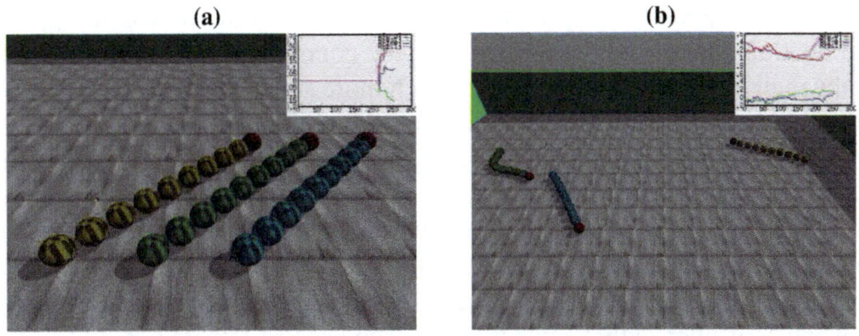

Video 8.7: Whole body movement with SLINGING SNAKE. Three SLINGING SNAKEs are placed in a square arena. **(a)** Initially they only move little but slowly more and more strong movements are observed. Nevertheless, due to the underactuated nature, the robot cannot be pulled straight along the ground. Eventually the learning rate is increased and the robots start to pick up the swaying movement of their body and enter a fast rotational mode. **(b)** The robots are slowed down or accelerated by collisions. The videos can be watched at `http://playfulmachines.com`.

ing robots, which cause heavy perturbations. Nevertheless, the controllers quickly adapt to the new frequency or initiate the spinning again, see Video 8.7(b).

Let us take a more quantitative look at the experiments and the possible pitfalls. For that we place a single robot on a flat ground without borders or obstacles. Hence there are no perturbations by the environment. Let us first consider an idealized condition without rolling friction. The robot again enters the spinning mode quickly and accelerates, see Video 8.8. The controller matrix clearly shows a rotational structure with increasing rotation angle as displayed in Fig. 8.15. While the rotation frequency keeps increasing, the whole body rotation frequency is decreasing steadily. Eventually the controller matrix reaches the period-4 cycle structure (diagonal elements are zero and off-diagonals are non-zero with opposite sign), which the robot cannot follow because of the prohibitively high frequency so that the robot stops rotating. As discussed in Sect. 7.2.1, this high frequency oscillation is an attractor and the parameter dynamics will not depart from there. We call this situation a high frequency

Video 8.8: Sweeping mode of SLINGING SNAKE with high frequency catastrophe. The initial phase illustrates very nicely how the robot enters the spinning mode. The frequency rises quickly and then is lowered again until the robot comes to a rest. It will not come out of this situation, see the corresponding data in Fig. 8.15. The video can be watched at `http://playfulmachines.com`.

catastrophe. Note the difference to the SEMNI case where the period-4 cycle still was productive in exciting a rocking mode. Probably, the difference is explained by the fact that in the SEMNI case all the controller had to do was to find the appropriate phase relation whereas in the present case the controller has to get into resonance with the many degrees of freedom in a much more subtle way.

Why is this happening? Once rotating the robot's motion is only little influenced by the actions, especially when considering only one time step as the forward model does. As the experiments show, the model degenerates quickly to a rotation matrix with high rotation angle with opposite sign as that of the controller matrix. This is reflecting the fact that the robot can not follow the high frequency oscillations requested by the controller. This leads to a kind of "misunderstanding" between model and controller so that, once the robot comes to rest, there is no 'good' sensor data received that would correct the model and thus also the controller.

If we, however, change the conditions and include a small rolling friction the high frequency catastrophe is avoided. In the experiments, the initial period is taking a bit longer, but after about 2 min the robot enters as before the oscillatory mode with a lower maximal frequency. Then we observe a slow fluctuation of the rotation frequency until a stationary value is reached, see Fig. 8.16. We then manually decelerated the robot until it is at rest. The robot is able to enter again a spinning behavior very quickly and a similar evolution of the frequency is observed until it again enters a stationary value.

8.4 ROCKING STAMPER

The second real robot we want to present here is the ROCKING STAMPER. It is yet another "crazy" robot that can demonstrate interesting properties of homeokinetic control. It consists of a bowl-like trunk with an actuated inverted pendulum on top. A somewhat similar robotic construction, called "Stumpy," was developed at the lab of Pfeifer, see [76, 77]. They have shown that the robot performs self-stabilization and displays behavioral diversity while controlled by very simple and fixed control structures—the exploitation of embodiment.

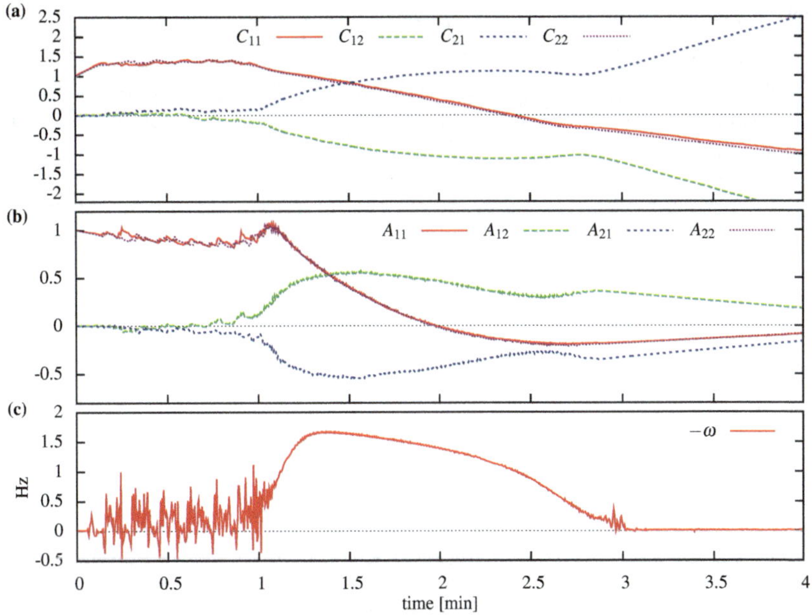

Fig. 8.15: Sweeping mode of SLINGING SNAKE without rolling friction. **(a)** Controller matrix C; **(b)** Model matrix A; **(c)** Rotation frequency of the robot around the $z-axis$ (perpendicular to ground plane) ω. The controller takes up the swaying of the body after about 1 min. The spinning speed rises and lowers again (c) until the robot comes to rest. The model matrix (b) and the controller matrix (a) develop into a period-4 cycle matrix at about 2.5 min (and then with even higher frequency). As soon as the robot comes to rest the parameters diverge since no motion can be initiated. See also the corresponding Video 8.8. Settings: $\varepsilon_c = 0.03$, $\varepsilon_A = 0.03$, $\mu_{\text{roll}} = 0.0$, $\gamma_h = 0.1$ (Sect. 15.1.1.4).

Similarly, our controller will automatically make use of the particular embodiment from scratch and excites different behavioral modes. Will mainly focus on the establishment of a sensorimotor coordination in a very inaccurate and non-trivial setting. Furthermore we will study the effects of changes in the sensor characteristics and even the failure of sensors, see also [41].

8.4.1 Hardware

The ROCKING STAMPER consists of a bowl-like trunk with a pole mounted on it that is driven by two motors ($m = 2$) in orthogonal directions, see Fig. 8.17 and 8.18. The robot is equipped with four infrared (IR) sensors ($n = 4$), two mounted at the front end of the trunk looking down and slightly sideways and two at the back, which can be flapped up and down. They measure the distance to the ground or to the wall. The sensor value changes depending on the pose of the trunk in a

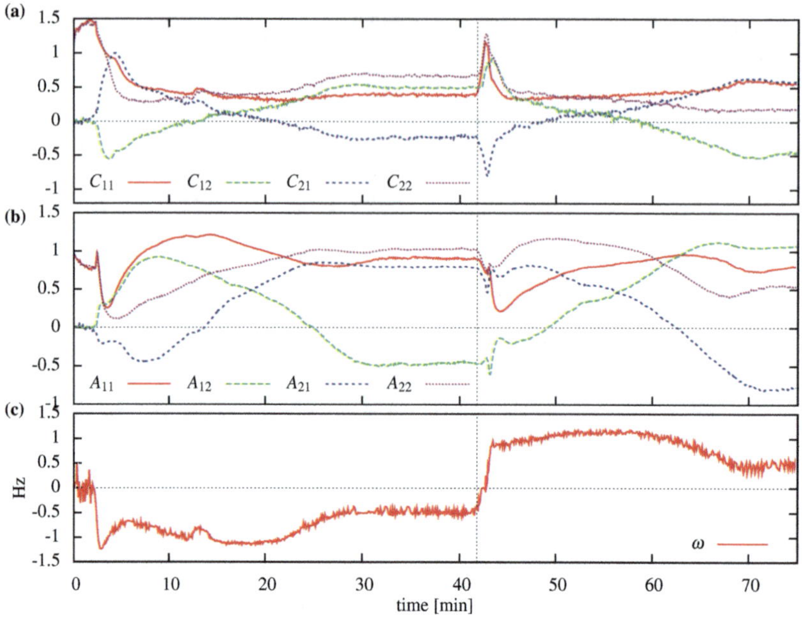

Fig. 8.16: Behaviors of SLINGING SNAKE with rolling friction. The same as in Fig. 8.15, but here with rolling friction and a lower learning rate. The controller matrix has also a rotational structure but with a lower maximal rotation angle. The frequency fluctuates but enters a steady state at about 30 min. At 42 min the robot was decelerated by hand. After a short resting period the robot enters a spinning mode again. Settings: $\varepsilon_c = 0.02$, $\varepsilon_A = 0.01$, $\mu_{\text{roll}} = 0.005$, $\gamma_h = 0.1$.

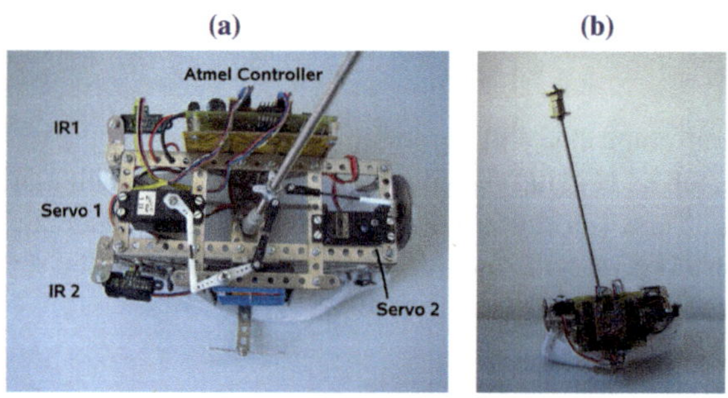

Fig. 8.17: The ROCKING STAMPER, a pole driven bowl-shaped robot. **(a)** Close view from the top; **(b)** Side view. Here only two sensors are mounted.

nonlinear fashion. The motor commands dictate the nominal positions of the servo motors, which determine the angles of the pole relative to the trunk, as illustrated in Fig. 8.18.

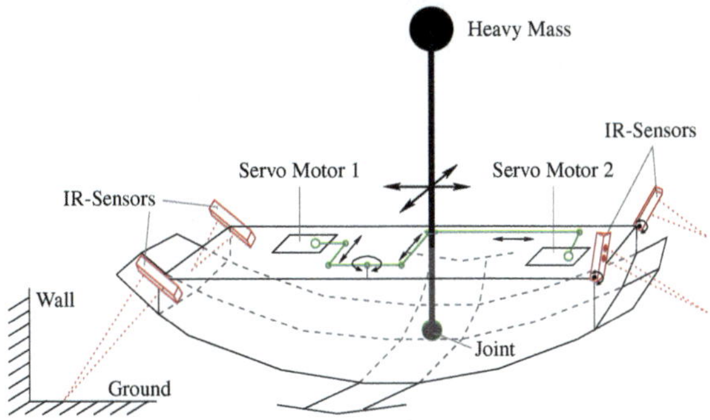

Fig. 8.18: Schematic diagram of the ROCKING STAMPER. The pole is moved relative to the trunk by the two servo motors. This causes the trunk to tilt so that the IR sensors to measure a different distance to the ground/wall. The two sensors mounted at the front are fixed and the two at the back can be manually flapped up and down.

The robot is equipped with a battery pack and an embedded controller board holding an Atmega32 processor. Thus the robot could operate autonomously. In the current implementation, however, the on-board hardware is only used to establish a connection to a workstation, which runs the actual control algorithm for practical reasons. From a performance point of view it is absolutely feasible to run the control algorithms on such an embedded processor, especially for a low-dimensional system.

The robot can basically rock in two dimensions. The swing in the direction front-back is more easily achieved because the body has a smoother curvature compared to the sideways direction. By suitable combination of both rocking dimensions the robot can perform a locomotion behavior. Another mode of operation occurs if the pole is moved quickly around the center position, such that the massive weight stays almost at a fixed position, while the trunk oscillates.

8.4.2 Experiments

When connecting the body to our controller we observe after a short time rocking (oscillatory) as well as a walking-like behavior. The latter is caused by a rotational mode of the pole with suitable phase shift. The emergence of these modes is a direct consequence of the homeokinetic paradigm. In fact in these modes the controller can—based on the current sensor values—evoke the maximum change in the sensor values in one time step. The modes preferred by the physical systems, e. g. oscillations at the eigenfrequencies, are excited for two reasons: First, the physical system stabilizes against perturbations so that a better prediction is possible and second these modes comply with the requirement of high activity. We are tempted to say that the controller develops a "feeling" for the body.

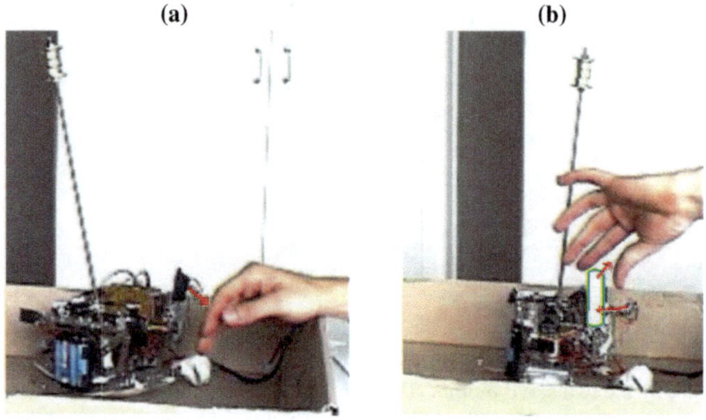

Video 8.9: Sensitivity and tolerance to sensor failure. The two infrared sensors at the back are mounted to the trunk with hinges so that their pitch can be changed during the experiment. **(a)** The robot is perturbed by moving the hand into the sensor range. The robot reacts immediately by moving the pole to the opposite side. **(b)** One of the sensors gets disabled and enabled by a paper hull. The robot adapts to the new situation quickly and reenters a rocking mode. The videos can be watched at `http://playfulmachines.com`.

In Video 8.9 the behaving robot can be observed. It is demonstrate that the robot remains sensitive to its sensors (Video 8.9(a)). The hand of the operator is moved into the infrared sensor range such that its values are changed and an immediate reaction of the robot follows. Also a quick adaptation to new situations is observed, for instance by disabling a sensor or changing the sensor setup during the experiment, see Video 8.9(b). In both cases the robot finds quickly back to an active rocking behavior. Sometimes the robot enters a locomotive behavior that is achieved by a cyclic movement of the pole, see Video 8.10. These experiments are interesting in that, despite the extremely nonlinear and non-deterministic behavior of the mechanical system, the controller learns to produce a motion which probes the possibilities of its body in a more or less controlled manner.

Let us now have a close look at the data to understand what happens. To make it easier we use the case where only the two front infrared sensors are used as depicted in Fig. 8.17. After initialization, we at first have subcritical values for the feedback strength of the sensorimotor loop so that the influence of the noise is damped and we observe only small fluctuations of the pole position. The learning dynamics increases the values of the controller parameters C and therefore the pole movements become stronger. Eventually the bifurcation point (Sect. 7.1.2) is reached and an (irregular) oscillatory motion sets in. In Fig. 8.19 the behavior, reflected by the sensor readings, and the parameter adaptation are displayed. It is insightful to take a closer

Video 8.10: Walk-like behavior of the ROCKING STAMPER. The robot rocks in a way that slow forward locomotion occurs. The first three frames (from left to right) show one full swing. The remaining ones show how the robot travels. Frames times relative to the first frame: 0.5, 0.9, 3.2, and 18 sec. The video can be watched at `http://playfulmachines.com`.

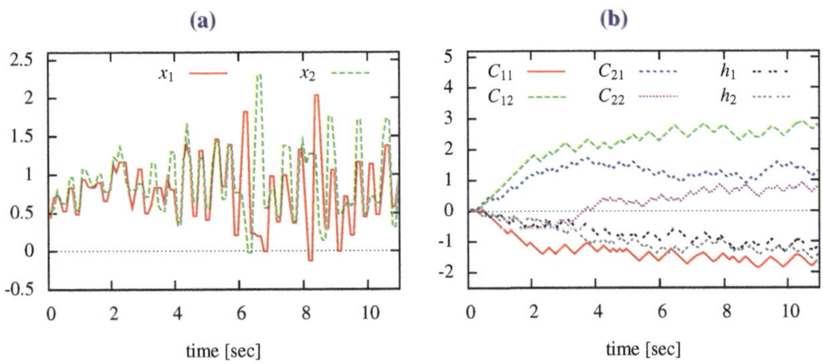

Fig. 8.19: The ROCKING STAMPER starts to rock. **(a)** Sensor values from left (x_1) and right (x_2) infrared sensor. **(b)** Controller parameters C over time. The sensitization is seen in the increase of $|C_{ij}|$. The controller matrix evolves to a structure that maps the difference of both sensor values to servo 1 (y_1) ($C_{11} \approx -C_{21}$) and the sum of both sensors to servo 2 (y_2) ($C_{21} \approx C_{22}$). The bias terms h_i adapt such as to compensate for the positive average of the sensor values.

look at the controller matrix. It indeed develops to a filter for the sensor values, such that both degrees of freedom are separated. The first motor neuron receives the difference between both sensor values, which corresponds to the sideways pitch of the robot. This is exactly the direction influenced by the first servo motor. The second motor neuron received the sum of both sensor values, which fits to front-and-back direction induced by the second servo motor. To summarize, the controller autonomously adapts from a unit initialization to a specific structure suitable for the particular system under control and manages to excite various modes of behavior at the natural frequencies of the hardware.

In order to demonstrate the environment related nature of the emerging behaviors we put the performing robot into a corner where the infrared sensors measure the distance to the wall, which is much shorter and also has a different characteristics, see Fig. 8.18. As a consequence the robot became calm for a short time. Then the parameters adapt to the new situation, so that again an oscillatory behavior sets in, see Fig. 8.20. The same adaptation scenario occurred when the robot was manually moved away from the corner.

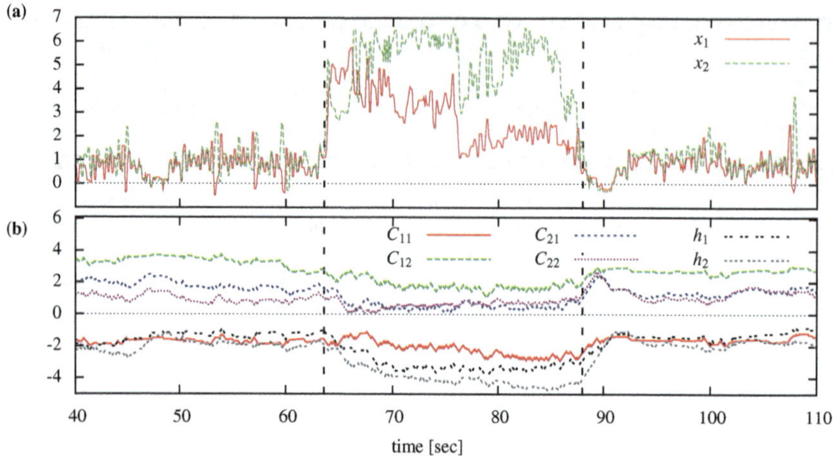

Fig. 8.20: Environment sensitive behavior of the ROCKING STAMPER. Evolution of **(a)** sensor values from left and right infrared sensor, and **(b)** parameters of the controller. Until second 64 we observed rocking (oscillatory) motion with a short break around second 48. Then the robot was manually set into a corner. The infrared sensors measure much shorter distances since they see the walls instead of the ground. At second 88 the robot was pulled back into free space. After each change of the environment the robot was calm for a while (low sensor fluctuation) and probed the new environment, however, after a short time the robot rocked again.

8.5 BARREL

The simultaneous learning of both model and controller from the time-loop error is also the source of creativity in handling unexpected situations. For illustration we consider the BARREL again, see Sects. 3.2 and 7.3.3. The robot was seen to produce interesting rolling modes and other behaviors, which however more or less rely on the fact that getting a barrel to roll is quite "cheap" since this is the most natural motion when lying on a surface.

8.5.1 Creativity in New Situations

What happens if the BARREL is put upright on the surface? This is a completely new situation, which does not fit very well into the picture. The problem is, that the internal axes are now horizontal so that the sensor values are equal to zero (apart from some small noise) and thus there is no reaction of the sensors to the motor actions. Nevertheless, if the approach disposes of some intrinsic creativity it should find a way out towards new activity. Video 8.11 shows one possible development in such a situation. After some time the motors start a kind of jiggling motion, which is essentially largely amplified sensor noise. But as soon as the BARREL starts to

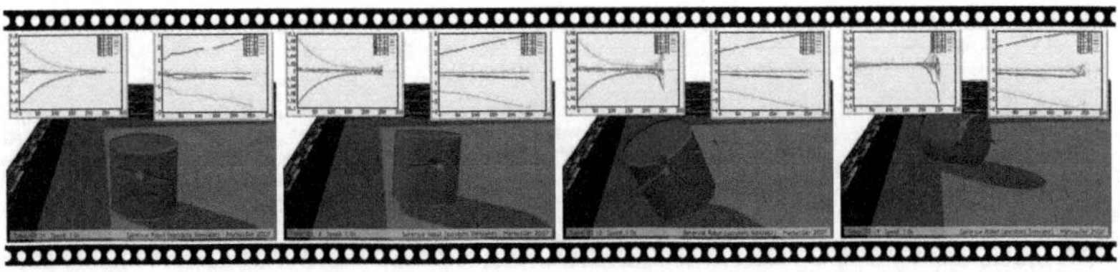

Video 8.11: Creativity in unexpected situations. The BARREL was put into an upright position by an external force about 10 seconds ago. Internal axes are horizontal now so that the sensor values, the inclination of the internal axes, are zero apart from some small sensor noise. The forward model does not get any reliable information in this situation so that rapid forgetting sets in. This is counteracted by the controller, which increases weights so that small perturbations are amplified. Note that the parameters are quite shaken at the beginning of the clip due to the moving of the barrel. After some time the motion of the internal weights become so strong that the BARREL is tossed over. The dynamics of the parameters of the model and the controller can be followed in the panels at the left and right upper corners, respectively. The panels depict the course of the parameters in a time window of 250 steps corresponding to about 10 sec. Note that the scales of the panels change, in particular the model parameters change by two orders of magnitude. The rapidly oscillating parameters are the bias terms of both model and controller. The video can be watched at `http://playfulmachines.com`.

give a definite reaction by whipping a little bit, the learning system synchronizes and amplifies the swaying motion of the body. As a consequence, the BARREL is falling over and now is free to go into a rolling mode again.

The reason for the strong amplification of the sensor noise is the decay of the forward model, which does not receive any reliable information. The controller counteracts this by increasing its weights. The amplification of the sensor noise in these situations can be enhanced with the `creativity` parameter introduced in Sect. 15.1.1.3 (p. 264), which makes use of the degenerated forward model.

This is not yet the full story. One can, out of this impasse situation, also observe the emergence of a new mode, a kind of precession of the BARREL which requires a high degree of sensorimotor coordination, as presented in Video 8.12. Nevertheless, the mode is quite stable and can last over a long time, about five minutes real time in this particular case. This mode is a truly emerging phenomenon that is produced through the interaction of the learning dynamics with the specific embodiment in the given physical situation. Note again, that both controller and forward model receive nothing but the inclinations of the internal axes so that the physical state of the body remains widely obscure to the "brain." In particular, there is no information about being put upright.

Experiment 8.2: Precession mode BARREL. Creativity in unpredictable situations.

In the simulation Homeokinetic control of BARREL(upright) the robot is in an upright position. In Video 8.11 the controller gets negative diagonal elements such that we initialized here with $C_{11} = C_{22} = -0.1$ and $C_{12}, C_{21} \in [-0.1, 0.1]$ randomly chosen. Observe the parameters evolution with the GUILOGGER and remember that you can speed up or slow down the simulation. If the robot manages to get free you can try to put it up again using the external forces (<Ctrl>+Mouse).

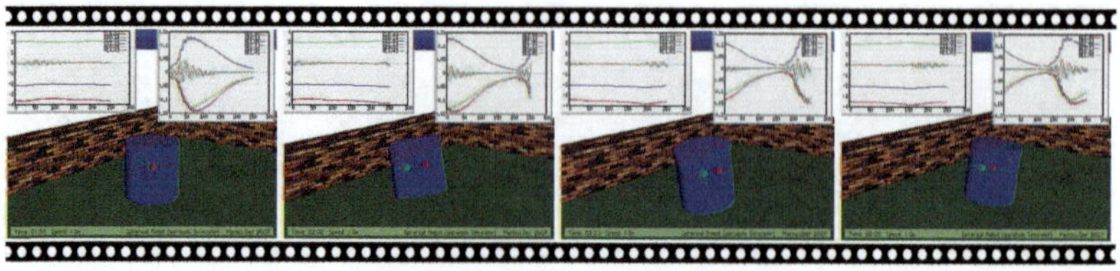

Video 8.12: Emergence of nontrivial modes — Precession of the BARREL. Out of the upright position there are different modes that can emerge. In the video you see the emergence of a precession mode which lasts for more than five minutes. Parameter and model dynamics are depicted in the panels in the right and left upper corners (swapped w. r. t. Video 8.11). Note that the scales of the panels change. The video can be watched at `http://playfulmachines.com`.

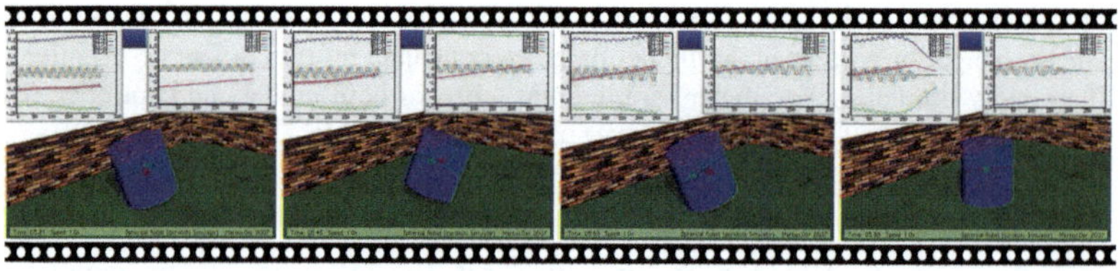

Video 8.13: Decay of nontrivial modes. One special feature of our approach is that modes do not last forever, instead the simultaneous learning of model and controller most often leads to a slow change of the parameters so that the system leaves the mode. In the present case, the precession mode, after lasting for about five minutes decays spontaneously. The parameter changes are most prominently in the diagonal elements of both the model (left) and the controller (right panel) depicted by the *red* and *purple lines* (which almost coincide). The video can be watched at `http://playfulmachines.com`.

The transient nature of emerging behaviors is nicely demonstrated by these experiments. For instance, Video 8.13 shows the decay of a precession mode after existing several minutes. The behavior of the parameters of both the controller and the forward model strongly suggests a systematic tendency towards the decay of the mode. This phenomenon is also observed in higher dimensional systems and is seen as a decisive advantage over the common dynamical system approaches which usually try to associate behaviors (or cognitive states) with stable attractors, instead of transients. In the attractor case exists not natural way out of a behavior in contrast to the transient dynamics.

8.5.2 Creativity Using the Real BARREL

A real physical counterpart to the BARREL was recently constructed by Wolfgang Rabe and colleagues at the University Leipzig, for what we are very grateful. Here we will present our first results with the robot. It consists like the simulated version

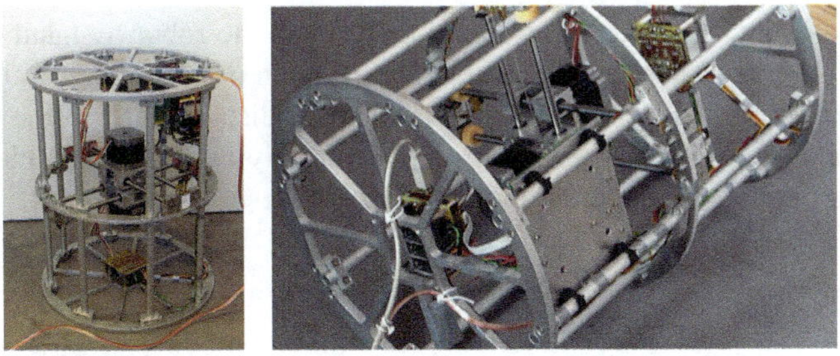

Fig. 8.21: Real BARREL. The robot was built by Wolfgang Rabe and Frank Güttler, Universität Leipzig.

of a cylindrical hull and two massive weights that are moved by step-motor (actually the motors are the weights) along sliders. Obviously, the parts cannot penetrate each other, such that the axes are offset, see Fig. 8.21.

The motor values control the positions of the motors along their axis. The robot is equipped with a 3-axes acceleration sensor (see Sect. 8.1.1 for details), where we use only the $x-y$ components along the slider axes. The sensor values are an overlap of the acceleration caused by the earth gravitation and the accelerations caused by the movement of the robot itself, see also Sect. 8.1.1. The former provides a very similar measure to the simulated orientation sensors, whereas the latter also provides some feedback from quick movements of the motors.

We connected our controller to the robot and put the barrel upright on the floor. In contrast to the soft ground in the simulation, where a strong mass movement causes a small tilting of the robot, the physical floor is very hard, such that no feedback

(a) (b)

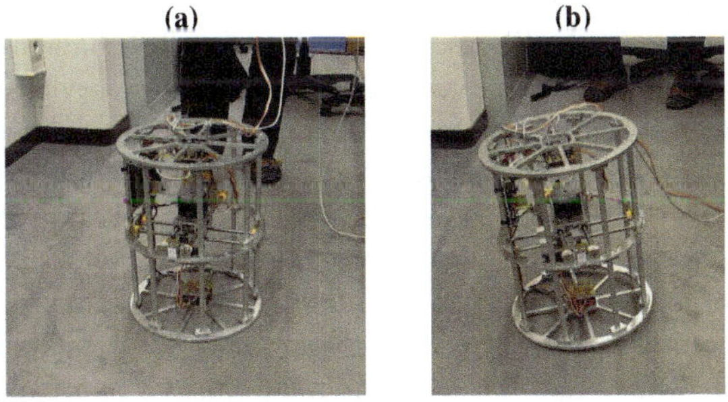

Video 8.14: Creativity in unexpected situations. (a) Robot with flat lower surface. The floor is very hard in comparison to the floor in the simulations above, such that the robot cannot get any feedback from the actions. If we push the robot by hand it starts to whip along the lower rim. **(b)** A small ring is added at the bottom of the robot, such that it can easier get into motion. The robot performs similarly to the simulated one. Note, the maximal amplitude of the moving weights was manually adjusted such that the robot does not flip over. The videos can be watched at `http://playfulmachines.com`.

of the actions is provided. In this situation we push the robot by hand a little bit, see Video 8.14(a). When we change the construction of the robot a little bit, such that the standing surface is not flat but has a small distance ring in the center, the robot starts to enter the precession mode more easily as shown in Video 8.14(b). The corresponding sensor data is plotted in Fig. 8.22.

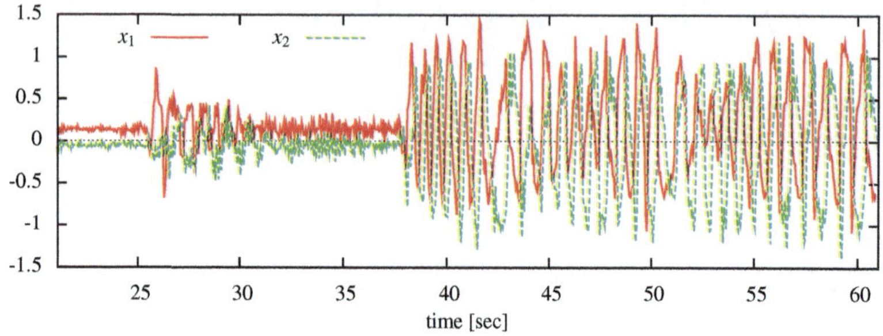

Fig. 8.22: Real BARREL in precession mode. Sensor values (acceleration sensors) during the experiment, see Video 8.14(a). The robot is slightly pushed at sec. 26 and a bit stronger at sec. 37. The oscillations are amplified and a precession mode with different intensity is observed.

8.6 Summary

We have considered the homeokinetic control of various real and simulated robots. They possess all a vastly different morphology, but have in common to be compliant to physical forces. In all cases active and smooth behavior emerged, which can be characterized as playful. Even with the rather crazy devices like the ROCKING STAMPER or the SEMNI we find the emergence of coordinated behavior most prominently at the natural frequencies of the body. We cannot stress enough that there are no specific goal, just the minimizing of the TLE. By now it should be clear that the seemingly oversimplified architecture of the controller is in fact sufficient for the control of many robots, due to the fast synaptic dynamics and the exploitation of the embodiment. The latter implies an export of computation complexity to the interaction of the brain-body system in the spirit of morphological computation [125, 132, 133], see also [131].

Chapter 9
Model Learning

Abstract: This chapter discusses several aspects concerning the simultaneous learning of controller and internal model. We start with discussing the bootstrapping dilemma arising in this context and the consequences of insufficient sampling. It appears that homeokinetic learning solves these problems naturally, which we illustrate in several examples. Further, we extend the implementation of the internal model by a sensor-branch. This is seen to increase the applicability of the homeokinetic controller because it allows for situations where the sensor values are subject to an action-independent dynamics. The extended model is prone to an ambiguity in the learning process, which can lead to instabilities. The problem can be resolved if the time-loop error is used as an additional objective for the model learning.

Internal models of the interaction with the environment form an essential part of many natural and artificial systems. For instance the human motor system that has been investigated for a long time, see e. g. [187, 188], is supposed to possesses a variety of internal models. Among them are so called forward models that predict the outcome of actions and inverse models that infer the required actions from desired states. This is supported by recent experiments that provided increasing evidence of the widespread use of predictive models in the human brain, see [35] for a short review. It seems quite clear that, once internal models are established, they can be used for planning ahead by means of internal (or mental) simulation of the sensorimotor dynamics and for detecting unexpected situations.

In our setting the forward model is given by (Eq. (3.6) in Sect. 3.1)

$$x_{t+1} = M(x_t, y_t) + \xi_{t+1}. \tag{9.1}$$

which relates the current motor signal y_t (*efference copy*, in biological terms) and the current sensor values x_t (*afferences*) to the estimated sensory feedback $M(x_t, y_t)$ (*corollary discharges*) and is thus in direct correspondence to the computational models of the human motor system. As in the human brain, our forward model is adaptive in order to cope with changes in the motor system and it seems obvious to use the sensory discrepancy ξ between estimated and real feedback as a source of learning.

R. Der and G. Martius: The Playful Machine, COSMOS 15, pp. 183–200.
springerlink.com

Internal models are also one of the prerequisites for a robot to become a cognitive system. The general aim of conventional robotics approaches is to learn the internal model as complete as possible. This is of particular interest for planning of action sequences for navigation or reaching movements in structured environments. Typically the internal model would be trained by trying random actions or action sequences. However, in high dimensional systems, such as complex robots or even humans where the individual degrees-of-freedom are not independent of each other, a purely random exploration is infeasible because of the curse of dimensionality well known from statistical learning theory.

Alternative approaches to complete internal representations are the attempts to build an internal model by a physics-simulation engine like the one used in our simulator. In this way the physical simulation runs in parallel on a computer [21]. This reduces successful modeling in principle to a question of resources. Although quite successful in recent applications, the approach faces both practical and conceptual difficulties. In many robotics systems, the practical problem is how to find the mapping from reality to the internal representation. Our simple BARREL, for instance, is described in the simulation by 8 differential equations and collision equations (contact with other objects), a dynamical system capable of all kinds of dynamical patterns reaching from pure fixed point attractors to non-trivial orbits and even chaos. Learning these differential equations from scratch by way of examples would require to get the robot into all kinds of dynamical patterns by appropriate control commands. However, we have seen that it is already difficult to get it into a high velocity rolling mode by open loop control. This leads us to the conceptual difficulty—in order to get the system by classical control paradigms into such complex modes, we have to already know the dynamics of the system quite well, far beyond the stage at which the learning would have led us so far.

The role of the forward model in the homeokinetic learning context differs in several respects from conventional robotics approaches. In particular, the aim is actually not to represent the full spectrum of future outcomes of current actions, but instead to provide a qualitative information on the current mode of behavior. Comprehensive competencies are substituted by flexibility, which is realized by a rapid learning strategy on simple structures. The "normal" task of the forward model consists in giving a coarse directional and/or qualitative information on the landscape of the prediction error. The reduction of the competencies of the model is even a favorable component supporting the explorative nature of the homeokinetic learning procedure. We will consider this specific effect of the interplay between model and controller in the next section.

In the most simple setting of the homeokinetic controller, we use a forward model, which predicts the next sensor state only, based on the motor signals (efference copies). There are situations where this oversimplified model is not sufficient and may lead to a runaway effect of the controller parameters. A more general definition of the internal model includes a sensor branch and helps to avoid the divergence problem. However, it also introduces an ambiguity problem due to the independent dynamics in the world, which we will treat in Sect. 9.2.

9.1 Cognitive Deprivation and Informative Actions

In this section we want to focus in particular on the consequences of the simultaneous learning of the controller and the internal forward model, which faces among others the so-called cognitive bootstrapping problem or learning paradox [16, 53, 56, 158]. Starting at a "do nothing" and "know nothing" initialization of the controller and the self model, respectively, the robot does not have any information about the structure and dynamics of its body, so that the forward model has to learn this from scratch. However, in order to learn effectively, the controls have to be informative so that the forward model is provided with the sensorimotor patterns necessary for its improvement. On the other hand, these actions require a certain knowledge of the reactions of the body—information is acquired best by informative actions.

This bootstrapping situation in principle reappears on all stages of the developmental process. We focus here, as before, on the low-level feedback loops to achieve self-exploration of the physical properties of the body. In this context the controller should issue just those actions that contribute most to the information gain of the model, since random actions are not suitable to effectively explore the behavior space in high-dimensional systems. First, we will see that a controller eliciting only actions confined to a subspace causes a deprivation of the forward model. Second, homeokinetic control will be seen to naturally produce informative actions that cause a quick recovery of the model.

In the following we give a simple example for illustration.

9.1.1 *Demonstration by the* TwoWheeled

Let us consider the TwoWheeled (Sect. 6.4.1) to demonstrate the effects of deprivation by restricted motor commands and the rapid recovery if the homeokinetic learning is switched on.

The robot has two motors actuating the wheels and two sensors measuring the wheel velocities. To illustrate the effect we have to restrict the controller in some way. We use here our standard controller $K(x) = g(Cx + h)$ but restrict C as

$$C = C^{\text{restr}} = upp^{\top} \text{ with } p = \frac{1}{\sqrt{2}}\begin{pmatrix}1\\1\end{pmatrix} \tag{9.2}$$

and $u = 1.1$ so that the sensorimotor loop is slightly supercritical. The matrix C is a projector into the space spanned by p, hence both wheels receive the same input and the robot will perform only straight forward and backward motion. The controller is fixed at the beginning of the experiment (no learning dynamics by putting $\varepsilon_c = 0$). In order to let the robot go forward and backward the bias h was modulated with a sine signal. As expected, we observe a degeneration of the forward model[1] see Fig. 9.1,

[1] As usual a small damping is added to the learning rule, see Eq. (9.4).

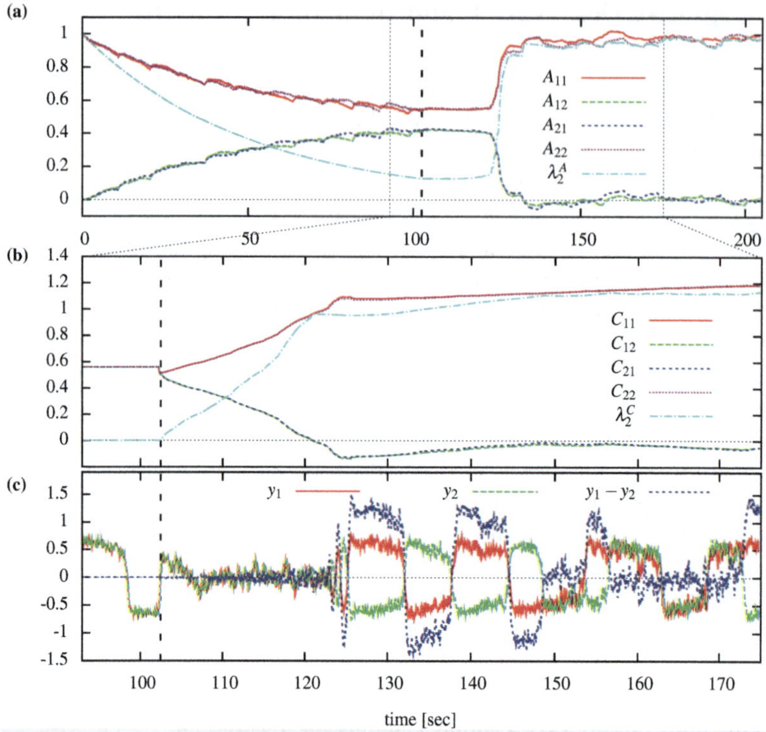

Fig. 9.1: Cognitive deprivation and recovery in the case of a TwoWheeled. The first 103 sec the robot was controlled by a restricted controller Eq. (9.2) with sine modulated h (oscillating forward/backward). After that the learning of the controller was activated, marked with a dashed line. **(a)** Model matrix A and its smallest eigenvalue λ_2^A. The matrix degenerates until the controller is activated and becomes then approximately a unit matrix; **(b)** Controller Matrix C with smallest eigenvalue λ_2^C; **(c)** Motor commands for left and right wheel y_0, y_1, and the steering value $y_0 - y_1$. The robot performs mainly rotational actions after the activation of learning. Parameters: $\varepsilon_c, \varepsilon_a = 0.01$, $\gamma_d = 0.0002$, square root of error (Sect. 15.1.1.6), update rate 100 Hz.

which converges towards a projector on the vector p. Eventually, the learning of the controller was activated by setting $\varepsilon_c = 0.01$ and the external modulation of the bias h was switched off. After a short recovery phase, the robot starts to move again, now controlled by the homeokinetic learning dynamics. Interestingly, the controller issues mainly motor commands lying in the subspace orthogonal to p. This means that the robot is rotating as if it aims at exploring the mode previously neglected. This exploration continues for about 40 sec until the forward model is converged to the true relationship and therewith both straight and rotational modes occur with about equal probability. The behavior can be seen in Video 9.1 and the parameter dynamics is displayed in Fig. 9.1. Both effects, the deprivation and the rapid recovery by homeokinetic learning can be understood analytically in the following way.

Video 9.1: Deprivation of internal forward model and recovery shown with the TwoWheeled. The controller was first restricted to only straight driving. The model degenerates as shown in Fig. 9.1. After the learning is switched on (robot turns green) the robot mainly rotates at the spot for a while before driving around normally. The video can be watched at `http://playfulmachines.com`.

9.1.2 Deprivation Effect

Let us study first how the forward model is effected if the controller elicits only a limited spectrum of actions. As a reaction to this restriction the forward model undergoes a process that we call cognitive deprivation. The term "cognitive" is used because the model reflects a certain ability of the robot to perceive and represent the reactions of the world in the current situation so cognition in a minimalist sense.

Consider the learning rule for the simple forward model $M(x,y) = Ay$. Including a small damping term γ_d, the learning rule reads, see Eq. (4.5),

$$\frac{1}{\varepsilon_A}\Delta A = \xi_{t+1} y_t^\top - \gamma_d A\,. \tag{9.3}$$

For the sake of demonstrating the deprivation effect it suffices to consider the most simple case of a linear world with response matrix $A^{(w)}$ and noise ς so that the true world dynamics is

$$x_{t+1} = A^{(w)} y_t + \varsigma_{t+1}$$

and the prediction error in Eq. (9.3) is $\xi_{t+1} = \delta A y_t + \varsigma_{t+1}$ with $\delta A = \left(A^{(w)} - A\right)$. With independent and zero-mean noise we obtain, in the average over a time window of length T, the following average gradient rule (the noise is assumed to average to zero)

$$\frac{1}{\varepsilon_A}\langle \Delta A \rangle - \delta A P - \gamma_d A \tag{9.4}$$

where

$$P = \frac{1}{T}\sum_{t=0}^{\top} y_t y_t^\top \tag{9.5}$$

is the covariance matrix of the controller outputs, assuming $\langle y_t \rangle_{t=0}^{\top} = 0$ for simplicity. The learning dynamics is stationary at $\Delta A = 0$ so that

$$A = A^{(w)} \frac{1}{P + \gamma_d \mathbb{I}} P\,.$$

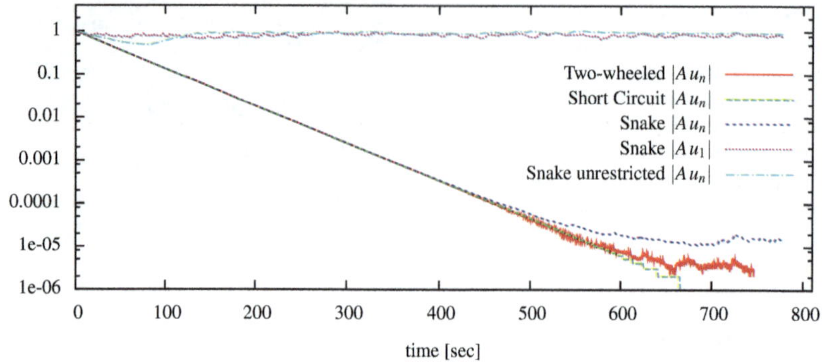

Fig. 9.2: Progressive deprivation of the forward model when a robot is driven by a restricted controller. The controller is restricted as in Eq. (9.9). The eigenvector in the nullspace of the controller matrix C is denoted by u_n and the eigenvector belonging to the largest eigenvalue of C is u_1. Plotted is the projection of the forward model matrix A onto these directions in logarithmic scale. Robots: TwoWheeled—2 Degrees of Freedom (DoF), Short Circuit—10 DoF, Snake (planar)—15 DoF (Sect. 9.1.4). Short Circuit is the idealized world condition $x_{t+1} = y_t + \varsigma_t$. There is a clear exponential decay for $|A\,u_n|$ expressing the deprivation of the forward model, except in the case the controller is not restricted (learned normally) (*dash-dotted cyan*). The projection on the largest eigenvalue is not effected by the deprivation (*dotted magenta*). Parameters: $\gamma_d = 0.0002$, $\varepsilon_A = 0.01$, update rate 100 Hz.

From this equation we see that A has the same nullspace as P. The matrix $\frac{1}{P+\gamma_d \mathbb{I}}P$ produces finite projections into the nullspace of P (due to γ_d) and thus the original nullspace of P is carried over to A. In this way the model degenerates if the controller is restricted in the sense that it does not produce outputs y covering the whole space.

Note that also a nonlinear controller in the form $y = g(Cx)$ will produce a non-empty nullspace if the matrix C is not of full rank, such that the above proof also holds. The progressive deprivation of the forward model in our setup for a restricted controller is shown in Fig. 9.2 for several robots. In all cases an exponential decay of the model in the restricted direction is observed. To conclude, if not all dimensions of the action space are probed into from time to time, then the forward model will become inaccurate for the unvisited subspaces.

9.1.3 Homeokinetic Learning and Bootstrapping

The degeneration of the forward model in the unexplored subregions of the state-action space is a severe problem for any autonomous, self-learning system. On a formal level, the challenge for the controller is to detect this deprivation of the forward model and to issue motor commands which provide the necessary sensor-action pairs (x_{t+1}, y_t) to correct the deficiency. However, in complex systems this procedure faces again the curse of dimension, since the sampling costs, if using uninformed sampling, are growing exponentially with the dimension of the system.

On the contrary, all a simple internal model can learn is the behavior on low-dimensional manifolds of the full space corresponding to certain modes of the system. Hence, the challenge is not to completely explore the unknown but to find out about different latent modes of the system that could be explored by the right "informative" actions of the controller. Fortunately, homeokinetic learning provides us, so to say for free, with just that feature, at least in a rudimentary way, as will be demonstrated by the following arguments.

9.1.3.1 Bootstrapping as a High Temperature Effect

Let us consider now the TLE as introduced in Eq. (5.9) of Sect. 5.1.1

$$E = \xi^\top \frac{1}{LL^\top} \xi \,,$$

and ask for the consequences of homeokinetic learning for the deprivation effect. In order to discuss this in detail, we use polar matrix decomposition writing $L = \Lambda U$ with U an orthogonal and Λ a symmetric matrix, respectively, i. e. $U^{-1} = U^\top$ and $\Lambda = \Lambda^\top$. In order to keep the argument simple, we assume for the moment that there is just one dimension heavily deprived, i. e. only the eigenvalue λ_n of Λ is closed to zero and the others are around one. Using that Λ is symmetric, we may decompose it as

$$\Lambda = \sum_{i-1}^{n} \lambda_i u_i u_i^\top \,, \tag{9.6}$$

where the λ_i are the (real) eigenvalues with the corresponding (real) eigenvector u_i. The effect of the eigenvalues in E is only reflected by the matrix $LL^\top$ so that the TLE in the leading term is given by

$$E \approx \frac{\left| u_n^\top \xi \right|^2}{\lambda_n^2} \,. \tag{9.7}$$

This shows that the TLE is increasing rapidly with growing deprivation (shrinking λ_n). Moreover, the error is proportional to the projection of prediction error ξ onto the deprived direction. Since the forward model has strong deficiencies in the deprived direction this will be large. Using the terminology introduced in Sect. 5.3.4 we may say that the deprivation is heating up any region of the error landscape that corresponds to the deprived dimensions. Given the picture of learning as a diffusion on the fluctuating error landscape, homeokinetic learning is expected to drive the system out of the deprived regions already by virtue of the statistical effects.

9.1.3.2 Effects on Learning

The consequences for the learning process can be worked out even more directly. Either by means of Eq. (9.7) or by starting with the symmetric representation of the general learning rule, see Eq. (5.17) that provides the update rule for any parameter p of the controller. We immediately obtain the learning rule as

$$\Delta p = \varepsilon_c \sum_{k=1}^{n} \frac{\left|\xi^\top u_k\right|^2}{\lambda_k^3} \frac{\partial \lambda_k}{\partial p} . \tag{9.8}$$

The expression shows how the deprived dimensions (with small λ) dominate the learning step. Updates in parameter space are driven by exactly those dimension that are most damped in the current dynamics of the states. Furthermore, via the quickly adapting parameters of the system, the eigenvectors of the deprived mode will change themselves very rapidly with a direct effect on the updates given by Eq. (9.8).

We may now provide some speculative arguments why, as stated above, this expression supports the self-amplification of latent modes of the system. Actually, at first sight, the update is dominated by the sum of all small eigenvalues of the Jacobian matrix of the system and especially by those that are subject to heavy fluctuations, caused either by a lack of regularity in the dynamics or more importantly by the deficiencies of the forward model in the deprived dimensions. This leads, besides the temperature effect, to further fluctuations of Δp realizing a more or less random search in the deprived subspace. This will lead nowhere in a high dimensional system. However, if one of the latent modes of the **embodied** system is getting excited even very weakly, the prediction error will reflect this regularities, such that the parameters dynamics can systematically amplify these modes and the general tendency of the homeokinetic learning to increase the eigenvalues can prevail. Admittedly, the argument is speculative but we are convinced that there is a general tendency to follow that scheme of bootstrapping of low-dimensional modes since it is observed in many ways in the practical applications.

An excellent example of the deprivation effect and the intrinsic bootstrapping was already presented, namely the BARREL in an upright position, see Sect. 8.5. There the deprivation was imposed onto the system by putting it into this physical situation. Since no sensor reactions are observed the Jacobian degenerates in all dimensions. The bootstrapping is first dominated by random actions, but as soon as the robot's body starts to slightly seesaw, the systematics is picked up and amplified. Let us study the effect in more detail starting with a high-dimensional example.

9.1.4 Planar SNAKE

Let us now consider the planar SNAKE as depicted in Fig. 9.3 with 16 segments and 15 joints. Each joint has one degree of freedom that is actuated by a servo

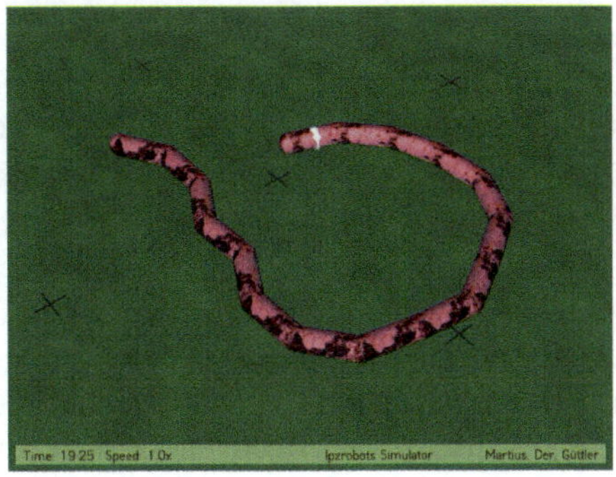

Fig. 9.3: Planar SNAKE with 16 segments and 15 degrees of freedom. In this experiment it is fixated at the first segment with a joint (white ring) and has no friction with the ground.

motor. A motor value of zero corresponds to a straight configuration and -1 and 1 are associated with the fully deflected positions at $\pm 90°$ off center. The robot is underactuated, meaning the power of the motors is not sufficient to move the joints independently of each other to any position. This effect is especially strong at the joints in the middle of the robot. In order to change the joint angle, both halves of the body have to be moved.

We assume again an artificial controller restriction and investigate the bootstrapping after the release of the restriction in this high-dimensional case. We define the direction of restriction by the vector $p \in \mathbb{R}^{15}$, with $|p| = 1$, where $P = pp^\top$ is the projection matrix into the subspace to be avoided. The restriction of the controller is achieved by

$$C_t = (\mathbb{I} - P)(C_{t-1} + \Delta C)(\mathbb{I} - P)^\top \quad \text{and} \quad h_t = (\mathbb{I} - P)(h_{t-1} + \Delta h)\,. \tag{9.9}$$

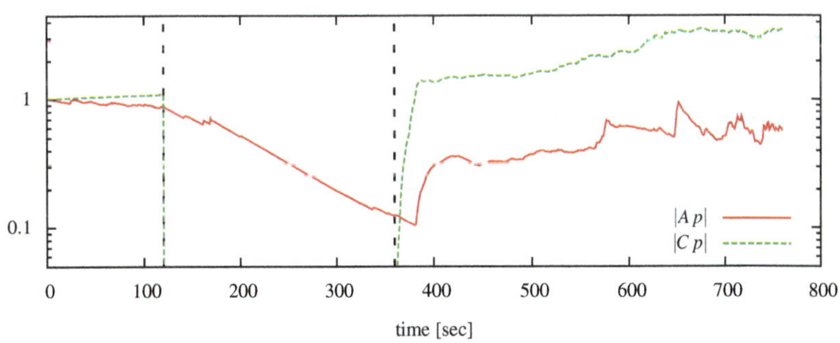

Fig. 9.4: Cognitive deprivation and bootstrapping with the SNAKE. The vector p is pointing in the deprived direction. After 120 sec the restriction of the controller (9.9) was switched on and at second 360 the restriction was released again. A quick recovery is observed. Settings: $\varepsilon_C = \varepsilon_A = 0.01$, $\gamma_d = 0.0001$, update rate 100 Hz, square root of error (Sect. 15.1.1.6).

Essentially the restriction is enforced after the parameter update each time step. Hence, the adaptation of parameters C and h is still performed, but the components corresponding to the p-space are projected away. In this example we use $p_i = \frac{1}{\sqrt{15}}(-1)^i$, which corresponds to a zigzag posture of the robot. In Fig. 9.4 the evolution of $|Cp|$ and $|Ap|$ are displayed, which are the lengths of the projections of C and A into the space of P. After 120 sec the restriction of the controller was switched on by constantly applying Eq. (9.9). One can see an exponential decay of the forward model in the direction of p. At second 360 the restriction was released and the controller matrix C quickly recovers (shows a large projection $|Cp|$ into the deprived space). The effectiveness of thereafter elicited actions is shown by the rapid increase of the projections of the model matrix A.

9.1.5 SPHERICAL *Robot in a Basin*

Let us now consider another case where both deprivation and bootstrapping occur in a way that is generic for homeokinetic learning. We want to take a closer look at experiments with the SPHERICAL in a circular basin that was already described in Sect. 8.2.3. Here we will focus on the development of the forward model and the resulting actions. Let us shortly recall the behavior of the robot. After some initial period we found the robot to roll roughly at a constant height for extended periods of time. In these periods we observe a deprivation of the forward model. In

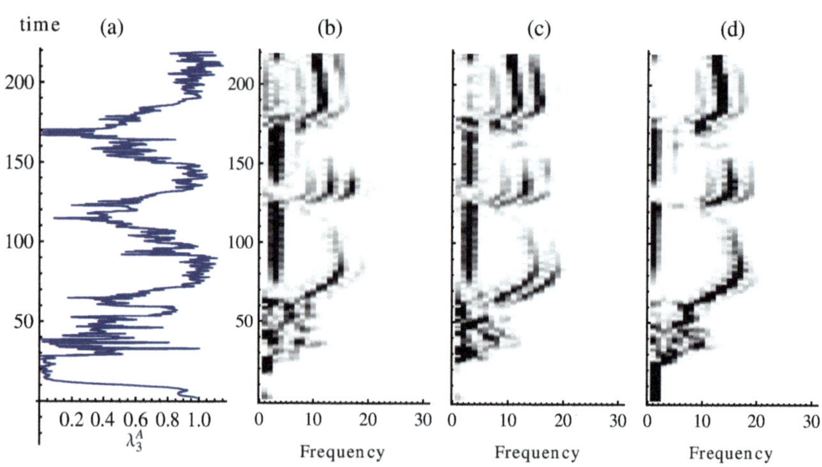

Fig. 9.5: Behavior of the SPHERICAL in a basin. Note that the time axes are from bottom to top. **(a)** Time evolution of λ_3^A, the smallest eigenvalue of the model matrix A; **(b-d)** Time evolution of the power spectrum of the three sensor values over time. Each row is the power spectrum of a certain time window. Consequent rows are overlapping. Darker pixels stand for more energy. High frequency means here high velocity of the robot. Parameters: $\varepsilon_C = \varepsilon_A = 0.1$, update rate 50 Hz.

Fig. 9.5, the behavior of the robot is characterized using the power spectrum of the sensor values and the smallest eigenvalue of the forward model. The initial phase is of the same nature as on the flat surface, i. e. from time 0–80 one can see self-explorational modes, where different frequencies are probed. Then a stable rotational mode emerged (time 80–110 and 140–170), which is the circulation in the basin at a constant height, exhibited over many laps. The circulation mode is seen in the power spectrum at the low frequency excitation. The high frequency excitations are the movements of the axes of the robot due to the rolling and precession motion. One can see that the value of the smallest eigenvalue λ_3^A of A decreases while the robot stays in one mode of behavior, see Fig. 9.5(a). The smallest eigenvalue is a measure for the degeneracy of the forward model, which deprives due to the restriction to a specific mode of behavior. At time 115 the forward model matrix is close to a singularity and the bootstrapping of new actions sets in. This period effectively explores the deprived subspace such that the smallest eigenvalue of A is seen to increase rapidly. The same starts again at time 140 and so forth.

In summary, the behavior of the SPHERICAL in the basin shows that the cognitive deprivation effect occurs naturally in a behaving robot once a stable mode of behavior persists for a longer time. However, with increasing deprivation homeokinetic learning is found to induce an exploration of the deprived dimensions so that the internal model is refreshed and eventually new modes of behavior can emerge.

9.1.6 Discussion

Model learning of high dimensional systems always faces the curse of dimensionality. The random sampling of actions is not very promising in such a setup. Additionally, in the closed loop setup with simultaneous learning of forward model and controller, it is well possible that substantial subspaces in the sensor-action space are not visited. We have shown analytically that a restricted controller leads to a deprivation of the model. However, we found that the homeokinetic approach provides a natural solution for this problem. The controller generates explicitly motor commands that are directed into the unknown regions of the sensor-action space. More details may be found in [43].

9.2 Extending the Forward Model

Most of the considerations done so far are based on $M(x,y) = Ay + b$ so that the forward model, as given by the matrix A and the vector b, reflects only the correlations between motor and sensor values and some offset. However, this is often quite unrealistic since the new sensor values may as well directly depend on the previous ones. In the following we will define the extended model and discuss possible issues and formulate the extended learning rules. Before that we will look at an example where the simple forward model causes problems.

9.2.1 Getting Misled by the Forward Model

Consider for example the SPHERICAL, see Sect. 8.2, when it is rolling down a slope. The sensors of the robot measure the orientation of the axes with respect to the ground. In this case the sensory dynamics is not caused by the robot actions, but rather by an environmental circumstance. The simple forward model cannot capture this, because it assumes a motor driven dynamics. A similar effect occurs if the SPHERICAL is equipped with light internal masses compared to the weight of its hull and is placed on level ground. In the experiments, after the robot acquired some rolling velocity, the sensory dynamics is again dominated by the current motion and is influenced only little by the current actions, due to the large inertia effects. In that hardware configuration the learning dynamics can become unstable, such that a diverging error value is observed. The divergence can happen even with comparably

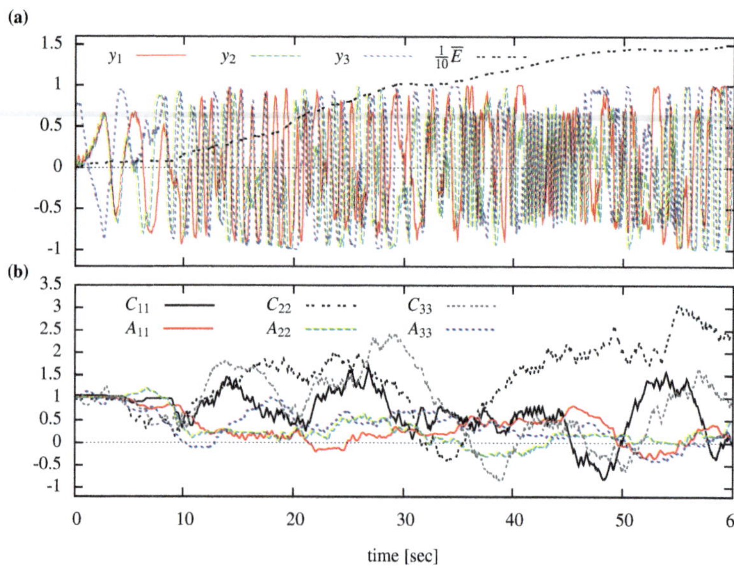

Fig. 9.6: Divergent error value for the SPHERICAL with light internal masses. Despite the low learning rates $\varepsilon_c = \varepsilon_A = 0.01$ the parameters fluctuate heavily and the forward model collapses. An irregular behavior is observed. **(a)** Motor commands and averaged error ($\overline{E}$) (scaled to fit in the plot); **(b)** Diagonal elements of controller matrix C and forward model matrix A. Parameters: $\varepsilon_c = \varepsilon_A = 0.01$, update rate: 100 Hz, weight of internal masses: $1/2$ of hull-mass.

low learning rates of $\varepsilon_c = 0.01$, in contrast to the normally used 0.1. Figure 9.6 illustrates the irregular behavior of the robot and the parameter evolution. The high error values are the result of the intrinsic dynamics of the sensor values when the body is rolling, which cannot be captured by the simple forward model.

9.2.2 Including the Sensor Branch

A simple extension of our forward model $M(x,y) = Ay + b$ (4.2) reads

$$M(x,y) = Ay + Sx + b \tag{9.10}$$

which includes the sensor-to-sensor mapping by the new matrix S. The sensorimotor map (dynamics model) is now

$$\psi(x) = Ag(Cx+h) + Sx + b \tag{9.11}$$

and the Jacobian is simply

$$L = AG'C + S\,, \tag{9.12}$$

where $G'_{ij} = \delta_{ij} g'_i(Cx+h)$ (as before). Due to the matrix S the forward model M represents the dynamics in the world. While the learning rules for the parameters A and b remain the same, see Eqs. (4.5, 4.6), S is updated as

$$\Delta S = \varepsilon_A \xi x^\top \tag{9.13}$$

In the homeokinetic learning rules, we have to invert L in order to compute χ and v, see Sect. 5.2.1 (p. 84). In most practical applications, S can not be singular, since it connects two closely related vectors. Hence, the inversion should be no problem in the mathematical sense, even if the sensors outnumber the motors, which is a common cause of divergences in the case without S. Still, the inversion in the full sensor space may be troublesome if the number of sensors is very high. In this case we propose to use generalized pseudoinverses as a convenient approximation, simply in order to reduce numerical effort, i. e. to reduce matrix inversions to the motor space ($m \times m$). These technicalities are dealt with in Sect. 15.4 (p. 277) so that we do not have to go into these details here.

9.2.3 Ambiguity of the Extended Model

Including the sensor branch introduces additional parameters for modeling the relation between the input and output vectors. This may lead to ambiguities, detailed in Sect. 9.A, with serious impacts on the model learning and therewith on the controller as well. In order to understand what happens, we can consider the model as the combination of two competing branches, namely the motor branch (Ay) and the sensor branch (Sx). Depending on the degree of fluctuations in the respective parts, the updates may be of different nature. In typical cases, the update ΔS of the sensor branch is very steady since S connects the two closely related state vectors x_t and x_{t+1}, which are often approximately mapped to each other by a unit matrix. On the contrary, A describes the correlation between motor and sensor values that may heavily change in the behaving robot so that actually A is fluctuating much more than S. Therefore, given the ambiguity, the sensor branch will dominate the learning steps so that the motor branch may be modeled very badly. This is rather disastrous for the homeokinetic learning procedure so that we have to resolve this ambiguity in a suitable way. In the following we will propose one of many ways to deal with this. More options may be found in [102].

9.2.4 Increasing Awareness of Causality

On the more conceptual level the ambiguity can be resolved by differently interpreting the sensor values. One extreme is to consider the sensor values to be mostly determined by the intrinsic dynamics of the world and the other extreme is to assume the sensor values are fully determined by the actions of the agent. The latter is actually the approach we choose if the simple forward model (4.2) without the sensor branch is used. The true correspondence is typically somewhere between these extremes, however it cannot be determined from the perspective of the agents in the closed loop setup. In order to estimate the consequences of its own actions appropriately the first extreme is not very productive since then the agent does not "believe" in the effectiveness and/or causality of its actions.

With active agents, we argue that it is reasonable to postulate a substantial bias towards the awareness of causality between actions and sensations. This can be achieved by trying to represent the causal effects of the action preferably with the motor branch (given by the matrix A). Very useful in this respect is to derive the learning rule for A not based on the prediction error $\|\xi\|^2$, as done so far, but directly on the TLE (5.9). In this way, the implicit dependence of the TLE on the A matrix is included into the learning dynamics, which was neglected before. Why would this be helpful in resolving the ambiguity? The TLE increases the sensitivity of the network to its inputs. In this way the matrix A is driven to grasp most of the correlations between motors and sensors. The gradient descent of the TLE w. r. t. to A yields

$$\Delta A = \varepsilon_A \chi^\top \frac{\partial L}{\partial A} v + \varepsilon_A \chi^\top \frac{\partial \psi}{\partial A}, \tag{9.14}$$

where we use $\xi = x_{t+1} - \psi(x_t)$. In our setting (9.11), the update rule for A is obtained explicitly as

$$\Delta A = \varepsilon_A \chi \hat{y}^\top, \tag{9.15}$$

where

$$\hat{y} = y + \rho_A G'Cv \tag{9.16}$$

with the additional factor ρ_A that allows to weight the influence of the new term. In comparison to the usual delta rule for learning A ($\Delta A = \varepsilon_A \xi y^\top$, Eq. (4.5)) we have two modifications, one is the replacement of ξ with $\chi = \left(LL^\top\right)^{-1}\xi$ and the other one is the shift of the controller output by the term $\rho_A G'Cv$. What is the meaning of the latter? In the model without S we have[2] $G'Cv = \eta = A^{-1}\xi$, which is the reconstructed error ξ at the input of the model. The fact that now the inverse of A is featuring in the update rule will increase the sensitivity of the model to its motor inputs.

This solution of the ambiguity problem consists in strengthening the connections in the motor branch independently of the actual causal relation. Remember, that the controller is generally adapted to make the internal representation of the sensor dynamics match the actual sensor dynamics (especially if the extension in Sect. 15.1.1.4 is used). If the internal model is adapted such that an external dynamics is explained in terms of its own actions then the controller tends to produce those actions that are coherent with this model.

9.2.5 Experiment with the SPHERICAL *Robot*

In Sect. 9.2.1 we have seen that the control of the SPHERICAL with small internal masses can be a problem if the simple forward model is used. In the following experiment the extended forward model (9.10) is used and the robot will be seen to behave smoothly. The extended model is able to make better predictions already before stable modes are found. This is especially true with systems that are only partially controlled by the actions due to a substantial momentum of the physical system, like the currently considered SPHERICAL.

The evolution of the behavior and the parameters during the experiment are depicted in Fig. 9.7. In contrast to the earlier experiment (Fig. 9.6) without the extended model, the error now falls to a low value. The robot remains in a certain rolling mode for a while until the behavior is changed. At theses points in time the error goes up and is then decreased again. A more detailed description of the behavior and a video is located in Sect. 8.2 (p. 162). Let us here take a closer look at the parameter dynamics. Even though the S matrix was initialized with zero we observe in the beginning that S captures a substantial part of the dynamics, see Fig. 9.7(d). Then the robot is controlled so that one of the internal axes is the rotation axis (behaviors A-C). During this time the model matrix A is clearly seen to deprive (smallest eigenvalue λ_3^A decreases), because the movements of the mass along the rotation axis have only little effect on the sensor value due to gyroscopic and inertia effects. At the same time the controller is increasingly destabilized (Fig. 9.7(c)) until the behavioral mode becomes unstable and a different mode is found. The model

[2] We assume normal inversion of the matrices or the sequential strategy for the pseudoinverse, Eq. (15.48).

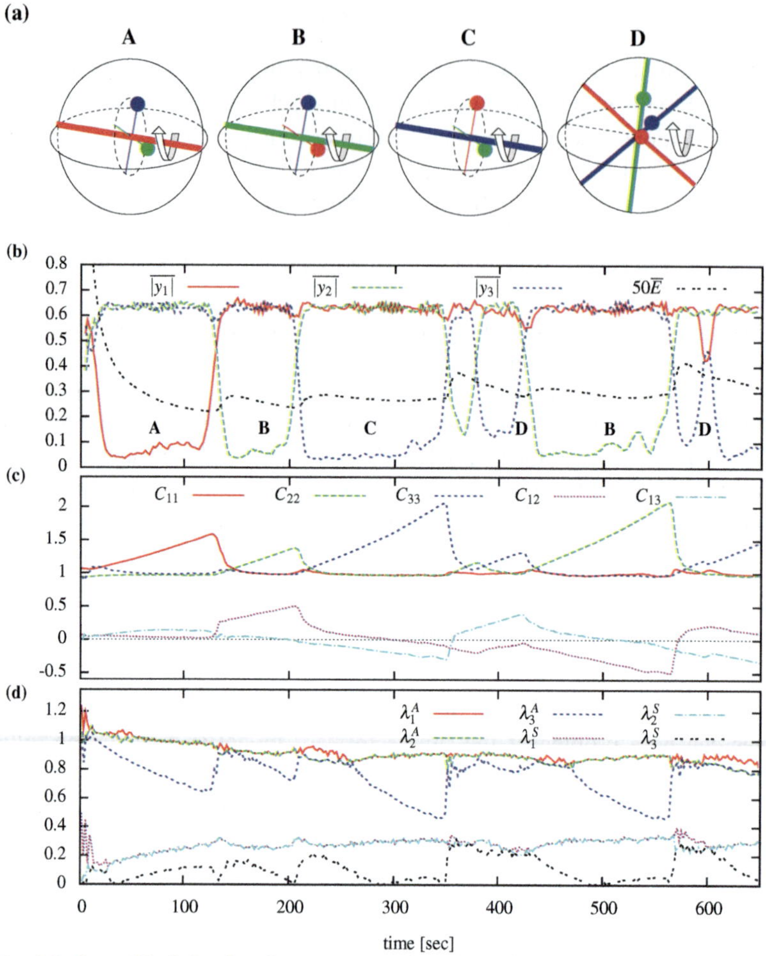

Fig. 9.7: Smoothly behaving SPHERICAL using the extended forward model. The TLE has a low value and the behavior is smooth but still diverse. **(a)** Sketch of four typical behaviors (**A-D**). **(b)** Amplitude of motor commands and the error (scaled for visibility) are averaged over 10 and 30 sec. Corresponding behaviors are indicated with letters **A-D**. **(c)** Diagonals and two non-diagonal elements of the controller matrix C. **(d)** Eigenvalues of the forward model matrices A and S. Parameters: update rate $100Hz$, $\varepsilon_c = \varepsilon_A = 0.1$, $\gamma_d = 0.0001$, $\rho_A = 0.1$.

is seen to recover and the game starts anew. Thus we have yet another example of the deprivation and bootstrapping effect.

To summarize, the extended forward model leads to a decreased prediction error, resulting in a smooth and stable parameter dynamics. Using the TLE also for learning the motor branch of the internal model increases the awareness of causality which is important for homeokinetic learning.

9.3 Summary

The learning of internal models in high-dimensional systems faces the curse of dimensionality. The random sampling of actions is not very promising in such a setup. Additionally, in the closed loop setup, with simultaneous learning of internal model and controller, it is well possible that substantial subspaces in the sensor-action space are not visited, such that the model degenerates. However, we found that the homeokinetic learning provides a natural solution for this problem. The controller generates explicitly motor commands that are directed into the unknown regions of the sensor-action space, see for instance the experiment with the SNAKE robot.

We also took a look at an extended forward model, that uses apart from the motor values also previous sensor values for prediction. Since it models also the dynamics in the world it is called forward model. The new approach creates an ambiguity in the learning such that possibly unfavorable configurations are reached. A simple solution for this problem is to learn parts of the internal model (the motor branch) using the TLE. The extended forward model enables the homeokinetic controller to handle systems with large inertial effects and to better cope with different sensor types.

Appendix 9.A Ambiguities in the Sensorimotor Loop

The occurrence of ambiguities in the sensorimotor loop are obvious if the sensor branch is included since we have now more parameters (matrix elements of A and S) than equations for finding the correct matrices. As a demonstration let us consider the case of a linear world like

$$x_{t+1} = A^{(w)} y_t + S^{(w)} x_t + \varsigma_{t+1} \tag{9.17}$$

with some zero mean noise ς, $A^{(w)}$ representing the response of the world on the motor values y and $S^{(w)}$ the direct influence of previous sensor values on the new ones. Using the model from Eq. (9.10)

$$x_{t+1} = Ay_t + Sx_t + b + \xi_{t+1} \tag{9.18}$$

with $b = 0$ for simplicity, the learning steps are (omitting the time indices)

$$\Delta A = \varepsilon \xi y^\top \quad \text{and} \quad \Delta S = \varepsilon \xi x^\top . \tag{9.19}$$

The effect is already obvious if we use the most simple case of a linear controller[3] as

$$y = Cx$$

so that

$$\xi_{t+1} = (\delta AC + \delta S)x_t + \varsigma_{t+1} = \delta P x_t + \varsigma_{t+1},$$

where $\delta A = (A^{(w)} - A)$, $\delta S = (S^{(w)} - S)$, and $\delta P = (\delta AC + \delta S)$. In the average over time the noise ς vanishes and we obtain the prediction error $E^{\text{pred}} = \xi^\top \xi$ as

$$2\left\langle E^{\text{pred}} \right\rangle = x^\top \delta P x + \left\langle \varsigma^\top \varsigma \right\rangle . \tag{9.20}$$

We have to consider this over a set of states x spanning the whole state space, so that E^{pred} is minimal only if $\delta P = 0$, i. e.

$$AC + S = A^{(w)}C + S^{(w)} \tag{9.21}$$

which has arbitrarily many solutions for A and S.

[3] The considerations are also true for more complex controllers.

Chapter 10
High-Dimensional Robotic Systems

Abstract: This chapter contains many applications of homeokinetic learning to high-dimensional robotic systems. The examples chosen for investigation and proposed as experiments to the reader comprise various robots ranging from dog-like, to snake-like up to humanoid robots in different environmental situations. The aim of the experiments is to understand how the controller can learn to "feel" the specific physical properties of the body in its environment and manages to get in a kind of functional resonance with the physical system. In order to better bring out the characteristics of homeokinetic learning in these systems, we use a kind of physical scaffolding, for instance suspending the HUMANOID like a bungee jumper, putting it in the RHOENRAD, or hanging it at the high bar. Interestingly, in all situations the robots develop whole-body motion patterns that seemingly are related to the specific environmental situation: the DOG starts playing with a barrier eventually jumping or climbing over it; the SNAKE develops coiling and jumping modes; we observe emerging climbing behaviors of a HUMANOID like trying to get out of a pit; and wrestling like scenarios if a HUMANOID is encountering a companion. Eventually, in our robotic zoo all kinds of robots are brought together so that homeokinesis can prove its robustness against heavy interactions with other robots or dynamical objects. Essentially this chapter provides a phenomenological overview and invites to play around with numerous simulations to see the "playful machine" in action.

We start now with applying the algorithms of homeokinetic learning developed in the previous chapters to physical systems of higher dimensionality. In Video 10.1 we metaphorically sketch the challenge of self-organizing robot behavior. The video shows three robots of very different morphology but with the same number of motors and proprioceptive sensors (joint angles), reporting the results of motor actions back to the brain. Each of these robots has a brain of its own but the brains are identical in structure and are learning by the same homeokinetic learning rule. The robots are "thrown" into the world and left to themselves. We already know that they will come to activity by the homeokinetic learning. However, what will the emerging motion patterns look like?

The challenge for true self-organization is the emergence of embodiment-specific motion patterns, i. e. that each robot develops an individual way for self-actualization depending not only on its morphology but also on its interaction with the environment encountered in the course of time. This is the "acid test" any self-organization paradigm of embodied robot development has to pass. Unfortunately, it

R. Der and G. Martius: The Playful Machine, COSMOS 15, pp. 201–221.
springerlink.com

is difficult to develop an effective measure for these features so that an assessment will always depend on subjective factors. This can be helped a little by designing experiments so that the specific features can develop.

In the experiment shown in Video 10.1 as in many of the examples of this chapter, we use the standard setting of a one-layer network for both the controller and the model together with the basic learning rules as given in Sect. 5.2.2 (p. 85), Eqs. (5.22, 5.23). We will indicate explicitly in the experiments which learning rule is being used.

Video 10.1: The challenge for self-organization. Emergence of embodiment-specific behavior in robots with different morphology but identical brains. The video can be watched at `http://playfulmachines.com`.

Initialization

In most experiments, the robot is initialized in a "do nothing" state so that in this phase the feed back strength of the sensorimotor loop is subcritical and the robot will stay in a resting position. The parameter dynamics, in this situation will increase the feedback strength little by little so that after some time the robot starts moving its limbs in a more or less spontaneous fashion. However, in the initialization one has to avoid the "dark side" problem sketched in Sect. 5.1.4 (p. 81). In the experiments described in this chapter we have explicitly taken care of that problem. More details of how to get a convenient initialization may be found in our cooking recipe in Sect. 15.2.

If the starting phase is chosen in that way it is advisable to add a little sensor noise. Otherwise, if for instance the robot is lying on the ground it is forced by gravity to total rest so that the sensor values are constant. In this situation the forward model will very rapidly learn to produce the exact sensor values (motor values being also constant) so that the time-loop error goes down to zero and learning stops altogether. Once the feedback strength exceeds a certain critical value so that the robot starts moving, the model will start producing a residual error in a natural way due to the extremely complicated dynamics in these high-dimensional physical systems. Then, the noise may be switched off so that the state-parameter dynamics becomes a purely deterministic dynamical system in several hundred dimensions.

Spontaneity

One of the interesting observations made in that setting is the seemingly random nature of the motion patterns although the system is deterministic. This is explained by the fact that the physical dynamics takes place in a low-dimensional subspace of the extremely high-dimensional state-parameter space. In this way what we see is not random, but spontaneous, namely the result of the hidden dynamics in the residual dimensions.

Nevertheless, in particular in long-time experiments, it is often advisable to inject some (very) weak noise into the sensors together with a weak damping of the parameter dynamics.[1] This avoids running into "pathological" regions of the state-parameter space. Speculatively, with several hundred dimensions, the dynamics of the combined system is prone to running into "very strange" attractors (possibly much different from what we know so far by dynamical systems theory). In experiments with real robots, some kind of sensor noise and even actuator noise (that we do not use in the simulations) is present in a natural way so that its use in simulations is not unrealistic. But again, we emphasize that the intended self-actualization and playful behavior is to be a result of the interaction between the deterministic inner dynamics and the uncertainties of the external (to the brain) world. In the experiments of this chapter, the uncertainties are essentially a result of the properties of the body—a complex mechanical system getting fed energy into by the actuators.

10.1 Underactuated and Compliant

We have seen in Chap. 8 how the homeokinetic controller realizes what we call body inspired control. We have seen there that the controller, quite by itself, develops a kind of feeling for the body and manages to induce whole body movements by forces of minimal strength, getting into a kind of functional resonance with the physical dynamics of the mechanical system. The SLINGING SNAKE in Sect. 8.3 (p. 168) gives a striking example of that scheme: the robot is very underactuated and has a poor sensor information measuring nothing but the velocity of the head over ground. Given the highly complex physical system of many degrees of freedom actuated solely by the head it is very surprising that the controller manages to develop a highly regular whole body motion by slinging the passive parts of the body.

We are now going to consider a certain parallel of that situation in systems with more motors and sensors but with the same setting of heavy underactuation and poor

[1] Of course, you are free to try strong noise, too, checking the robustness of homeokinetic learning against external perturbations.

sensor information. The latter is only obtained from the joint angles and this is of course not enough to understand the physical situation of the robot. In such a setting, what will the emerging motions look like? Will there be a relation to the body and will the controller manage to excite collective motions of the system, i. e. whole body motions in tight correlation with the sensor values?

Let us consider two experiments that demonstrate in a particular way how the controller is being inspired by a highly compliant body.

DOG *Fixed in the Air*

In the first experiment we consider the DOG, a dog-like robot, which is like all robots in this chapter, simulated in our LPZROBOTS environment. The DOG consists of boxes and capsules connected by 12 actuated hinge joints, see Sect. 16.4.1. In this first experiment we fix the trunk of the robot so that the legs may move freely. In order that the "brain" may feel the embodiment we use the case of an underactuated system where the torques are so weak that large angular values can only be reached by getting the legs into a swinging motion, exploiting the inertia given by the mass of the legs. Initialization is in the standard way described above. In the underactuated system this means that the legs will hang down freely. The interesting point is that after some time the legs start swinging in different modes and temporarily we even observe an emerging coordination between the legs, see the Video 10.2 and/or do Experiment 10.1.

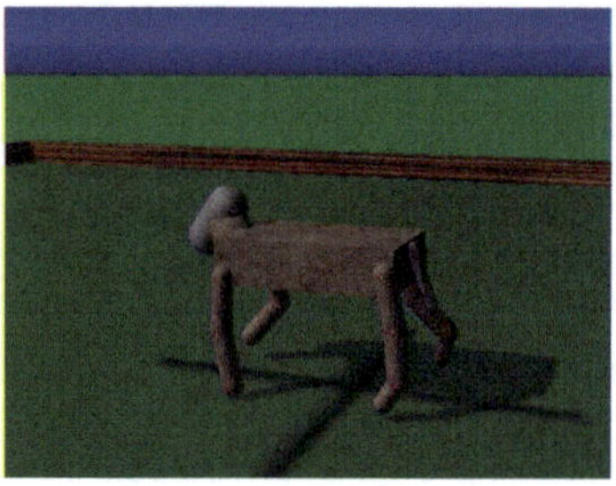

Video 10.2: Swinging legs. The trunk of the DOG is fixed so that the legs can move freely. The motors are so weak that the controller must learn to excite a resonance mode. There are no physical cross couplings between the legs. Nevertheless, there are some frequency and phase correlations observable. The video can be watched at `http://playfulmachines.com`.

This means that the legs swing at about the same frequency, with phase relations being constant over some time. We observe both in-phase and anti-phase correlations in the leg motions much in the way necessary for different gaits. The coordination is only transient but that there is a coordination is seen best by its breakdown: the swinging of the legs stops almost simultaneously, which can happen only if there is a coordination between individual legs.

However, this is highly nontrivial since there is no direct information exchange, neither physical nor between motor neurons, the only way of establishing correlations being the emergence of cross channel couplings in the matrix C of the controller. This is what is happening temporarily at least, corroborating our general

Experiment 10.1: Swinging legs. The heavily underactuated dog.

Start the simulation DOG with swinging legs where the robot is fixed in the air. You can drop the robot by pressing `X`. Default values of the parameters are set such that you will observe after some time coordinated motion patterns. You may wish to change the following parameters: `forcefactor` (this is a relative quantity multiplying the force of the motors, default value is 1), `gravity` (default is -9.81), and the learning rates for controller and model (`epsC, epsA`) (default is 0.1). Use the GUILOGGER and MATRIXVIZ for monitoring the parameters (x, y, C, h).

understanding of homeokinetic learning as a process of making each motor neuron sensitive to the inputs by all sensors and this can be realized best if there is some correlation between the sensor inputs.

At a more speculative level we may also argue that the coordinated swinging motion is emerging because our parameter dynamics preserves the symmetries of the physical system so that, with a convenient initialization (and isotropic sensor noise, if any at all), any motions can only arise from spontaneously breaking these symmetries. As we have seen in Sect. 5.3.5 (p. 96), oscillatory motions are very natural in this context. Moreover, due to the explorative character of the full dynamics, the system is not observed to establish and stabilize a specific mode. Instead all modes are of a transient nature, so that we observe the playful realization of several such collective modes.

The Suspended HUMANOID

The next example we are going to consider is the underactuated HUMANOID. Like the DOG, our HUMANOID consists of boxes and capsules but has human-like proportions, see Video 10.3. The motor power and sensor configuration is similar to the DOG but we will not fix the trunk of the robot in the air. Instead, we consider the case that the robot is suspended like a bungee jumper so that its legs are just able to reach ground. This is realized in the simulator by an emulated spring producing an upwards directed force linearly depending on the position of the trunk. The position and strength of the spring can be defined conveniently, see Experiment 10.2.

There is nothing spectacular happening in this scenario, see Video 10.3. Different from the DOG, however, there is now a physical coupling between all degrees of freedom mediated on the one hand by the collisions with the ground and, on the other hand, by inertia and gravitation that become effective if ever the robot or any of its parts changes its position or inclination. How can we see that the controller is "feeling" its body? Again, this is a controversial question that can not be answered in an objective way. Of course, it would be very interesting to have a measure for the whole body movements. Predictive information of the sensor values might be a convenient solution, as could be concluded from the results obtained with a chain of wheeled robots with decentralized control [40] but we do not have concrete results yet. Other methods for using information theory as proposed for instance in [98, 135, 140, 155] could help here in order to quantify the emerging coordination between the parts of the body.

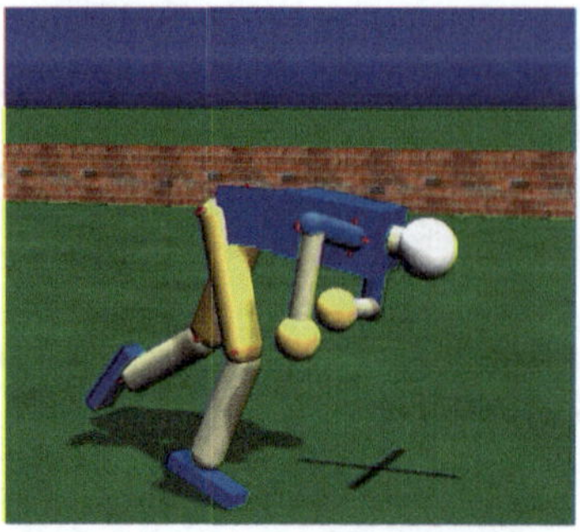

Video 10.3: Suspended HUMANOID. The robot is pulled upwards by a simulated spring such that the robot can freely move, but still interacts with the ground. This is a kind of scaffolding situation in order to give the brain time to accustom to the body. The video can be watched at `http://playfulmachines.com`.

Experiment 10.2: Bungee jumping.

Start the simulation Bungee jumping. Default values of the parameters are set such that you will observe after some time the robot taking all kinds of whole body motion patterns. You may wish to change the following parameters: `forcefactor` (this is a relative quantity multiplying the force of the motors, default value is 1), `gravity` (default is -9.81), learning rates for controller and model (`epsC, epsA`) (default is 0.1). The robot is lifted by a spring force (bungee) that you can adjust with `bungeestrength` (relative quantity, default 1) and `bungeeheight` that specifies the height of the spring origin in multiples of the body height. You may apply additional forces to the center of the trunk by using <Ctrl>+<left mouse button>. Monitor the parameters (x, y, C, h).

Visual inspection can tell us a little about the nature of the motion patterns, in particular if changing the physical parameters like motor power, gravity, or the length of the spring giving the robot more ground contact or less. It is readily seen that the behavior changes and one may well see that the changes are by trend taking account of the modified conditions.

These experiments reveal more interesting results in the course of time, so it is worthwhile to have the simulations run for several hours since then the more typical patterns are emerging. In these simulations it is also advisable, as discussed above, to inject a little sensor noise and include a small damping into the controller parameter dynamics. As already mentioned, you are invited to try strong noise, too, checking the robustness of homeokinetic learning against external perturbations.

The freely moving robots offer a good opportunity to test the potentials of the more complex brain architecture, using multilayer networks for both model and controller, as implemented in the `soml` algorithm, see Sect. 15.3. The point is that the multilayer architecture needs coherent learning signals over a longer period of time. This is not so well guaranteed if the robot is subject to rapidly changing conditions, as is the case already if the robot is in full ground contact without any suspension. It is interesting to observe how the robot can be prepared to the latter situation by having it suspended for a longer time, dropping it only afterwards. This can be considered as a kind of scaffolding, giving the multilayer controller network time to accustom to the specifics of the body mechanics in a less challenging setting. Interesting experiments in that direction have for instance been reported in [90]. An experiment is prepared in Experiment 10.3.

Experiment 10.3: Multilayer networks for forward model and controller.

Start the simulation Multilayer networks: Bungee. Both controller and forward model are now networks are two layers. You can change the number of hidden units with `numcontrolunits` and `nummodelunits` (default 17). The changes will become effective after restarting the robot by pressing `r` in the graphical window. See Experiment 10.2 for further details. Use the MATRIXVIZ for monitoring the parameters (x, y, $C1$, $C2$, and h).

10.2 DOG and HIPPODOG

Let us now study two dog-like robots that differ in an essential construction element of the body. The DOG is equipped with a large massless box fixed to its back (not visible in the simulations), preventing it from falling over. The HIPPODOG on its part has a spherical trunk so that it may roll over when falling down. Our interest is in the difference in the motion patterns developed by these two different mechanical structures under homeokinetic control.

In most of the experiments the gravity is reduced to about one half of the earth gravity (-9.81) much in the sense of a scaffolding.

The DOG

Above, we considered the DOG in the air, but the main interest is of course the behavior of the dog when on the ground. Nevertheless, in our experimental settings, see Experiment 10.4, we often let the dog for some time fixed in the air so that the "brain" can feel the motions of the unperturbed legs as in the experiments above, much in thc spirit of a scaffolding of the development process. After that we let the dog fall to the ground. However, the normal settings is to start the dog directly on the ground. In that case it is useful to initialize the controller matrix C and model matrix A as scaled unit matrices so that (i) there are no cross correlations between the channels (made up of the motor and the respective proprioceptive sensor), and (ii) the feedback strength is already close to the critical point so that the amplification of noise (actually the prediction error) is already substantial. In this way, initial irritations resulting from the physics (falling down) are already sufficient to "get things going." If not, injection of a little sensor noise or a mechanical shake up (use mouse buttons) will do.

In order to illustrate the emerging sensorimotor coordination we use an environment consisting of three concentric squares with barriers of increasing height. In a typical run, see Experiment 10.4, after its initial phase of getting into activities the

Experiment 10.4: DOG surmounting barriers.

Start the simulation Dog Inside Barriers, which starts with the DOG fixed in air. You may drop it by pressing `x`. The default setting of the arena is three barriers of heights $1/3$, $1/2$, and 1. You may change the distance and height ratios by the relative parameters `barriersheight` and `barriersdistance`. You may change the parameters as described in Experiment 10.1.

DOG rather soon surmounts the innermost barrier and then lingers around for quite some time. It may return into the inner barrier and/or will eventually face the next barrier, which has a height of about half the dogs clearance. In the interplay with this barrier, it eventually will surmount that barrier, see Video 10.4 for an example.

When caught on the barrier, a more cautious, probing behavior is emerging. Video 10.5 shows that the DOG is keeping its body low most of the time so that it has good contact with the barrier. Please note again that the only information the robot has on its body and the contact with the obstacle is given by the sensors measuring the joint angles. Contact with the barrier is felt only by the perturbations of the sensorimotor correlations caused by the physical forces. The preference of the forward direction in the motion is interpreted as a result of the anatomical asymmetry.

Video 10.4: DOG "playing" with barrier. The DOG video demonstrates the playful manner of exploring the world in the homeokinetic learning scenario. Note that the robot has nothing but its joint angle sensors as source of information about its body and the environment. The DOG is protected by an invisible box on its back preventing it from falling over. The video can be watched at `http://playfulmachines.com`.

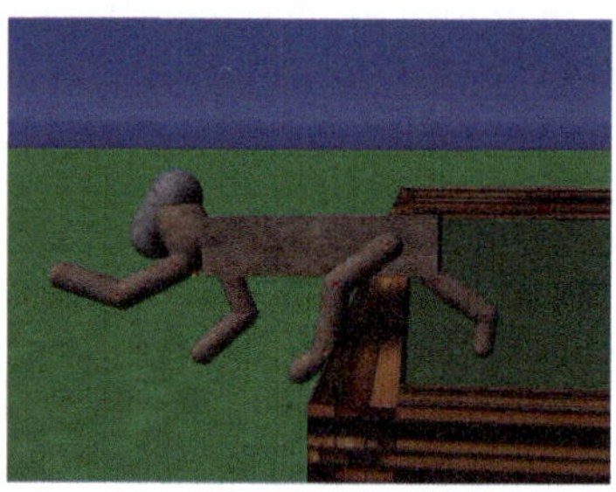

Video 10.5: Dog climbing over a barrier. When encountering a barrier, the DOG can behave in many different ways since there is no goal prescribed. However, more often than not, the robot manages to climb over the barrier in a seemingly goal oriented way. In the video, the DOG has acquired a rather cautious behavior slowly probing different possibilities of interacting with the barrier. After some time the left hind leg is swung onto the barrier and after several minutes it climbs out completely. The video can be watched at `http://playfulmachines.com`.

Experiments with the HIPPODOG

In another artifact, the HIPPODOG, we have modified the DOG by giving it a spherical trunk, omitting the artificial stabilizing box. The spherical shape of the body is sufficient that the robot, after falling over, manages to get back to its "working" stance from nearly any situation. Thus one may leave the robot to itself in the same way as its "ancestor" that was protected by the invisible box. In the course of time we observe similar behaviors as described with the DOG. However, due to its higher mobility, the HIPPODOG has more the tendency to reach a very active, jumpy regime. In particular, getting back to its feet is not realized, as a real dog might do, by rolling over, keeping the legs stretched so that they are out of the way. Instead, the robot gets back to its feet by heavily agitating its legs catapulting itself up by ground contact with its feet, exploiting the high ground-foot friction.

Video 10.6: The HIPPODOG — The effect of the anatomy. The HIPPODOG with its spherical trunk is developing more active and jumpy motion patterns due to its specific anatomy. The video can be watched at `http://playfulmachines.com`.

Experiment 10.5: The HIPPODOG.

Start the simulation HIPPODOG . Choose arena and parameters as in Experiment 10.4.

In conclusion, let us emphasize again that the robot has no goal or aim to surmount the barriers so that often it returns instead of moving further ahead. The point we want to make is that the observed motion patterns may be described as activities in close dynamical contact with the objects like the barrier. This leads to an increased probability for seemingly dedicated actions like jumping over the barrier as in Video 10.4.

10.3 HUMANOID

Let us consider the HUMANOID again. We think that the HUMANOID lends itself best to qualitative visual inspection because of its similarity to humans. As humans, we are particularly trained to observe and "classify" other human beings by their body language, an ability that is thought useful for our purpose of getting qualitative understanding of the diversity of behaviors emerging under homeokinetic control.

The emerging behaviors largely depend on the environmental conditions, as encountered for instance on level ground, in a cluttered environment, or if together with other robots where often a wrestling-like behavior is emerging, see below.

Like in the previous experiments, the gravity is reduced to about one half of the earth gravity (-9.81) in order to provide some scaffolding. In the experiments it is interesting to change the gravity parameter, observing the capability of adapting to new physical situations.

Starting from Scratch

Let us study first the initial development of the robot behavior in the case of the HUMANOID.

We study here two situations, the first one, starting on level ground, gives the robot more freedom although the motions are constrained due to the gravity pulling the robot to the ground. In the second situation the HUMANOID is falling into a narrow pit. In the following experiments the motor forces are strong enough to partly counteract the constraints. As usual, we have chosen the slightly subcritical initialization of the controller without any cross channel correlations, see above. In the first situation, see Video 10.7, the robot develops smooth and slow motions. Already during the first minutes an increase of the behavioral spectrum is observed. Later on one often observes patterns that look like the intention to get up, repeating several times. In Video 10.8 the behavior in the second situation is shown, where many different behaviors emerge due to the much stronger constraints given by the geometry of the pit.

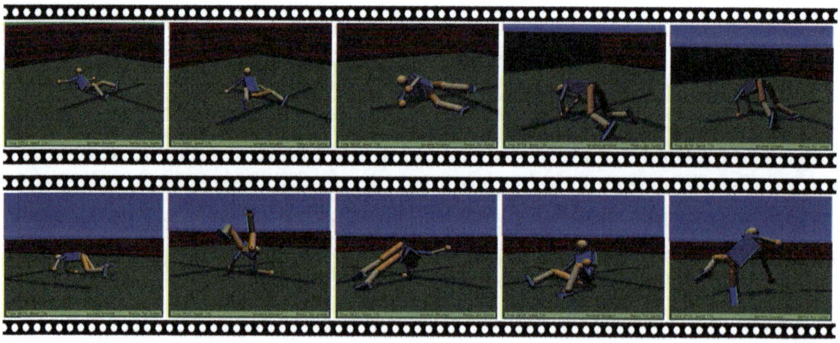

Video 10.7: Initial development of the HUMANOID (I). Top: From scratch. **Bottom:** 6 min later. If the robot is started initially on level ground, it develops rather smooth and slow motions, increasing steadily its behavioral spectrum. The videos can be watched at `http://playfulmachines.com`.

Video 10.8: Initial development of the HUMANOID (II). If started in the pit, the robot develops increasingly complex motion patterns related to that situation. Interestingly, these patterns keep repeating in the course of time, getting more pronounced and lasting over a longer time. The video can be watched at `http://playfulmachines.com`.

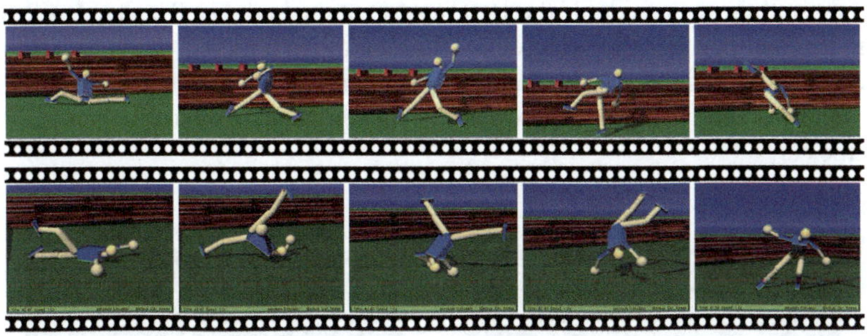

Video 10.9: Later developments. In the later development of the HUMANOID on level ground we observe more active behaviors. Interestingly, one often observes patterns that look like the intention to get up, these patterns often repeating several times. The bottom video shows another motion pattern like they often emerge in the course of time. In the experiment the gravity was chosen half the earth gravity in order that the robot can move more freely. This is necessary in order to get enough feedback from the body. The videos can be watched at `http://playfulmachines.com`.

High Bar and Rhönrad - Feeling the Body

The following two applications are to illustrate the emergence of what we call a feeling for the body. The examples studied in Sect. 10.1 (p. 203) have already demonstrated how in strongly underactuated and compliant systems the homeokinetic controller manages to excite whole body modes despite the minimal control forces. That setting was chosen so that the combined state-parameter dynamics has some time to settle into coherent modes, i. e. that the entanglement described in Sect. 5.3 can become effective. Otherwise, the underactuated and suspended situation is not

Experiment 10.6: HUMANOID

Start the simulation HUMANOID. The robot is not connected to a spring so that it falls down freely. You can change the arena with the parameters `arenasize` and `arenaheight`. Press `r` to restart the robot at the initial position. You may wish to change the following parameters: `forcefactor` (overall factor for strength of motors, default 1), `relforce` (ratio between strength of arms to rest, default is 1), `gravity` and the learning rates for controller and model (`epsC`, `epsA`) (default is 0.1). There are many more other parameters as seen in console output. In a second simulation you can try the HUMANOIDin a rescue setup similar to (Video 10.8). Use the GUILOGGER and MATRIXVIZ to monitor all parameters (x,y, C, h).

very realistic with respect to real robotics applications and, moreover, the situation is largely unconstrained so that the emerging behavior patterns are very diversified.

Let us therefore consider two different situations where the physical constraints are more dominant without restricting too much the independent development of behaviors.

HUMANOID at the High Bar

The first setting we consider is the HUMANOID at the high bar. The hands are fixed to the bar, but they can slide sideways. Clinging to the high bar, the HUMANOID is forced into an upright position while still having much freedom to sway and swing. The motor forces are again quite weak so that for instance the robot is not able to chin the bar.

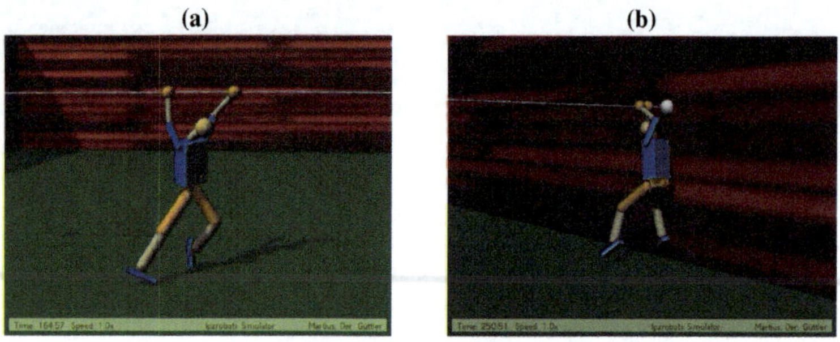

Video 10.10: HUMANOID exercise at high bar. Underactuated robot under physical constraints: motor forces are quite weak so that for instance the robot is not able to chin the bar. The observed motion patterns express the emerging cooperativity of body components. The videos can be watched at `http://playfulmachines.com`.

Experiment 10.7: HUMANOID at high bar

Start the simulation High Bar where the humanoid is hanging on a fixed high bar. The robot can move along the bar and of course swing axially. You may change the height of the bar with the parameter `barheight`. The other parameters are the same as in Experiment 10.6.

As Video 10.10 shows, ambitious expectations are only partly met. On the one hand we observe several motion patterns that look the way a very unathletic human might behave. On the other hand, there is no swinging up observed as might be possible by gaining enough momentum even in this weak force setting. The potential reason is seen if the bar is chosen so high that the legs can not reach the ground. What happens in this situation is that the robot most of the time is hanging in a stretched posture without much trying to get into an oscillatory motion, as it would be possible with the given motor forces?[2] Our preliminary explanation for that effect leads us back to the one-dimensional sensorimotor loop analyzed in Chap. 6. There, the system was shown to go into a hysteresis oscillation due to the entanglement with the bias dynamics. Why does this not happen here? The explanation is found in the strong torques exerted by the gravity on the joints. Strong external forces change the hysteresis properties of the sensorimotor loop and thereby the entanglement between state and parameter dynamics. Probably this effect is strong enough here to damp away the oscillations altogether—the robot keeps staying in the elongated posture.

Does this knock out the claimed universality of our approach? One might argue so, in particular since this phenomenon plays a role in many other situations as well (see below), if the underactuation is substantial. However, there is a way out. The point is that we are working here still with the forward model that relates sensor and motor values simply by the matrix A. But this is completely wrong if there are strong external forces. We expect that a convenient inclusion of these forces into the model will improve the situation but we have no concrete results yet.

The RHOENRAD

The Rhönrad, a purely German item as it seems, is a large wheel of human size that can be moved by shifting the center of gravity, see Fig. 10.1. Video 10.11 shows the robotic counterpart RHOENRAD realized in our simulator, see Experiment 10.8.

Fig. 10.1: The Rhönrad in reality. The wheel is maneuvered by body movements shifting the center of gravity. Image included with permission by V. Hümpfner.

This construction was suggested by the BARREL in the first place since the latter is also driven by an internal mechanism shifting its center of gravity. However, the conditions for the controller are much more involved here. The BARREL disposed of sensors measuring the inclination of the axes and there were only two degrees of freedom for maneuvering the internal weights. The Rhönrad does not have any sensors measuring the rotation angle of the wheel or any of its physical coordinates, and shifting the center of gravity is done by an intrinsic mechanism involving now the 15 degrees of freedom of the HUMANOID. Being underactuated (weak motor

[2] Please remember that the only sensor information is from the joint angles so that the robot has no information on its inclination, say. This makes the task of getting into a whole-body swinging motion much more difficult.

forces for the joints) the information on the position of the wheel can be obtained by the different effects of the gravitation and possible inertia forces at the joints, but this is of course an extremely intricate relation. Nevertheless, let us quite naively connect our standard `sox` algorithm or the multilayer `soml` algorithm to the robot and look what happens.

What can we expect? In the ideal case we expect to see a similar scenario like with the BARREL or with the SPHERICAL in the spherical basin, that the complex internal mechanism is guided by the controller in such a way that the most natural modes, rolling in that case, are excited.

Our investigations are in a very preliminary state, Video 10.11 showing some very first results. Of course what you see there is still far away from any stable rolling mode, but the motions of the HUMANOID may be interpreted as getting in some resonance with the motions of the wheel. This behavior is strongly depending on the compliance of the HUMANOID that can be set in the experiments by changing the motor-power factor. We also think that the RHOENRAD is a good opportunity to test the multilayer `soml` algorithm of the homeokinetic controller since the whole system is not prone to running into deadlocks by mechanical or environmental constraints. In this scaffolded situation it is possible to play with physical parameters like the gravity, masses, and motor forces in a very free way.

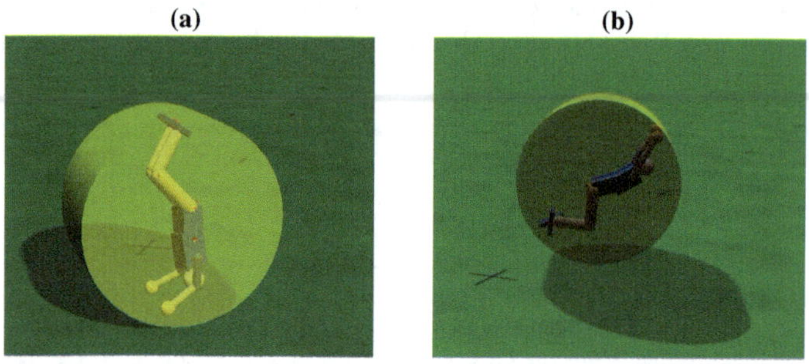

Video 10.11: HUMANOID in Rhönrad. The wheel can be moved by shifting the center of gravity. The only sensor information is from the joint angles so that the robot does not have any information on the physical coordinates and orientation of the wheel. The videos can be watched at `http://playfulmachines.com`.

Experiment 10.8: RHOENRAD

Start the simulation RHOENRAD. Here the multilayer version (`soml` algorithm) of the controller is used. Choose the parameters as in Experiment 10.6. Occasionally it is helpful to push the wheel (<Ctrl>+<mouse buttons>)

Fighting

The behavior in highly dynamic environments is one of the most challenging problems in robotics. The dynamics can be created by independent objects but it is even more interesting to consider the interaction between different robots of the same or different kind. The most interesting point in these scenarios is that the robots develop a much higher variety of behavioral modes than in the case of a static environment. We demonstrate this in particular by two humanoid robots in a narrow arena where contact between the two is nearly unavoidable, see Videos 10.12 and 10.13. Partly, we use also a weak force acting between the robots so that the probability to come into contact is slightly increased.

Video 10.12: Fight, Fight, Fight. If robots come into close contact they may excite each other to complex motion patterns. This is amazing since the only way of "feeling" each other is by the perturbations in the sensorimotor dynamics caused by the physical forces exerted in the interaction. Adherence is due to normal friction but essentially also by a failure of the ODE physics engine, which after heavy collisions produces an unrealistic interpenetration effect, which acts like a special gripping mechanism. So, the "fighting" is a truly emerging phenomenon not expected before we observed it. The video can be watched at `http://playfulmachines.com`.

Although without any aim of fighting each other the robots develop scenarios like in a real wrestling. How does this fit into the general paradigm of homeokinetic learning? We may, in a rather coarse argument, root this behavior back to the general property of homeokinetic learning as a compromise between increasing sensitivity while still trying to "keep things under control." Loosely speaking, the sensitivity increasing effect (see Sect. 5.3.2) makes the robot react to collisions caused by the opponent, the reactions being moderated by the confinement effects as discussed in Chap. 5. Additionally, due to a bug in the ODE physics engine, a heavy collision can lead to an interpenetration acting like a gripping mechanism. In this way the robots are directly influencing each other in a very intensive way. Under homeokinetic learning the interaction between the robots is determined by being sensitive but controlled, which may help understanding the behavioral variety. You can try it yourself by doing Experiment 10.9.

(a) (b)

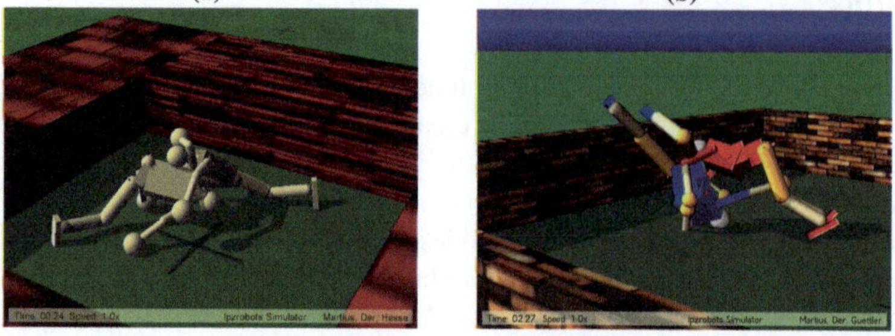

Video 10.13: More Fighting. In the left video the robots come into contact by the narrowness of the arena, in the second one there is a small force attracting the robots towards the center. The videos can be watched at `http://playfulmachines.com`.

Experiment 10.9: Fighting

Start the simulation Fighting. There are two HUMANOIDs in the arena, which you can modify as in Experiment 10.6. In case the robots penetrate the walls or get stuck otherwise, remember to press `r` to relocate them (the controller remains unchanged). In order to increase the interaction of the robots there is a force that moves them to the center. You can change its strength with `centerforce` (default 1). For other useful parameters see Experiment 10.6.

10.4 Snakes — Adaptation and Spontaneity

Snake-like artifacts have already been considered in [41]. However, those "snakes" have been of a simple construction with only a hinge joint connecting the flat segments. The experiments described here use capsules as geometric primitives connected by a universal joint realizing two degrees of freedom. Angles are constrained so that a joint can not freely rotate. When in free space, the emerging motions depend very much on the friction with the ground and the constraining effect of the gravity. The snakes display a large variety of behaviors that are difficult to classify. The reason is that the physical constraints on the motion are not so specific as for instance in the DOG or HUMANOID case. Again, it is interesting and helpful for the development of different behaviors to play with the physical parameters like the gravity.

The situation is different however, if the snake is in a narrow vessel with diameter of about two segment lengths so that the motions are heavily constrained, see Video 10.14. Under this condition the snake displays typical modes repeatedly in a loose succession. In the videos you find the coiling mode with the snake coiling into the vessel while still rotating around its axis. In this way the snake can be active while still being in good agreement with the physical constraints given by the vessel. The most interesting effect however is in the fact that sometimes the snake manages to (nearly) escape from the vessel in a rather spectacular way, see Video 10.15. The experiment has been done without any noise so that the state-parameter dynamics is deterministic and this sudden change to a completely new mode of behavior is to be considered as purely spontaneous.

Video 10.14: The SNAKE adapting to its environment. Under strong physical constraints, the homeokinetic learning finds a way to keeping the robot active while taking account of the geometry of the vessel. The video can be watched at `http://playfulmachines.com`.

Video 10.15: How the SNAKE may manage to get out of the pit. The first picture in the sequence shows the snake in a seemingly relaxed situation. However, as the video shows there are tensions building up in the body and after some time it suddenly crunches into a tight bundle of which it unrolls with very high velocity. By the inertia effects it manages to nearly jump out of this very deep pit despite the quite weak motor forces. The video can be watched at `http://playfulmachines.com`.

(a) (b)

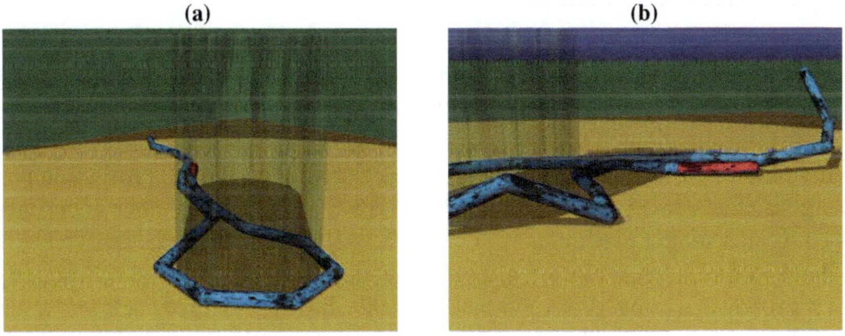

Video 10.16: Emergence and decay of collective modes. The video sequence demonstrates nicely the transient nature of the emerging modes. In the first part of the behavioral sequence **(a)** the SNAKE goes into a collective mode with the segments rotating in a coherent manner generating a kind of strangling effect by the interaction with the glassy cylinder. This mode, after prevailing for several minutes, is seen in **(b)** to break down rapidly giving way to the emergence of new motion patterns. The videos can be watched at `http://playfulmachines.com`.

The experiments with the SNAKE are also well suited for demonstrating another feature of homeokinetic control related to the contingency of all behavioral patterns. The homeokinetic objective (minimization of the time-loop error) does not make any specific prescriptions on the kind of behaviors to emerge. Nevertheless, we have seen that there are clear preferences corresponding to vast plateaus or valleys in the error landscape, which act as attractors for the state dynamics, see Chap. 5. However, we have seen also that the state dynamics is always entangled with the parameter dynamics. This entanglement effect is kind of macerating these attractors so that fixed motion patterns are converted into transients. A simple example of that kind is the frequency drift described in Sect. 7.2.2 (p. 134). In the high-dimensional systems the transient nature of emerging behaviors is quite prominent. The interesting point is the long-lived nature of the transients. We give one clear example in Video 10.16, being part of an experiment where the "strangling" mode was excited and existed for several minutes, decaying rapidly afterwards.

Experiment 10.10: SNAKE

Start the simulation Snakes. Initially there is one SNAKE, falling into a circular pit. You can change the height of the pit with `pitheight`. By pressing `i/I` you can add/remove a snake inside the pit and with `s/S` add into/remove from the surrounding. You may wish to change the following parameters: `motorpower`, `noise`.

10.5 Self-Rescue Scenario

The results obtained so far suggest a potential practical application. In the scenarios of this chapter we could observe that the robotic objects manage to get free from very intricate situations. The first clear indication of that property was the surprising ability of the BARREL to get into new modes of behavior like the precession mode emerging after being put upright, see Sect. 8.5 (p. 178). Many other experiments have revealed similar properties.

We therefore suggest to use our homeokinetic controller as a kind of rescue system if a robot, driven by a conventional controller, has maneuvered itself into a situation it can not cope with. The idea is to switch in such a situation to our homeokinetic controller, which will spontaneously generate new movements that are contingent but in correspondence with the specific properties of the body and its interaction with the environment.

Consider for instance the scenario where a robot was fallen into some pit or ditch. We have presented in Video 10.8 above already a scenario of the robot in the pit in its initial phase. Video 10.17 presents another situation somewhat later in the development. As demonstrated by the video, the robot repeatedly shows climbing patterns, so that it is nearly escaping from the deep pit. Another example is the snake in the vessel, see Video 10.15.

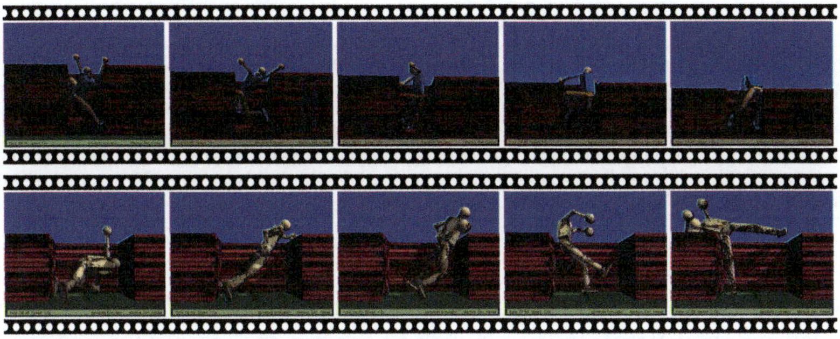

Video 10.17: Self-rescue scenario with the HUMANOID. After falling into a narrow pit, the robot develops several alternative motion patterns adapted to the new situation. After some time one often observes the emergence of climbing like behavior patterns. The patterns are emerging without any goals as a consequence of the sensitive but active interplay between the robot and the specific environment. Nevertheless, they may help the robot to get out of this impasse. The videos can be watched at `http://playfulmachines.com`.

10.6 World of Playful Machines

As already seen in the robot wrestling scenarios, the interaction with other robots is a productive element for the individual development. In a multi-agent arena we observe that each of the agents develops a kind of behavior of its own, which is emerging from the interplay of its body with the environment, see Video 10.18 and/or do Experiment 10.11. Behaviors are largely modified by the interactions with the obstacles in the arena but mainly with other agents. The latter kind of interactions makes this setting a highly dynamic environment.

Video 10.18: The world of playful machines. The video is an example of a robotic world, all robots being driven by homeokinetic learning. Such simulations can run over many hours, producing an unforeseeable sequence of motion patterns, even in the purely deterministic case. In this video, the most interesting agent is the ARMBAND that seemingly tries to get into a rolling mode. After heavy collisions it falls over, but in the later development it gets upright again by another collision. After that it starts rolling and jumping again. The video can be watched at `http://playfulmachines.com`.

Experiment 10.11: The world of playful machines

Start the Play World simulation. There are no robots at the beginning, but you can add them yourself. You may change the arena as in Experiment 10.6. Choose the controller type by: `>multilayer=0` for `sox` algorithm and `>multilayer=1` for `soml` algorithm. Then you can add the robots by pressing: `b` for SPHERICAL, `s` for SNAKE, `u` for HUMANOID, `a` for SLIDER ARMBAND, `d` for DOG, `x` for HEXAPOD, and `l` for LONGVEHICLE. Pressing <Shift>+X (where X is one of the above keys) a robot of the respective type will be removed. The other useful parameters are: `noise` (default is zero!), `damping`, learning rates `epsC`, `epsA` and so on, see the numerous parameters displayed on the command line.

In continual interaction, the robots with homeokinetic learning develop more complex motion patterns. In the simulations one can run the system for hours and days without running into dead locks. In principle, the system can run fully deterministic so that the whole scope of complex patterns and their development is defined by the initial conditions of the physical system and the initialization of the controller and forward model. In the simulations it is however sometimes necessary to inject a little sensor noise combined with a weak damping of the parameter dynamics in order to get rid of the determinacy by the initial conditions.

The multi-robot scenario is also an ideal possibility to test the multilayer `soml` algorithm that comes with the LPZROBOTS simulator. In preliminary single-robot experiments using the `soml` algorithm we observed a kind of channeling towards a behavior spectrum of restricted complexity. This could be avoided by the collisions with other robots, which, so to say, provide novel input from the body of the robot. We hope that in this way the playful unfolding of motion patterns, although still diversified, transient, and contingent by nature, is directed towards more pronounced and more stable behavioral primitives.

10.7 Discussion

Let us start the discussion with noting the differences in dimensionality between low- and high-dimensional systems. While for instance in the BARREL we have 8 degrees of freedom[3] (DoF) and 6 parameters of the controller (4 matrix elements of C and 2 of the bias term h), the SPHERICAL disposes of 9 + 11 so that the dimension of the state-parameter space jumps from 14 to 20. However, in the case of the SNAKE there are, beside the physical degrees of freedom, $m \times n$ parameters for the controller matrix C. With 10 joints with two motors and two sensors each, there are already 400 parameters (plus the bias terms) of the controller running freely on a time scale comparable to that of the physical system. Additionally, there are the $n \times m$ matrix elements of the A matrix. When using the multilayer network (`soml` algorithm) with just two layers, the number of parameters easily exceeds 1000.

However, despite the jump in dimensionality, there is no principal difference between the high- and the low-dimensional systems. So, we should also be able to relate the observed motion patterns back to the embodiment of the agents. At first sight this seems very difficult if not impossible. There are exceptions, though, seen

[3] The BARREL is a rigid body with 6 DoFs (3 translational + 3 rotational) plus the two positions of the internal masses.

by the emergence of collective modes where all DoF are cooperating. Particular examples of that kind are given by the SNAKE under certain environmental conditions, such as being in the vessel where the so-called coiling mode develops—a whole-body motion involving all DoF in a highly coordinated way. The emergence of the mode largely depends on the specific environmental situation, which can be considered as a kind of physical scaffolding.

At first sight, a collective mode like in the SNAKE seems rather an exception than the rule. We have made also the experience that many roboticists assess the emerging motion patterns as random. Our attitude is different for several reasons. Firstly, the assessment of the emerging patterns may seem obscured by the goal-focused view in robotics, concentrating on special properties or skills like reaching, grasping, walking, jumping, or swimming. Instead, homeokinesis, while not caring about particular needs, may help us to a fresh view on robotics. Namely letting robots unfold their potential behaviors in a playful way and generating motion patterns reminding of young children playing, acrobatics, or modern dancing to give a few examples.

A second argument is seen in the fact that the development of the behaviors is completely deterministic (switching off any sensor noise). The assessment of patterns as being random usually is motivated by the assumption that similar behavior patterns can be achieved by just using random actions. Even if this might work in specific cases, there is a fundamental difference to the homeokinetic behaviors, the latter being controlled in a deterministic way by a low-complexity controller. This has far reaching consequences. For instance, if ever the robot comes back to a behavioral situation encountered before, its behavior can be repeated in a deterministic way.[4] This is a good starting point for behavior based architectures, one of the future research objectives.

A third point is made by the observations of many different high-dimensional systems that all show quite a restriction to a rather small subset of the full behavioral variety, in particular if the systems are under physical scaffolding (suspended HUMANOID, SNAKE in the vessel, and many further examples treated above) where the system can run for a very long time without getting itself into situations where any DoF is being blocked in a severe way. In these systems, one often has the impression that the combined dynamics has settled into a very low-dimensional attractor living in the phase space of several hundred dimensions. Involuntarily, sometimes in the experiments one gets an impression of the full complexity by the parameter runaway effect generating behaviors that are much more complex (seemingly random) than the ones in the "sound," homeokinetically controlled regions of the behavior space.

We are convinced by this and many other observations, that in the considered high-dimensional systems there is a self-induced dimension reduction similar to the one known from natural systems; and that the emerging patterns, although they are many and varied, are in a definite way related to the physical properties of the system. It would of course be very interesting to make this statement quantitative, but we have no results yet.

---

[4] This would require to have stored controllers in a long term memory tagged with a context stamp.

Chapter 11
Facing the Unknown — Homeokinesis in a New Representation*

Abstract: Homeokinesis has been introduced and analyzed in the preceding chapters on the basis of the time-loop error (TLE). This chapter presents an alternative approach to the general homeokinetic objective by introducing a new representation of the sensorimotor dynamics. This new representation corrects the state dynamics for the predictable changes in the sensor values so that the transformed state is constant except for the interactions with the unknown part of the dynamics. The single-step interaction term will be seen to be identical to the TLE so that the learning dynamics is not altered. However, besides giving an additional motivation for the TLE, this chapter will extend the considerations to the case of several time steps and will eventually consider infinite time horizons making contact with the global Lyapunov exponents and chaos theory.

Coordinate transformations, in particular changing to a co-moving coordinate system, are helpful tools in physics and engineering in order to keep considerations as simple as possible and to concentrate on the essential features of a process.[1] Nature has also developed many strategies of that kind. When you are moving your eye while reading these lines, the images are moving on the retina. Nevertheless you are not under the impression that the world is moving. Instead, one observes a constancy of the space representation, which can only be achieved by the remapping in the brain on the basis of the motor signals sent to the eye. Quite generally, space constancy for a moving subject is one of the most important ingredients for feasible information processing. It is realized through the remapping of visual information in a cortical network situated between the occipito-parietal and frontal cortex [137]. This system necessitates the integration of multiple sensor and motor modalities in different frames of reference such as the object, head, and body centered reference frames based on the current motor behaviors [178].

Correcting for changes in perception that are caused by actions of the agent itself gives the information processing a stable background. Concentration on the essentials is a decisive evolutionary advantage since a moving object, be it predator or prey, can be perceived without perturbations caused by the motions of the agent it-

[1] A simple example is given by children playing ball on a train. Trajectories of the ball look quite different for the outside and the inside observer. As compared to the world-centered coordinate system, trajectories look much easier in the coordinate system moving with the train.

R. Der and G. Martius: The Playful Machine, COSMOS 15, pp. 223–234.
springerlink.com

self. In this chapter we will study temporal remapping that will be seen to result in a new, co-moving frame of reference, which focuses the perception on the unknown challenging a productive interplay with the uncertainties of the real world.

11.1 Interaction Representation of Sensorimotor Dynamics

Remapping in time aims at transforming the state vector x_t in such a way that, in the new representation, the state dynamics is invariant under the systematic part of the original dynamics. We are starting in this section with the transformation over a single time step since this will be sufficient for most of our applications, the more general definition will be given in Sect. 11.2.

Let us consider our time series of sensor values $x_t \in \mathbb{R}^n$, $t = 0, 1, \ldots$ produced by the behaving robot. Let us assume as before that we have established a map ψ acting as a time series predictor so that

$$x_{t+1} = \psi(x_t) + \xi_{t+1} \tag{11.1}$$

where ξ is the prediction error.

Let us now try to find a new state dynamics which is invariant under the predictable behavior of the robot, i.e. if the trajectory (in state space) is reducible to the effect of ψ alone. The new representation we want to introduce aims at a state constancy with respect to a fixed reference time t. Let us denote the states in the new representation by $\tilde{x}$, put $\tilde{x}_t = x_t$ and define the next state $\tilde{x}_{t+1}$ as

$$\tilde{x}_{t+1} = \psi^{-1}(x_{t+1}) \tag{11.2}$$

stipulating for the moment that the inverse ψ^{-1} of the map ψ, defined by

$$\psi^{-1}(\psi(x)) = x\,, \tag{11.3}$$

exists for all states x in the domain of ψ. Note that both state variables x and $\tilde{x}$ in Eq. (11.2) are at the same time step indeed. Obviously, the definition Eq. (11.2)

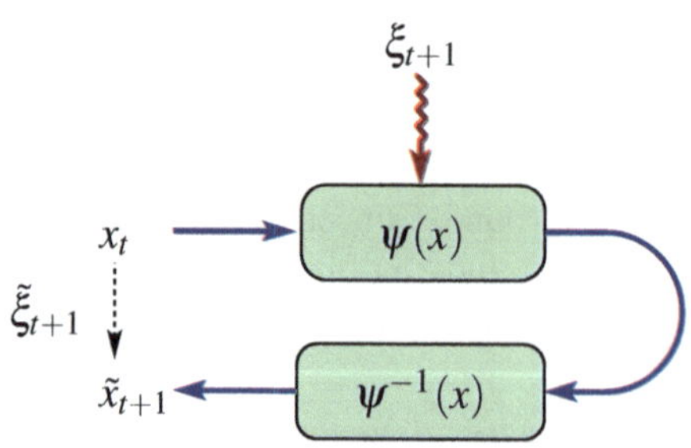

Fig. 11.1: Interaction representation — one time step. Starting from x_t, the dynamics goes one step ahead through the dynamics of the robot in its environment as described by the map ψ plus perturbations given by ξ. The new state is then projected back by means of the inverse of the dynamical map ψ. The result is the new state in the interaction representation, its difference to the previous state is the interaction term.

yields the desired result that $\tilde{x}_t = \tilde{x}_{t+1}$ if $\xi = 0$, i. e. that the dynamics is given by ψ alone.

The definition is worked out further in Sect. 11.B (p. 233) but we will introduce here a more handy representation. By using Eq. (11.1) in Eq. (11.2) we obtain by means of a Taylor expansion[2]

$$\tilde{x}_{t+1} = x_t + \frac{1}{L(x_t)}\xi_{t+1} + O\left(\|\xi\|^2\right) \tag{11.4}$$

where

$$L(x) = \frac{\partial \psi(x)}{\partial x} \tag{11.5}$$

is the Jacobian matrix of ψ, see Sect. 5.6 (p. 78). We use the notation $\frac{1}{L}$ for the inverse matrix L^{-1} as before. In case L is not invertable a pseudoinverse is to be used, see Sect. 5.G. Actually this expansion is helpful only if the noise ξ is very small. However, we can use an effective state in the Jacobian matrix to make the higher order terms in Eq. (11.4) disappear, see Sect. 5.B (p. 100). Hence, using $\tilde{x}_t = x_t$ (t is the reference time) we may write the dynamical law for the transition $\tilde{x}_t \to \tilde{x}_{t+1}$ as

$$\tilde{x}_{t+1} = \tilde{x}_t + \tilde{\xi}_{t+1} \tag{11.6}$$

where we **define** (omitting the subtleties with the effective state)

$$\tilde{\xi}_{t+1} = \frac{1}{L(x_t)}\xi_{t+1} \tag{11.7}$$

as the *interaction term* that plays the role of the noise in the new representation.

What is the relation of the interaction representation to the original representation of the time series? Comparing Eq. (11.7) with Eq. (5.8), we find that the interaction term $\tilde{\xi}_{t+1}$ featuring in the new representation is identical to the time-loop (reconstruction) error v_t, i. e. (using $\tilde{x}_t = x_t$ by definition)

$$\tilde{x}_{t+1} - \tilde{x}_t = v_t \, . \tag{11.8}$$

Moreover, once in the interaction representation, the prediction $\tilde{x}^{pred}_{t+1}$ of the next state $\tilde{x}_{t+1}$ is just given by

$$\tilde{x}^{pred}_{t+1} = \tilde{x}_t$$

[2] The derivative of the inverse function $\psi^{-1} : \mathbb{R}^n \to \mathbb{R}^n$ is obtained by taking the derivative of $x = \psi^{-1}(\psi(x))$ on both sides. Using the chain rule we find

$$\mathbb{I} = \frac{\partial}{\partial u}\psi^{-1}(u)\bigg|_{u=\psi(x)} \frac{\partial}{\partial x}\psi(x) \quad \text{so that} \quad \frac{\partial}{\partial u}\psi^{-1}(u)\bigg|_{u=\psi(x)} = \left(\frac{\partial}{\partial x}\psi(x)\right)^{-1} = (L(x))^{-1} \, .$$

By means of that we obtain the Taylor expansion in terms of ξ as

$$\psi^{-1}(\psi(x)+\xi) = x + \frac{\partial}{\partial u}\psi^{-1}(u)\bigg|_{u=\psi(x)} \xi + O\left(\|\xi\|^2\right) = x + L^{-1}(x)\,\xi + O\left(\|\xi\|^2\right) \, .$$

since $\tilde{\xi}_{t+1}$ represents the irreducible effect of the perturbations on the dynamics of the system. In view of Eq. (11.8) we may also say that **the prediction error in the interaction representation is identical to the reconstruction error (time-loop error) in in the original dynamics.**

The intriguing picture of the new representation is the stationarity of the state in the transition $\tilde{x}_t \to \tilde{x}_{t+1}$ if the model ψ describes the transition $x_t \to x_{t+1}$ exactly ($\xi = 0$). A dynamics comes in only if there is a discrepancy between the internal representation ψ and the true dynamics of the brain-body system ($\xi \neq 0$). Metaphorically speaking, the new dynamics focuses on the interaction between the internal representation and the true dynamics by taking away the knowable influences of the agent's own actions on the development of its sensor values. This is why we call Eq. (11.6) the interaction representation of the sensorimotor dynamics. We will give further details and justifications in the following.

11.2 Extending the Time Horizon

The interaction representation was introduced above for a single time step. This is sufficient for many applications, in particular if the correlations across time are not too important. From our experience with the time-loop error (TLE), it is obvious that the single time-step approach is creating already many highly sophisticated motion patterns in a variety of brain-body systems. The interaction representation of course can be defined over many time steps and this is what we are going to consider now. We will start with some preliminaries, defining orbits forward and backward in time in a deterministic dynamical system, applying this subsequently to the definition of the interaction representation of a time series. We think that this is an interesting new representation that might be of interest for time series analysis and prediction in general so that we will keep our considerations at a general level.

As the next step, we will apply the interaction representation to the problem of self-learning robot control by giving the general learning rule based on minimizing the interaction term (now over arbitrarily many time steps). Eventually, we are going to extend the time horizon to infinity in order to make contact with classical chaos theory. In this way we will recognize general homeokinetic learning as a flow of Lyapunov exponents of the model dynamics given by ψ.

11.2.1 Preliminaries: Orbits in Dynamical Systems

Let us consider now a **deterministic**, time discrete dynamical system defined by iterating a map $\psi : \mathbb{R}^n \to \mathbb{R}^n$, producing a time series x_t, $t = 0, 1, \ldots,$ as

$$x_{t+1} = \psi(x_t) \ . \tag{11.9}$$

This is quite general but we may consider it as the systematic part of the time evolution of the sensorimotor loop. We can describe trajectories or forward orbits of our system by introducing the maps $\psi^k : \mathbb{R}^n \to \mathbb{R}^n$ defined iteratively by

$$\psi^k(x) = \psi\left(\psi^{k-1}(x)\right) = \psi(\psi(\ldots\psi(x)))$$

together with $\psi^1(x) = \psi(x)$ and $\psi^0(x) = x$. In particular, starting from some state x_0 given at time $t = 0$, the state at time t is $x_t = \psi^t(x_0)$. In the same way we can also define trajectories backward in time assuming for the moment that the function ψ is invertible, i.e. that there exists a map $\psi^{-1} : \mathbb{R}^n \to \mathbb{R}^n$ so that $\forall x$ in the domain of ψ

$$\psi^{-1}(\psi(x)) = x\,.$$

Analogously to the forward orbit we define the k-fold iterate of ψ^{-1} iteratively by

$$\psi^{-k}(x) = \psi^{-1}\left(\psi^{-k+1}(x)\right) \tag{11.10}$$

so that we can define the backward orbit over k steps. As to notation we use the index k as the number of steps relative to some reference time, mostly t. In order to simplify notation we will assume $t = 0$ during the derivations and reintroduce the current value of t afterwards. Hence

$$x_0 = \psi^{-k}(x_k) \tag{11.11}$$

means the reconstruction of the present state from some state k steps later in the future. Of course, we can use this also in order to reconstruct states from the past based on the present one but we use this setting of the times since this is useful for the further developments. It is instructive to consider the linear case where $\psi(x) = Lx$ (with a constant matrix L), so that $\psi^{-k}(x) = L^{-k}x$ where L^{-1} is the inverse of the matrix L and $L^{-k} = \left(L^{-1}\right)^k$.

We can now follow, in a deterministic system, the propagation of a perturbation $\xi \in \mathbb{R}^n$ backward in time, i. e. consider in leading order [3] of the perturbation ξ

$$\delta x_0 = \psi^{-k}(x_k + \xi) - \psi^{-k}(x_k) = \psi_{x_k}^{-k}\xi \tag{11.12}$$

where

$$\psi_{x_k}^{-k} = \left.\frac{\partial \psi^{-k}(u)}{\partial u}\right|_{u=x_k}.$$

In Sect. 11.A we derive that in the same sense

[3] In order to keep derivations simple we assume here and in the following that the noise is infinitesimally small so that the higher order terms vanish. The leading order terms are then used in the sense of a definition for the general case or they may alternatively be improved by using effective states.

$$\delta x_0 = \psi_{x_k}^{-k} \xi = \frac{1}{L(x_0)} \frac{1}{L(x_1)} \cdots \frac{1}{L(x_{k-1})} \xi \tag{11.13}$$

where $k = 0, 1, \dots$ and $1/L$ denotes the matrix inverse L^{-1}, as always. The propagated perturbation δx_0 can, in leading order of ξ, also be obtained by iterating the auxiliary quantities δx_k as

$$\delta x_{k-1} = \frac{1}{L(x_{k-1})} \delta x_k \tag{11.14}$$

which has to be iterated starting with $\delta x_k = \xi$. This rule describes the step by step propagation of an infinitesimal perturbation ξ backward in time down to δx_0 by means of the deterministic dynamics.

11.2.2 Interaction Representation of a Time Series

Now let us return to the time series of sensor values $x_t \in \mathbb{R}^n$, $t = 0, 1, \dots$ produced for instance by a behaving robot. Let us assume as before that we have established a map ψ acting as a time series predictor so that

$$x_{t+1} = \psi(x_t) + \xi_{t+1} \tag{11.15}$$

where ξ is the prediction error. This extends the deterministic dynamical system given by Eq. (11.9) to the case of a more realistic time series with unpredictable (at the level of the dynamics model ψ) perturbations.

Our aim now is the derivation of a transformed state dynamics which is invariant under the predictable behavior of the robot, i.e. if the trajectory (in state space) is reducible to the effect of ψ alone ($\xi = 0$). The new representation we want to introduce aims at a state constancy with respect to a fixed reference time t, putting $t = 0$ again for the derivation. We define a new state $\tilde{x}_k$ by projecting later states x_k over k steps backwards in time back to the reference time, i. e.

$$\tilde{x}_k = \psi^{-k}(x_k) \ . \tag{11.16}$$

The reason for choosing this new representation of the state is seen if we follow the development of the state $\tilde{x}_k$ for $k \geq 0$ into the future. As shown in the Appendix, Sect. 11.B, Eq. (11.41), the dynamics of the transformed state $\tilde{x}_k$ is given by the iterative rule

$$\tilde{x}_{k+1} = \tilde{x}_k + \tilde{\xi}_{k+1}, \, k \geq 0 \tag{11.17}$$

with initial condition $\tilde{x}_0 = x_0$. The interaction term $\tilde{\xi}_{k+1}$ is, in leading order of the noise, obtained explicitly with $k = 0, 1, \dots$ as

$$\tilde{\xi}_{k+1} = \frac{1}{L(x_0)} \frac{1}{L(x_1)} \frac{1}{L(x_2)} \cdots \frac{1}{L(x_k)} \xi_{k+1} \ . \tag{11.18}$$

The new dynamics, represented by Eq. (11.17), deserves some discussion. The essential point is that the new state is constant in time indeed, i. e. $\tilde{x}_k = x_0 \ \forall k \geq 0$, if the model is exact i. e. $\xi = 0$. Moreover, the transformed noise is directly given in terms of the prediction error and the inverse of the Jacobian matrix.[4] We may consider the deterministic part of the dynamics as the behavioral component that is under complete control by the robot itself whereas the interaction term represents the interaction with the uncertainties of its body dynamics and the world. The transformed dynamics focuses completely on this interaction and this is why we want to call it the interaction representation of the sensorimotor dynamics.[5]

In the linear case where $\psi(x) = Lx$ we simply have $\tilde{\xi}_{k+1} = L^{-(k+1)}\xi_{k+1}$ and in the one step case, i. e. with $k = 0$ in Eq. (11.17), the interaction term is (setting the physical time back to t)

$$\tilde{\xi}_{t+1} = \frac{1}{L(x_t)}\xi_{t+1} \tag{11.19}$$

which is just the one step TLE v introduced in Eq. (11.8) above. The identification with the interaction representation may be seen as both a deeper foundation and a systematic basis for extending homeokinetic learning to a larger time horizon. Interestingly and most importantly for our approach, **the interaction term in** Eq. (11.18) **can directly be related to the global Lyapunov exponents (for $k \to \infty$) of the deterministic dynamical system generated by ψ.** This point will be further discussed in Sect. 11.3.

11.2.3 Relation to the Time-Loop Error

The interaction term in the interaction representation can be understood in various ways. A first point of interest is its relation to state reconstruction and the multiple-step time-loop error (TLE). Actually, the state in interaction representation is obtained by backward projecting a future state x_{t+k} to an earlier time, the reference time t, see Eq. (11.16), by means of the current knowledge on the process represented by the dynamical map ψ. In other words, the latter is used in order to reconstruct the state x_t from x_{t+k}, the state **observed** at time $t+k$. The difference $\tilde{x}_{t+k} - x_t$ (or its square) may be called the reconstruction error from time $t+k$ back to time t,

---

[4] Please note that $\tilde{\xi}$ depends explicitly on the reference time, but we omit this in our notation for the sake of simplicity whenever this does not lead to ambiguities.

[5] The notion is borrowed from quantum mechanics where the original Schrödinger equation with Hamiltonian H

$$i\frac{\partial}{\partial t}\Psi = H\Psi$$

is transformed by splitting $H = H_0 + H_{int}$ into the dynamics

$$i\frac{\partial}{\partial t}\Psi_{int} = H_{int}(t)\,\Psi_{int}$$

where $H_{int}(t) = e^{itH_0}H_{int}e^{-itH_0}$.

i. e. we define

$$E = \|\tilde{x}_k - x_0\|^2 = \left\| \psi^{-k}(x_k) - x_0 \right\|^2 \tag{11.20}$$

(putting $t = 0$ again).

The reconstruction error can be related to the interaction term in the interaction representation by iterating Eq. (11.17) backward in time (replacing iteratively $\tilde{x}_k$ with $\tilde{x}_{k-1} + \tilde{\xi}_k$ and using $\tilde{x}_0 = x_0$) to obtain

$$\tilde{x}_k - x_0 = \sum_{l=0}^{k-1} \tilde{\xi}_{l+1} \tag{11.21}$$

so that the k-step reconstruction error $E^{(k)}$ is given by the sum of the interaction terms collected between times t ($k = 0$) and $t + k$ as

$$E^{(k)} = \left\| \sum_{l=0}^{k-1} \tilde{\xi}_{l+1} \right\|^2 . \tag{11.22}$$

11.2.4 General Multiple-Step Learning Rule

The minimization of the TLE E from Eq. (11.22) can be realized by gradient descent so that we obtain a dynamics for any controller parameter p. Together with the system dynamics we obtain the following set of equations defining the self-referential robotic system (returning to the full time notation)

$$\tilde{x}_{t+l+1} = \tilde{x}_{t+l} + \tilde{\xi}_{t+l+1} \tag{11.23}$$

$$\Delta p_t = -\varepsilon \sum_l \tilde{\xi}^\top_{t+l+1} \frac{\partial}{\partial p_t} \tilde{\xi}_{t+l+1} \tag{11.24}$$

where $p \in \mathbb{R}^P$ is the set of the parameters (namely all C_{ij} and h_i in our standard setting).

In practice one uses a fixed time horizon k for the interaction dynamics. When at time t we use it as reference time, wait for the next states x_{t+l} for $l = 1, 2, \ldots, k$ to come in and find the interaction terms $\tilde{\xi}_{t+l+1}$ for each l by means of Eq. (11.18). Then, the parameters are updated according to Eq. (11.24). Subsequently t is increased by 1 and the procedure of finding the interaction terms is restarted.

11.3 Homeokinesis as a Flow of Lyapunov Exponents

Let us try now to formalize our results a little further by relating the homeokinetic learning rules to the global Lyapunov exponents of the model dynamics given by the map ψ. The considerations are kept a little sketchy in order not to burden the reader with too much technical details.

Our aim is first to relate the interaction term in the interaction representation to the separation vector between neighboring trajectories. The latter forms the basis for the global Lyapunov exponents of the dynamical system given by ψ. In fact, the usual (forward in time) definition of the Lyapunov exponent with k iterations rests on the separation vector for the propagation of an infinitesimal perturbation ξ_0 of the starting state x_0 into the future. In terms of the k-step Jacobian matrix

$$L^{(k)}(x_0) = \frac{\partial \psi^k(x_0)}{\partial x_0} = L(x_{k-1}) \ldots L(x_0) , \tag{11.25}$$

the maximal Lyapunov exponent for the perturbation vector ξ is given by

$$\lambda_{\max}(\xi) = \lim_{k\to\infty} \frac{1}{k} \ln \left\| L^{(k)} \xi \right\| . \tag{11.26}$$

There is a subtlety in that the maximal Lyapunov exponent depends on the vector ξ. We can circumvented this subtlety by recalling that in the general case the vector ξ is essentially noise, meaning that it will not be restricted (with probability one) to some subspace of the full phase space so that $\lambda_{\max}(\xi)$ is just the maximal Lyapunov exponent $\lambda_{\max}$ of the full Lyapunov spectrum. So, let us assume that is the case for our system, the extension to the general case being straightforward.

Now, let us consider a trajectory starting at x_0, generated by iterating k times the deterministic state dynamics, to generate the state $\hat{x}_k = \psi^k(x_0)$. Then, we retrace the dynamics back to $k = 0$ in two cases, one starting with $\hat{x}_k$, the other one with $\hat{x}_k + \xi_k$. The length of the separation vector δx_0 between these two trajectories is given for k steps in terms of

$$\tilde{\xi}_{k+1} = \frac{\partial \psi^{-k}(x_k)}{\partial x_k} \xi_k = L^{-1}(x_0) L^{-1}(x_1) \ldots L^{-1}(x_{k-1}) \xi_k = \left(L^{(k)}(x_0)\right)^{-1} \xi_k \tag{11.27}$$

Before taking the limit let us compare this to the separation vector forward in time. The difference between Eq. (11.25) and Eq. (11.27) is seen in the inversion of the Jacobian matrix $L^{(k)}$, taken at the starting state x_0. In a singular value decomposition of $L^{(k)}$ this means that the singular values are inverted so that in the limiting process we obtain instead of (11.26)

$$\lambda_{\min} = \lim_{k\to\infty} \frac{1}{k} \ln \left\| \frac{\partial \psi^{-k}(x_k)}{\partial x_k} \xi \right\| = \lim_{k\to\infty} \frac{1}{k} \ln \left\| \tilde{\xi}_{k+1} \right\| , \tag{11.28}$$

$\lambda_{\min}$ being the minimal Lyapunov exponent of the full Lyapunov spectrum with starting state x_0 (and not x_k).

Taking the logarithm of the interaction strength as objective for the gradient descent, we obtain the learning rule for the parameters p of the system as

$$\Delta p = \varepsilon \frac{\partial \lambda_{\min}}{\partial p} . \tag{11.29}$$

(We use finite k, suppressing the dependence on k since we tacitly assume that $k \to \infty$ eventually, where this dependence disappears.)

The message is that it is the lowest Lyapunov exponent of the full spectrum that is driving the change in the parameters, which for their part change the spectrum of the Lyapunov exponents. With the lowest Lyapunov exponent driving the process, it is expected that the learning dynamics generates a flow of the Lyapunov exponents so that the system is destabilized in all dimensions, i. e. it is driven more and more towards a chaotic regime. Just as in the case of the one-step time-loop error, there are confining mechanisms, for instance by the nonlinearities and the effects described in Sect. 5.3 (p. 89). In particular, being driven by ψ, the destabilization mechanism will loose its "grip" on the real system if the latter is driven too much towards chaos since ψ will not be able to represent the system in that regime. Another confining component is the temperature effect, see Sect. 5.3.4 (p. 95), since the chaotic regions in the error landscape will be of higher temperature than the more regular ones.

The solution of the conflict between sensitizing and confining tendencies is expected to produce a rich but controlled behavior of the robot just as with the one-step TLE. The advantage of the multi-step case is the explicit involvement of correlations over several time steps but the ascertainment of these expectations must be left to future research.

Appendix 11.A Proof of Eq. (11.13)

Applying, for any state u, the chain rule to

$$\psi^{-l}(u) = \psi\left(\psi^{-(l+1)}(u)\right)$$

yields the iterative rule

$$\psi_u^{-(l+1)} = \frac{1}{L\left(\psi^{-(l+1)}(u)\right)} \psi_u^{-l} .$$

With $u = x_k$ we may write

$$\psi_{x_k}^{-(l+1)} = \frac{1}{L\left(x_{k-(l+1)}\right)} \psi_{x_k}^{-l} \tag{11.30}$$

so that by iteration, starting with $l = 0$ and using $\psi_u^{-0} = \mathbb{I}$

$$\psi_{x_k}^{-k} = \frac{1}{L(x_{k=0})} \frac{1}{L(x_1)} \cdots \frac{1}{L(x_{k-1})} \tag{11.31}$$

Appendix 11.B Derivation of the Interaction Representation

Let us formulate at first the interaction representation for an arbitrary dynamical system given by the iterated map $\Psi : \mathbb{R}^n \to \mathbb{R}^n$

$$x_{t+1} = \Psi(x_t) \ . \tag{11.32}$$

Note that Ψ may be any map and will in general be the realization of a random function. The interaction representation rests on the assumption that, instead of Ψ, we have a simplified version ψ the dynamics of which we can treat more or less explicitly. In physics ψ often represents the dynamics of the isolated system while Ψ contains also all the interactions with the external world or between the constituents of the system. The idea is to find a new representation where this interaction features explicitly. This can be done by fixing a reference time t with state x_t. We set $t = 0$ during the derivations so that x_{t+k} is equal to x_k and reintroduce t afterwards. We introduce a new state $\tilde{x}_k$ by the backward orbit of x_k, defined in terms of the approximate map ψ, as

$$\tilde{x}_k = \psi^{-k}(x_k) \tag{11.33}$$

so that $\tilde{x}_k = x_0$ for all $k \geq 0$ if $\Psi = \psi$. We define the update step from any $k \geq 0$ to $k+1$ by applying $\psi^{-(k+1)}$ to both sides of the full dynamics given by Eq. (11.32), to obtain the update rule for the transformed state as

$$\tilde{x}_{k+1} = \tilde{\Psi}(\tilde{x}_k) \tag{11.34}$$

where

$$\tilde{\Psi}(u) = \psi^{-(k+1)}\left(\Psi\left(\psi^k(u)\right)\right) \tag{11.35}$$

for any state u. Obviously if $\Psi = \psi$ we have $\tilde{\Psi}(u) = u$ so that $\tilde{x}_{k+1} = \tilde{x}_k$. Hence, it is reasonable to write the rule as

$$\tilde{x}_{k+1} = \tilde{x}_k + \tilde{\xi}_{k+1} \tag{11.36}$$

where

$$\tilde{\xi}_{k+1} = \tilde{\Psi}(\tilde{x}_k) - \tilde{x}_k \tag{11.37}$$

is obviously zero if $\psi = \Psi$. $\tilde{\xi}_{k+1}$ may be seen as the residual interaction between the true and the model dynamics so that the new representation is called the interaction representation. These considerations are valid for any pair of functions ψ and Ψ.

Let us now consider more specifically the case that we have the explicit splitting of Ψ into a systematic and a (more or less) random part, i.e.

$$\Psi(x_k) = \psi(x_k) + \xi_{k+1}$$

The Linear Case

In the linear case these considerations become very simple. We have $\psi(x) = Lx$ so that $\psi^k(x) = L^k x$

$$\tilde{x}_k = \frac{1}{L^k} x_k$$

and Eq. (11.35) immediately yields

$$\tilde{x}_{k+1} = \tilde{x}_k + \tilde{\xi}_{k+1} \tag{11.38}$$

with

$$\tilde{\xi}_{k+1} = \frac{1}{L^{k+1}} \xi_{k+1} \,. \tag{11.39}$$

Obviously the state in interaction representation is constant for all k if the noise is zero, as it was intended.

The Nonlinear Case

We use $\tilde{\xi}_{k+1} = \tilde{x}_{k+1} - \tilde{x}_k$ so that

$$\begin{aligned} \tilde{\xi}_{k+1} &= \psi^{-(k+1)}(x_{k+1}) - \psi^{-k}(x_k) \\ &= \psi^{-(k+1)}(\psi(x_k) + \xi_{k+1}) - \psi^{-k}(x_k) \\ &= \psi^{-(k+1)}(\psi(x_k) + \xi_{k+1}) - \psi^{-(k+1)}(\psi(x_k)) \\ &= \Psi_{\psi(x_k)}^{-(k+1)} \xi_{k+1} + O\left(\|\xi\|^2\right) \end{aligned} \tag{11.40}$$

In leading order we may replace

$$\Psi_{\psi(x_k)}^{-(k+1)} \xi_{k+1} = \Psi_{x_{k+1}}^{-(k+1)} \xi_{k+1} + O\left(\|\xi\|^2\right)$$

so that in leading order

$$\tilde{\xi}_{k+1} = \Psi_{x_{k+1}}^{-(k+1)} \xi_{k+1} = \frac{1}{L(x_0)} \frac{1}{L(x_1)} \cdots \frac{1}{L(x_k)} \xi_{k+1} \,, \tag{11.41}$$

valid for $k = 0, 1 \ldots$. In particular,

$$\tilde{\xi}_1 = \frac{1}{L(x_0)} \xi_1$$

which is what we use most of the time.

Chapter 12
Guided Self-Organization — A First Realization

Abstract: We introduce here an in the following chapters *guided self-organization* as the combination of specific goals with self-organizing control. As a first realization we propose in this chapter the guidance with supervised learning signals. First, we investigate how these signals can be incorporated into the learning dynamics and present then a simple scenario with direct motor teaching signals. We find that the homeokinetic controller explores around the given motor patterns and thus may find a more suitable behavior for the particular body. Second, we transfer this into a teaching at the level of sensor signals, which is very natural in our setup. This mechanism of guidance builds the basis for higher level guiding mechanisms as discussed in Chap. 13.

This is the first of three chapters that focus on the generation of goal-directed behaviors from self-organizing control. For the combination of self-organizing and external drives we coined [106] the term *guided self-organization*, which was before only rarely used e. g. in nano technology [31] or swarm robotics [147]. Goal-oriented methods optimize for a specific task and require a prestructuring of the control problem in high-dimensional systems. Self-organization, on the other side, can generate coherent behavior and structure in the behavior space. Furthermore self-organizing systems show a great flexibility and high tolerance against failures and degrade gracefully rather than catastrophically [142, 143]. The perspective of guided self-organization is to obtain a system which unites benefits of both.

What can we expect from a *guided homeokinetic controller*? We have seen so far that a variety of behaviors emerged solely from the principle of homeokinesis. The process of self-organization has quickly structured the vast space of behaviors into a set that shows a coherent sensorimotor dynamics of the particular robot in its environment. We can hope to shape the self-organization process to produce specific behaviors within a short time if we can channel the exploration of the homeokinetic controller towards certain desired behaviors. This is especially important in high-dimensional systems where the self-organized search for potentially useful behaviors can take a long time.

Guided self-organizing differs in several aspects from typical ways to obtain goal-directed behaviors in autonomous robots. Most commonly optimization methods are used that adapt the parameters of the control program to achieve a specific

R. Der and G. Martius: The Playful Machine, COSMOS 15, pp. 235–242.
springerlink.com

goal, for example via supervised learning, reinforcement learning [166], or evolutionary algorithms [113]. In this setting the learning is expected to converge after many iterations and its final result is the goal-directed behavior. With guided self-organization the behavior develops during the interaction of the robot with its environment. The homeokinetic learning process may not converge, but instead causes different behaviors to be visited which increasingly satisfy the given goal more or less strictly. In this way goal-directed modes can be found which fit well to the particular robotic device.

On the methodological level we investigate three mechanisms of guidance in the following chapters. The first one allows for the incorporation of supervised learning signals, e. g. specific nominal motor commands. To make this possible we study the integration of problem specific error functions into the homeokinetic learning dynamics in the next section. Using a distal learning [82] setup we also study the use of teaching signals in terms of sensor values. The second mechanism, investigated in Chap. 13, consists in formulating relationships between motors. This will be proven to be an effective and simple way to introduce constraints into the system and facilitate the unsupervised development of specific behaviors. The third mechanism for guiding the self-organization is discussed in Chap. 14 and uses online reward signals to shape the emerging behaviors.

12.1 Integration of Problem Specific Error Functions

In this section we want to set the foundation for the combination of problem specific goals with the homeokinetic controller. The combination of self-organizing processes and additional constraints is, however, not trivial and the most complicated part is how to balance the two. The emergent properties of self-organizing systems are quickly lost if the appropriate working regime is left, but with the appropriate mechanisms there is hope to guide the system gently into a desired direction without loosing the benefits of a self-organizing dynamics.

Recall, that the learning dynamics of the homeokinetic controller was brought about by the gradient descent on the TLE, see Chap. 5. An additional goal may be specified in terms of a problem specified error function (PSEF) that is minimal if the goal is fulfilled, such that we can also employ a gradient method. The most simple approach to combine the TLE and such an PSEF would be a weighted sum of both functions. If we now perform a gradient descent on this sum than both error functions should be minimized, such that we minimize the TLE obtaining the self-organizing behavior and follow the goal. This, however, is likely to either bring the self-organizing system out of balance or to insufficiently follow the goal. The reason is that the size of the TLE varies over orders of magnitude, whereas the additional terms are often within one order of magnitude. Therefore we cannot find a fixed weighting factor for the PSEF that would cause it to show an effect in the average situation without destroying the self-organization process.

We will use a trick to scale the gradient on the PSEF in order to be compatible with the TLE. This is motivated by the natural gradient method, which states that for an arbitrary Riemann metric of the parameter space the steepest direction is given by the transformed gradient, which is obtained by multiplying with the inverse of the metric matrix as proven by Amari [4]. We use the metric defined by the matrix $J^\top J$, where J (Eq. 15.58 (p. 283)) is the Jacobi matrix of the sensorimotor loop in motor space. Thus, we can think of a 'scaling' matrix that modifies the gradient according to the sensitivity of the sensorimotor loop.

The PSEF is denoted by E^G and it must be functionally dependent on the controller parameters in order to derive a gradient. To avoid confusion we denote the TLE by E^{TLE}. The update rule for the controller matrix C can now given by

$$\frac{1}{\varepsilon_c}\Delta C = -\frac{\partial E^{TLE}}{\partial C} - \gamma\left(J^\top J\right)^{-1}\frac{\partial E^G}{\partial C}, \tag{12.1}$$

where $\frac{\partial E^{TLE}}{\partial C}$ is the homeokinetic learning rule given by Eqs. (5.25, 15.26) and γ is the guidance factor defining the strength of goal following. The entire update size is still controlled by the learning rate ε_c. For the update of the parameter h we apply the same procedure. More details may be found in [102].

12.2 Guidance by Teaching

Let us now investigate two concrete examples of problem specified error functions (PSEFs) that implement the guidance by supervised learning signals. We call them *teaching signals*, because they are given externally with respect to the self-organizing system. At first we will study *motor teaching signals* influencing the behavior of the robot by directly providing wanted motor patterns.

The reader might wonder why this is of practical value since it requires precise information about the motor setup and its dynamics. Even more reason for astonishment is given by the fact that precise motor pattern, once present, can be used to directly control the robot. Nevertheless, there are two aspects, which make the guidance indeed useful. The explorative character of the homeokinetic controller can be used to explore around the given motor pattern and thus find a more suitable behavior for the particular situation. More importantly, the direct motor teaching can be used by higher level guiding mechanisms as discussed in Chap. 13. These are again unsupervised and require much less information about the dynamics of the robotic device. After considering the implementation in more detail in the next section we will afterwards translate this guidance mechanism to teaching signals in sensor space.

12.3 Direct Motor Teaching

In order use motor teaching signals we define a PSEF, which penalizes the misfit

$$\eta_t^G = y_t^G - y_t \tag{12.2}$$

between motor teaching values y_t^G and actual motor values y_t (output by the homeokinetic controller). Similarly to the prediction error for the forward model (3.9) we find

$$E^G = {\eta_t^G}^\top \eta_t^G \,. \tag{12.3}$$

Using the gradient descent we get the additional update for the controller matrix C as

$$\frac{\partial E^G}{\partial C} = -G' \eta_t^G x_t^\top \,, \tag{12.4}$$

where G' is the diagonal matrix given by $G'_{ij} = \delta_{ij} g'_i(Cx+h)$. Similarly, for h we obtain $\frac{\partial E^G}{\partial h} = -G' \eta_t^G$. These additional terms are integrated into the final learning rule using Eq. (12.1). The guidance factor γ regulates the strength of the additional drive and has to be determined empirically. A small value of γ leads to a small influence of the teaching signal and results in a behavior that is mostly dominated by the original homeokinetic controller. For large values of γ the teaching signals are followed narrowly and few exploratory actions are performed, however, with the increasing danger to break down the self-organization.

12.3.1 Experiment

Let us evaluate this mechanism of guidance by using the TwoWheeled robot, see Sects. 6.4.1 and 9.1.1. We want to show that teaching signals can be used to specify a certain behavior and that the influence of the teaching can be conveniently adjusted using γ. For that let us consider two different motor teaching signals, which are subsequently applied. First the nominal motor values are given by a sine wave and then by a rectangular function with the same value for both motors, i. e.

$$(y_t^G)_i = \begin{cases} 0.85 \cdot \sin(\omega t) & t < 75 \\ 0.65 \cdot \mathrm{sgn}(\sin(\omega t)) & \text{otherwise}\,, \end{cases} \tag{12.5}$$

with $i = 1,2$ and $\omega = 2\pi/50$. The choice of the teaching signal has no deeper meaning. Note that the nominal motor values should not be too large because otherwise the controller is driven into the saturation region of the motor neurons, see Fig. 3.7 (p. 36). The fixed point of the sensor dynamics in the simplified world condition is

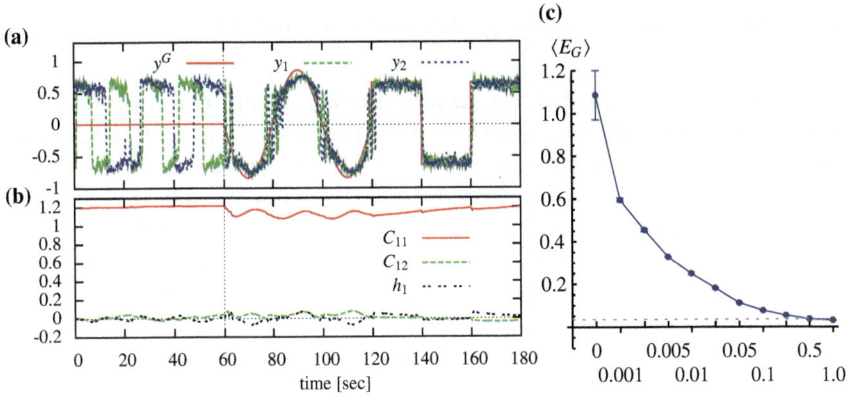

Fig. 12.1: TwoWheeled robot controlled with homeokinetic controller and direct motor teaching signals. **(a)** The teaching signals y^G (identical in both components) are followed partially by the motor values $y_{1,2}$ after teaching was switched on with $\gamma = 0.01$ at 60 sec. **(b)** Time evolution of the controller parameters affecting the first motor is shown to illustrate that only little changes are necessary, however, the adaptations do not vanish. **(c)** Average value of the PSEF E^G (for 5 experiments à 5 min) in dependence of γ (note the logarithmic scale). The noise level (*dotted gray line*) is reached at $\gamma = 1$. Parameters: $\varepsilon_c = \varepsilon_A = 0.1$, $\gamma = 0.01$ (a,b), update rate 100 Hz.

at $y \approx \pm 0.65$, see Sect. 6.3. This is a good mean teaching signal size, which was also used in Eq. (12.5). As a rule of thumb we recommend confining the motor teaching values to the interval $[-0.85, 0.85]$.

In Fig. 12.1 the produced motor values and the parameter dynamics are displayed for different values of the guidance factor γ. For a low value of γ the desired behavior is only followed by trend, whereas for higher values, e. g. $\gamma = 0.01$, the robot mostly follows the given teaching value with occasional exploratory interruptions. These interruptions cause the robot, for example, to move in curved fashion instead of strictly driving in a straight line as the teaching signals dictate. The exploration around the teaching signals might be useful to find modes which are better predictable or more active. It was shown in the theory of cognitive deprivation (Sect. 9.1) that a long performance of a single low-dimensional behavior can lead to the inaccuracy of the adaptive forward model. Thus, the explorative actions can supply the forward model with necessary sensation-actions pairs for complete learning.

The experiment demonstrated that motor teaching signals can be used to achieve a specific behavior. This result is not very surprising, because the system is very simple and the target behavior did not conflict with the homeokinetic principle (sensitive and predictable). However, it served as a proof of principle and showed that the balance between target behavior and remaining self-organized behavior can be adjusted using a single parameter.

12.4 Direct Sensor Teaching and Distal Learning

In this section we transfer the direct teaching paradigm from motor teaching signals (Sect. 12.3) to sensor teaching signals. This is a useful way of teaching because desired sensor values can be more easily obtained than motor values, for instance by moving the robot, or parts of the robot by hand. This kind of teaching is also commonly used when humans learn a new skill, e. g. think of a tennis trainer that teaches a new stroke by moving the arm and the racket of the learner. Thus, a series of nominal sensations can be acquired that can serve as teaching signals. Setups where the desired outputs are provided in a different domain than the actual controller outputs are called *distal learning* [46, 82, 163]. Usually a forward model is learned that maps actions to sensations (or more generally to the space of the desired output signals). Then the mismatch between a desired and occurred sensation can be transformed to the required changes of action by inverting the forward model. Another option is the use of an inverse model, which learns the mapping from sensations to actions. The main difference between inverse models and inverted forward models is the handling of noise, which is amplified by the inverted forward model and damped by an inverse model.

In our case a forward model is already at hand, namely, it is given by the internal forward model, see Eqs. (3.6, 4.2). Let the sensor teaching signal be given by x_t^G. The distal learning error is the misfit ξ_t^G between desired sensations x_t^G and actual sensations x_t, thus

$$\xi_t^G = x_t^G - x_t \; . \tag{12.6}$$

From Eq. (12.6) we can calculate the misfit η_t^G in motor space via the forward model M (3.6) in a linearized way by

$$\eta_t^G = A^+ \xi_t^G \; , \tag{12.7}$$

where $A = \frac{\partial M(x,y)}{\partial y}$. In our linear implementations of the forward model the parameter matrix A features explicitly. Now the update formulas for C and h from the direct motor teaching setup are used, see Eq. (12.4).

Alternatively this guidance mechanism can be obtained in a very similar way using the `harmony` extension as introduced in Sect. 15.1.1.4 if the distal learning error ξ_t^G is added to the normal prediction error ξ_t.

The application to the TwoWheeled driving robot, as it was done in the previous section, is trivial since the forward model consists essentially of a unit matrix. Among the previously considered robots the Spherical has a non-trivial relation between sensor and motor values, hence we will use it in the following experiment.

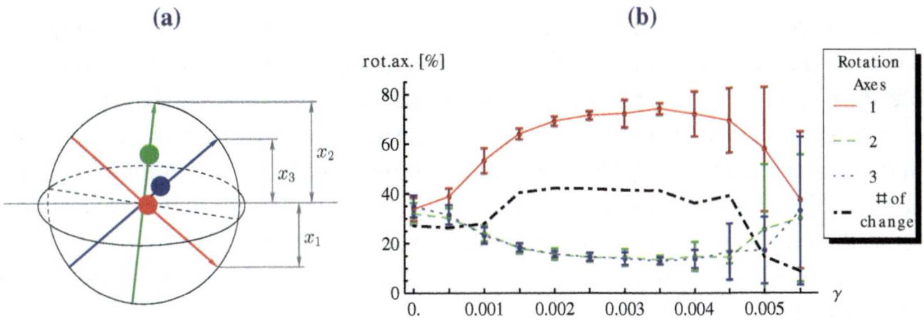

Fig. 12.2: SPHERICAL in a homeokinetic plus distal learning setup. (a) Illustration of the robot with its sensor values. **(b)** Behavior with the distal learning signal, Eq. (12.8). The plot shows the percentage of rotation time around each of the internal axes and the number of times the behavior was changed for different values of the guidance factor γ (no teaching for $\gamma=0$). The rotation around the red (first) axis is clearly preferred for non-zero γ. The mean and standard deviation are plotted for 20 runs each 60 min long, excluding the first 10 min (initial transient, no guidance). For too large values of the guidance factor the self-organization process is too much disturbed such that the robot gets trapped in a random behavior (*dash-dotted line*). Parameters: $\varepsilon_c = \varepsilon_A = 0.1$, update rate 100 Hz.

12.4.1 Experiment

In this experiment we use the SPHERICAL, as described in Sect. 8.2. For each axis we have one sensor value that is the z-component of the axis vector in world coordinates, which is illustrated in Fig. 12.2(a). We use the extended forward model as proposed in Sect. 9.2 with increased the awareness of causality, see Sects. 9.2.4 and 15.1.1.2.

The objective of this experiment is to show that a simple teaching signal in terms of sensor values can be used to effectively guide the behavior of the SPHERICAL towards rotations around the first internal axis. In such a mode of behavior the first sensor value remains at a low absolute value, see Fig. 12.2(a). This can be directly specified in the distal learning setting by putting

$$x_t^G = \begin{pmatrix} 0 \\ (x_t)_2 \\ (x_t)_3 \end{pmatrix}, \tag{12.8}$$

thus only the first component of the sensor value produces an error signal. To evaluate, we performed for different values of the guidance factor 10 runs each, with the robot on the flat ground. Each run was 60 min long.

The distal learning setup requires a well trained forward model. Therefore pure self-organization was used during the first 10 min of the experiment ($\gamma = 0$). As a descriptive measure of the behavior, we used the index of the internal axis around which the highest rotational velocity was measured at each moment of time. Figure 12.2(b) displays for different values of the guidance factor and for each of the axes the percentage of time it was the major axis of rotation. Without teaching there is no preferred axis of rotation. With distal learning the robot shows a significant preference (up to 75 %) for a rotation around the first axis. For overly strong teaching, a large variance in the performance occurs. This is caused by a destructive influence of the teaching signal on the homeokinetic learning dynamics. Remember that the rolling modes can emerge due to the fine regulation of the sensorimotor loop to the working regime of the homeokinetic controller, which cannot be maintained for large values of γ. The reader is invited to try the simulation described in Experiment 12.1.

Experiment 12.1: SPHERICAL with direct sensor teaching.

Start the simulation Spherical Robot - Direct Sensor Teaching The simulation is now running without guidance, i. e. $\gamma = 0.0$. The robot will start to roll and change the axis of rotation every so often.
Observe the change in behavior when you set the guidance factor to $\gamma = 0.003$. The robot changes to rotate around the red axis. You can change the teaching signal to another axis by `>axis=1,2,3`. Try different guidance strengths (parameter `gamma`) and observe the parameters as usual.

Why is it not possible with this method to force the controller to stay in the rotational mode around the first axis? The answer is rather intuitive: While the robot is in this rotational mode the teaching signal is negligible. However, the sensitization property of homeokinetic learning increases the impact of the first sensor, such that the mode becomes eventually unstable again. Again this may be considered as an advantage since the temporary breaking out avoids the cognitive deprivation effect (Sect. 9.1). Note, moreover, that the learning success in the current setting of controller and forward model could **not** be achieved by the distal learning alone, at least not with a constant learning signal.

To summarize, the direct teaching mechanism proposed in Sect. 12.3 allows one to specify motor patterns that are more or less closely followed, depending on the strength of integrating the additional drives into the learning dynamics. In this section we considered sensor teaching signals that were transformed into motor teaching signals using the internal forward model. We have shown that the SPHERICAL with the homeokinetic controller can be guided to locomote mostly around one particular axis, by specifying a constant sensor teaching signal at one of the sensors [104].

The supervised learning in terms of sensor signals is one step in the direction of imitation learning. Imitation learning [149] deals with how an autonomous robot can acquire behaviors from other agents or from humans. In this setting the robot can typically perceive the movement via a camera or perhaps via its joint sensors if the demonstrator was moving the parts of the robot's body. In the latter case the methods proposed here can be directly used. If only visual information is available the correspondence problem concerning the mismatch between the teacher's body and the robot's body needs to be resolved [34, 45, 145], which is not discussed here.

Chapter 13
Channeling Self-Organization

Abstract: Many desired behaviors are distinguished by a certain structure in the motor or sensor activity. In particular the phase relation between different motors or sensors capture a lot of this structure. We will now propose a way to embed these relations as soft constrains to the learning system, such that we break certain symmetries and let desired behaviors emerge. Starting from the guidance by teaching we introduce the concept of cross-motor teaching that allows to specify abstract relations between motor channels. First we study simple pairwise relations and shape the behavior of the TwoWheeled robot to drive mostly straight by a relation between both motor neurons. Then we will consider a high-dimensional robot—the Armband and demonstrate fast locomotion behaviors from scratch by guided self-organization.

Many behaviors, in particular for locomotion, have a particular structure with respect to the phase relation of the joints. We will now discuss how to provide the control system with information about these relations of the desired behaviors to influence the self-organization process. For that we will introduce soft constraints which break symmetries and reduce the effective dimension of the sensorimotor dynamics and thus guide the self-organization along a sub-space of the original control problem. In biological systems these kinds of constraints are often implemented on a low level of neuronal circuitry, e. g. pairs of antagonistic muscles are connected such that activity of one inhibits activity of the other on the level of α-motor neurons and inter-neurons in the spinal cord [128].

Inspired by this regulation we use the motor values of one motor neuron as a teaching signal for another motor neuron. We call this method *cross-motor teaching* and use *cross-motor connections* to describe which motor neuron receives a teaching signal from which other neuron. Note that despite the use of 'teaching signals' the algorithm is completely unsupervised, because the signals are generated internally. Before we go into the details let us briefly make a connection to the symmetry breaking and conservation discussion in Sect. 5.3.6. The self-organization progress preserves a high amount of symmetries of the physical system. As an example the TwoWheeled drives forward and backward and rotates clockwise and counter-clockwise equally often. The physical system (morphology of the body and interaction) is essentially symmetric with respect to forward-backward, left-right (lateral), and also straight-rotational behavior. To the contrary, the LongVehicle is lacking

R. Der and G. Martius: The Playful Machine, COSMOS 15, pp. 243–252.
springerlink.com

the forward-backward symmetry and, more importantly, also the straight-rotational symmetry because of friction and inertia. This is also reflected in behavior in that the robot is more driving straight than rotating, see Sect. 6.4.1.

Let us see how we can permanently break some of these symmetries with the following mechanisms.

To introduce this method we will first consider pairwise relations and then generalize to permutation relations.

13.1 From Spontaneous to Guided Symmetry Breaking

First we want to influence the controller to prefer a pairwise *in phase* or *antiphase* relation in the motor patterns [104]. Let us consider a particular pair of motors (r,s). We place a cross-motor connection from r to s and back, which means that the motor neuron s receives its teaching signal from motor neuron r and vice versa. In this way both motor neurons are driven to an in phase activity. The (internal) teaching signal is

$$\left(y_t^G\right)_r = (y_t)_s \qquad \text{and} \qquad \left(y_t^G\right)_s = (y_t)_r \,, \tag{13.1}$$

which is used in Eqs. (12.1–12.4).

Likewise, an antiphase relation can be expressed with $\left(y_t^G\right)_r = -(y_t)_s$ and vice versa. In this simple setup the cross-motor connections have either a positive or negative sign. For those motors i that are not part of a pair we need to set $\left(y_t^G\right)_i = (y_t)_i$, in order to produce no error signal in the framework of direct motor teaching, see Sect. 12.3. More complex connection setups are discussed later.

Experiment

To illustrate the concept we will consider the TwoWheeled, see Sects. 6.4.1 and 9.1.1. The robot has two motors actuated according to y_1 and y_2. Let the goal be to prefer straight driving. This can be obtained by an in phase relation between both motors following Eq. (13.1), i. e.

$$\left(y_t^G\right)_1 = (y_t)_2 \qquad \text{and} \qquad \left(y_t^G\right)_2 = (y_t)_1 \,. \tag{13.2}$$

For experimental evaluation we placed the robot in an environment cluttered with obstacles, see also Experiment 13.1

We performed, for different values of the factor γ, five runs of 20 min length. In order to quantify the influence of the cross-motor teaching we recorded the trajectory, the linear velocity, and the angular velocity of the robot. We expect an increase in linear velocity because the robot is to move straight instead of turning. For the same reason the angular velocity should go down. In Fig. 13.1 a sample trajectory

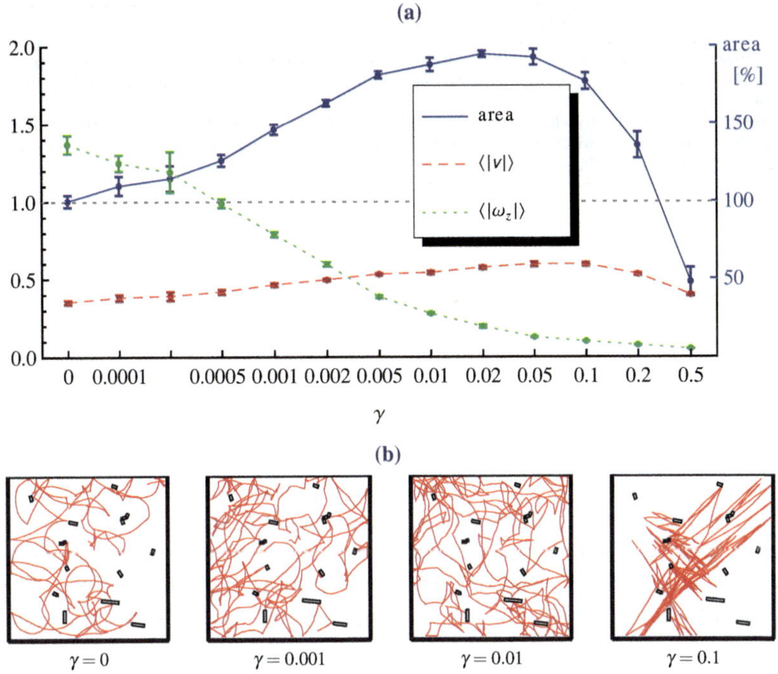

Fig. 13.1: Behavior of the TwoWheeled when guided to move preferably straight. **(a)** Mean and standard deviation (of 5 runs each 20 min) of the area coverage (**area**), the average velocity $\langle|\mathbf{v}|\rangle$, and the average turning velocity $\langle|\omega_z|\rangle$ for different values of the guidance factor γ. Area coverage (box counting method) is given relative to the case without influence ($\gamma=0$: 100%) (right axis). The robot is driving straighter and its trajectory covers more area for larger γ, until at large γ the teaching dominates the behavior of the robot. **(b)** Example trajectories for different guidance factors. Parameters: $\varepsilon_c = \varepsilon_A = 0.01$, update rate 100 Hz.

and the behavioral quantifications are plotted. Additionally, we plot the relative area coverage,[1] which reflects how much area of the environment was covered by the robot with cross-motor teaching compared to the original homeokinetic controller. As expected, the robot shows a distinct decrease in mean turning velocity and a higher area coverage with increasing values of the guidance factor. Note that the robot is still performing turns and drives both backwards and forwards and does not get stuck at the walls, as seen in the trajectory in Fig. 13.1(b). The properties of the homeokinetic controller, such as sensitivity and exploration, remain.

We have seen that a pairwise cross-motor teaching can be used to guide the self-organizing control to drive mostly straight in the TwoWheeled. The strength of this preference can be adjusted by the guidance factor. The algorithm is self-supervised and the only specific information that is given is the pair of motors to

[1] The area coverage of the trajectory is calculated using a box-counting method.

Experiment 13.1: TwoWheeled robot with cross-motor teaching.

The simulation Guided TwoWheeled drives straight starts with the robot in a cluttered environment (smaller than in Fig. 13.1) without guidance, i. e. $\gamma = 0.0$. The robot will equally probable move forward, backward and turn in both directions.
Observe the change in behavior when you set the guidance factor (gamma) to $\gamma = 0.02$. The robot starts to move more straight than before. After a while set $\gamma = 0.5$, which is too large for this application. The robot will now drive very straight but will get stuck at the walls.

be synchronized. Let us now proceed to a more flexible specification of cross-motor connections in order to deal with more complex situations.

13.2 Multiple Motor Relations

Now we want to consider a more general cross-motor connection setup where each motor has one incoming and one outgoing connection, such that there is still only one teaching signal per motor neuron [105]. The cross-motor connections can be described by a permutation π_m of the m motor neurons assigning each motor neuron a source of teaching input. The teaching signal is then given by (dropping the time index)

$$y_i^G = y_{\pi_m(i)} \qquad \text{for } i = 1, \ldots, m. \tag{13.3}$$

Additionally a sign function could be used defines whether the motors are supposed to be in phase or antiphase, but we do not need it in the following. The pairwise setup (Eq. (13.1)) is of course a special case of this notation. Note, that with a cyclic schema of connections also a group of motors can be synchronized.

13.2.1 Guiding Towards Directed Locomotion

Let us now consider a more complex robot—the Armband. The name comes from the German word *Armband* meaning a wristband or bracelet of a watch. We have considered a similar robot already in Sect. 6.4.3, which has additional slider-joints in the segments. The robot a locomotion behavior with a decentralized control setting, namely each joint was individual and independently controlled by a homeokinetic controller. In this section we will see that we can explicitly guide the robot to a directed and much faster locomotion in our normal control setting (one controller for the entire robot).

The Armband consists of a sequence of $m+1$ flat segments placed in a ring-like configuration, where subsequent segments are connected by hinge joints (in total m). The resulting body has the appearance of a bracelet or chain, see Fig. 13.2(a),(b). Each joint is driven by a servo motor and has a joint-angle sensor. The joints have a center position, which is such that the robot is in a perfectly circular configuration,

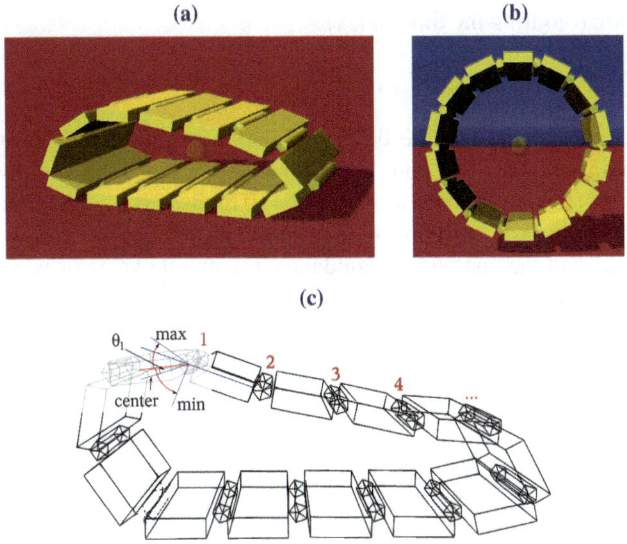

Fig. 13.2: The robot ARMBAND. **(a,b)** Screenshots from the simulation. The transparent sphere in the center marks the center of mass of the robot. **(b)** Configuration where all sensors are zero (joint center position). **(c)** Schematic view of the robot. The prismatic structures are hinge joints actuated by servo motors. All joints are equally constructed. The motor values and sensor values are defined in terms of joint deflection angles (θ_i) from the center (initial) position (see **(b)**). The motor values are scaled to the interval $[-1,1]$. The joint limits are: max$= 4/3 \cdot 2\pi/m$, and min$=-\pi/3$.

see Fig. 13.2(b) (angle of $2\pi/m$ with respect to a straight positioning). The motor values and sensor values are given in terms of joint angle deviations from the center, as displayed in Fig. 13.2(c). Note that the joints are highly coupled through the ring configuration. Therefore, an independent movement of a single joint is not possible. Instead it has to be accompanied by a movement of the neighboring joints and of distant joints.

Since the robot is symmetric there is by construction no preferred direction of motion, meaning that the robot controlled by the homeokinetic controller will equally probable move forward or backward. The robot cannot turn or move sideways, but it can produce a variety of postures and locomotion patterns.

With the method of cross-motor teaching we can help to break different symmetries, such that the robot is more likely to perform a directed motion. One possibility is to connect motors on opposite sides of the robot with a bias in clockwise or counterclockwise direction. For that we define the permutation, (used in Eq. (13.3)), as

$$\pi_m(i) = (i + k + \lfloor m/2 \rfloor) \mod m\,, \tag{13.4}$$

where $k \in \{-1, 0, 1\}$ and $\lfloor\cdot\rfloor$ denotes the truncation rounding (floor). We will only use positive connections, such that the sign function is not required. Thus, the teaching signals are (omitting the time index)

$$y_i^G = y_{(i+k+\lfloor m/2 \rfloor) \mod m} \qquad \text{for } i = 1,\ldots,m\,. \tag{13.5}$$

The choice of k depends on the desired direction of motion and on whether the number of joints m is even or odd. If m is even then $k=-1$ and $k=1$ are used for both directions (forward or backward) and $k=0$ represents a symmetric connection setup. In the latter case the robot will not prefer a direction of motion and the behavior is similar to the one without guidance. For an odd value of m, which is also used here, $k=0$ and $k=1$ need to be used for backward and forward motion.

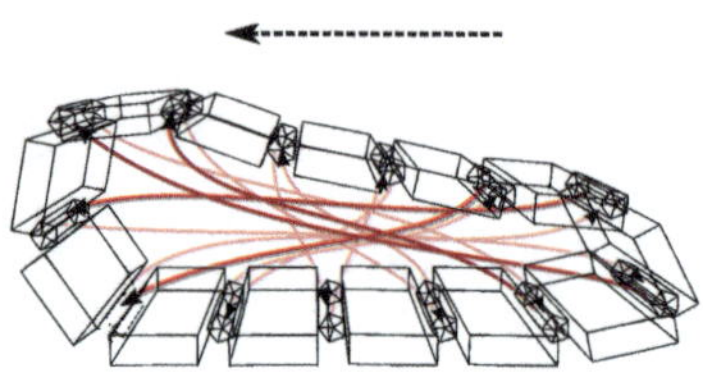

Fig. 13.3: ARMBAND with cross-motor connections. The arrows indicate unidirectional cross-motor connections, where the head points to the receiving unit. All links are equal, but for visibility reasons only four links are drawn boldly. For this connection setup the robot preferably moves leftwards.

In the following experiments the robot has $m=13$ motors. The motor connections for $k=1$ are illustrated in Fig. 13.3. Each motor connection is displayed by an arrow pointing to the receiving motor. Note that the connections are directed and a motor neuron is not teaching the motor neuron from which it is receiving teaching signals. For $k=0$ all arrows are inverted, meaning that for each connection the sending and receiving motor neurons swap roles.

To evaluate the performance we conducted, for different values of the guidance factor γ, 5 trials each 30 min long. In a first setting the cross-motor connections were fixed ($k=1$) for the entire duration of the experiment. Without guidance the robot moves equally to both directions but with comparably low velocity. This can be seen at the mean of the absolute velocity in Fig. 13.4(a). If the value of the guidance factor is chosen conveniently, we observed the formation of a locomotion behavior after a very short time and the robot moves in one direction with varying speed see Fig. 13.4(b) for 3 velocity traces. Note that this behavior requires all joints of the robot to be highly coordinated. We also observe a peak of high velocity after the first few minutes, which is followed by a dip before a more steady regime is attained. During this time the controller is going from a subcritical regime (at $t=0$) to a slightly supercritical regime.

The locomotive behavior can also be seen in Video 13.1 for a low value of guidance factor ($\gamma_s = 0.001$) and in Video 13.2 for a medium value of guidance factor ($\gamma = 0.003$). The average velocity of the robot increased distinctively with rising guidance factors, see Fig. 13.4(a). However, for excessively large values of the

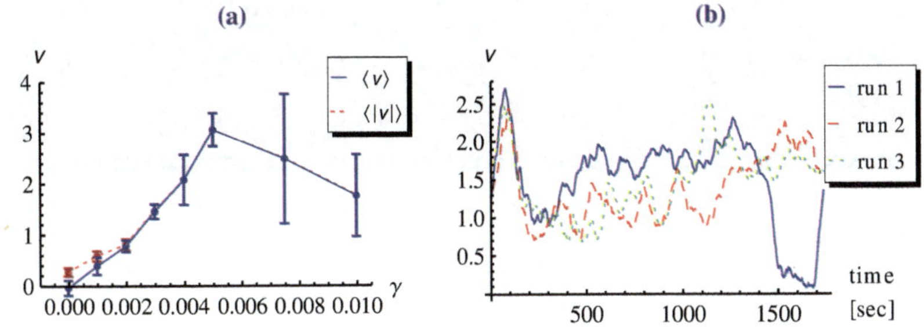

Fig. 13.4: Performance of the ARMBAND with constant cross-motor teaching. (**a**) Mean and standard deviation of the average velocity $\langle v \rangle$ and the average absolute velocity $\langle |v| \rangle$ of 5 runs for different value of the guidance factor γ. (**b**) Velocity of the robot $\overline{v}$ (averaged over 1 minute sliding window) for 3 runs at $\gamma = 0.003$. Parameters: $k = 1$, $\varepsilon_c = \varepsilon_A = 0.1$, update rate 100 Hz.

Video 13.1: ARMBAND learns to locomote — weakly guided. Behavior of the robot with cross-motor teaching and weak guidance ($\gamma = 0.001$). A slow locomotive behavior with different velocities is exhibited. Explorative actions cause the posture of the robot to vary in the course of time. The video can be watched at `http://playfulmachines.com`.

guidance factor the velocity goes down again. This occurs for two reasons: First, the cross-motor teaching has a too strong influence on the working regime of the homeokinetic controller and second the actual motor pattern of the locomotion behavior does not perfectly obey the relations between the motor values as specified by Eq. (13.5). In order to satisfy the constraints imposed by Eq. (13.5) all motor values need to be equal, which is of course not the case in the locomotion behavior.

In a second setup we changed the cross-motor connections every 5 min, i. e. k was changed from 0 to 1 and back. A value of $k = 0$ should lead to a negative velocity and a $k = 1$ to a positive velocity. To study the dependence on the guidance factor and to measure the performance we use the average absolute velocity $\langle |v| \rangle$ and the correlation of the velocity with the configuration of the coupling $\rho(v,k)$, see Fig. 13.5(a). Without guidance ($\gamma = 0$) there is, as expected, no correlation with the supposed direction of locomotion. For a range of values of the guidance factor we find a high total locomotion speed with a strong correlation to the supposed direction of motion. Note that the size of the correlation depends on the length of the intervals of one connection setting. For long intervals the correlation will approach one. In Fig. 13.5(b) the velocity of the robot is plotted for different runs with the same value of the guidance factor that was used in the previous experiment ($\gamma = 0.003$). We

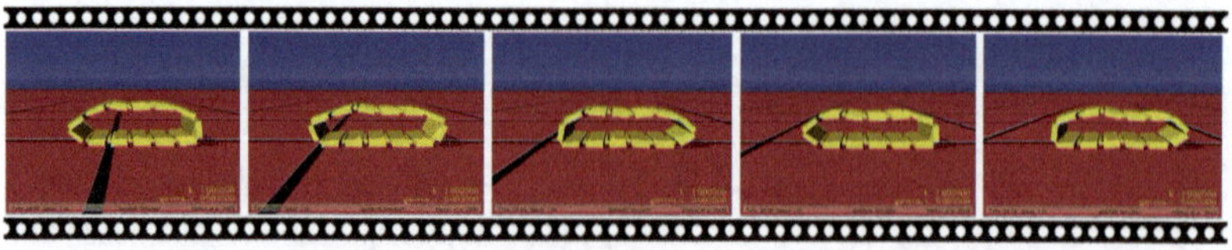

Video 13.2: ARMBAND quickly learns to locomote. Behavior of the robot with cross-motor teaching and medium guidance ($\gamma = 0.003$). Comparable fast locomotive behavior emerges quickly and is persistent. Nevertheless the velocity varies. Only small exploratory actions are takes, such that the posture is mainly constant. The video can be watched at `http://playfulmachines.com`.

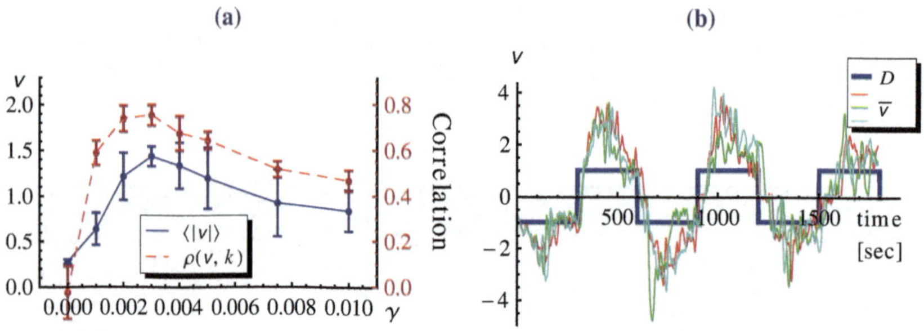

Fig. 13.5: Performance of the ARMBAND with changing cross-motor teaching. **(a)** Mean and standard deviation of the average absolute velocity $\langle |v| \rangle$ and the correlation $\rho(v,k)$ of the velocity with the configuration of the coupling for 5 runs with different values of the guidance factor γ. **(b)** Velocity (averages over 10 sec sliding windows) of the robot for 3 runs at $\gamma = 0.003$ and the target direction of motion $D = 2k - 1$ for better visibility. Parameters: $\varepsilon_c = \varepsilon_A = 0.1$, update rate 100 Hz.

observe that the robot changes the direction of motion shortly after the configuration of the coupling was changed, see Video 13.3 or do Experiment 13.2. In the latter the robot also has to overcome some barriers.

13.2.2 Scaling Properties

The locomotion of the robot is essentially influenced by the number of cross-motor connections. To study this we use again the fixed connectivity. In a series of simulations a number $0 \leq l \leq m$ of equally spaced cross-motor connections (Fig. 13.3) are used. With increasing l the robot starts to locomote earlier. Full performance is reached already if 8 out of the 13 connections are used, see Fig. 13.6(a).

In order to study the scaling properties of the learning algorithm we varied the number of segments m of the robot and thus the dimensionality of the control problem. The results are astonishing, see Fig. 13.6(b): The behavior is learned with the same speed also for large number (40) of segments. There is no scaling problem here

Video 13.3: ARMBAND changing the direction of motion. The behavior of the robot with cross-motor teaching if the connections are changed. The video starts with a fast locomotive behavior to the left ($k = 1$). At time 5:00 the couplings are changed ($k = 0$) and the robot slowly stops. A period of probing actions follows until a reversed locomotion starts to show up. The video can be watched at `http://playfulmachines.com`.

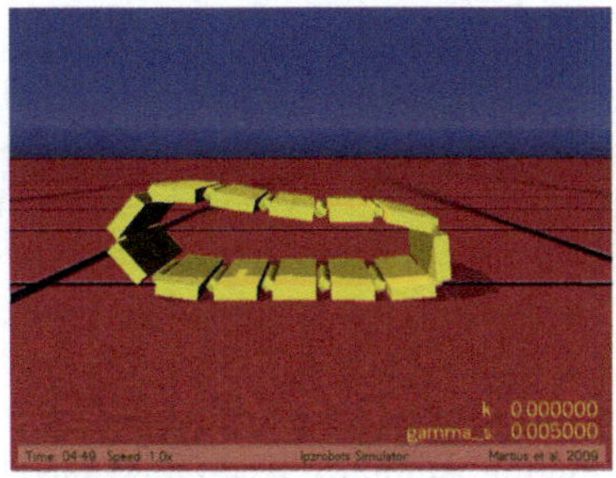

Experiment 13.2: Locomotion of the ARMBAND.

The simulation ARMBAND with cross-motor teaching starts in pause mode, so that the robot can be inspected. Press `<Ctrl>+P` to toggle the pause mode. At the beginning the guidance is switched off and the values of $\gamma = 0.0$ (`gamma`) and $D = 1$ are displayed. Observe the unguided behavior (speed up the simulation if impatient using `+`/`-`). Use the MATRIXVIZ (`<Ctrl>+M`) to observe the motor values y and the controller matrix C. Enable now weak guidance with:
`>gamma=0.001`
The robot will slowly start to perform a locomotion in one direction. Observe the change in behavior when you change the guidance factor to `gamma=0.005` and then the connection setup with `D = -1`. Keep in mind that a change in the guidance needs time to show up in the behavior. Therefore it is recommended to wait about 30 seconds after a change in parameters. There are also bars on the floor to overcome. If the robot gets stuck then speed up and observe the slow change in motor activity—the robot will manage eventually.

for the following reason. In the closed loop with an approximate feedback strength (self-regulated by the homeokinetic controller) the robot needs only very little influence to roll. The length of the robot can even help because other behavioral modes (e. g. wobbling) are damped increasingly due to gravitational forces. For the same reason, small robots are slower than medium ones. Large robots are again slower because the available forces at the joints become too weak.

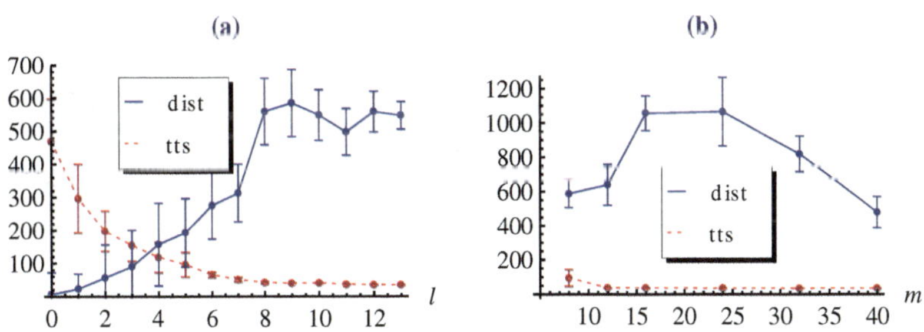

Fig. 13.6: Scaling of learning time and performance for different robot complexity. The plots show mean and standard deviation of the distance traveled by the robot ('dist' in units of 1 segment size) and of the time-to-start ('tts' in seconds) of 20 runs à 10 min ($\gamma = 0.003$). **(a)** Performance as a function of the number of cross-motor connections l (equally spaced around a robot with $m = 13$ joints). **(b)** Performance for different numbers of segments m (DoF) with full cross-motor connectivity ($l = m$).

The experiment illustrates that specific behaviors can be achieved in a high-dimensional robot by using cross-motor teachings. Cross-motor connections can break the symmetry between the two directions of motion such that a locomotion behavior is produced quickly. When the connections are switched during runtime, the behavior of the robot changes reliably.

The mechanism proposed here can also be transferred to sensor space using the direct sensor teaching (Sect. 12.4) instead of the motor teaching. One obtains a cross-sensor teaching analogously to the definitions given above. This can become useful, for example if a certain behavior is demonstrated by a human operator by activity moving the robot. In the case of the ARMBAND, one can imagine pushing the robot along the ground forcing it into a locomotion pattern. Based on the observed sensor readings, the correlations between the sensor channels can be determined and used as a basis for the construction of a specific cross-sensor teaching setup.

13.3 Summary

Starting from the guidance by teaching we introduced the concept of cross-motor teaching allowing for the specification of abstract relations between motor channels. There are no external teaching signals required, because the motor values are used mutually as teaching signals. The only specific information put into the system is the cross-motor relation. First we studied simple pairwise relations and shaped the behavior of the TWOWHEELED robot to drive mostly straight through the coupling between both motors. The couplings introduce soft constraints that guide the self-organization process to a subspace of the entire sensorimotor space and therewith the effective dimension of the search space for behaviors is reduced. This was demonstrated using a high-dimensional robot—the ARMBAND. With a simple cross-motor teaching the robot developed within a short time fast locomotion behaviors from scratch. The direction of motion was altered by a change in the connection setup. Remarkable is also the scaling property with respect to the dimensionality of the control problem.

Chapter 14
Reward-Driven Self-Organization

Abstract: In this chapter we investigate how to guide the self-organization process by providing an online reward or punishment. The starting point for the following considerations is that the homeokinetic controller explores the behavioral space of the controlled system and that those behaviors which are well predictable will persist longer than others. The idea we pursue in this chapter is to regulate the lifetimes of the transient according to the reward or punishment. The mechanism is applied to the SPHERICAL with two goals, fast motion and curved rolling.

This chapter focuses on the use of online reward or punishment to guide the self-organization process, see also [106]. The essential feature of the homeokinetic controller that we exploit is the exploration of the behavioral space of the controlled system, for instance the behavior of the BARREL that showed a systematic sweeping through the accessible frequencies of the sensor state reflected by rolling modes with different velocities (Sect. 7.3.3). In the case of the SPHERICAL with its three dimensional motor and sensor space we also observed a sweeping through a large set of possible behaviors (Sect. 8.2). In a setup where the robot can move freely, it will exhibit different slow and fast rolling modes around different axes.

Before introducing the new mechanisms, let us recall an important feature of homeokinetic control, namely that well predictable behaviors persist longer than others, which we called the temperature effect see Sect. 5.3.4. Due to this effect the well predictable behaviors are also quickly found because badly predictable ones are left quickly. Translating this into the case of reward and punishment, we want that rewarded behaviors persist longer than punishment ones and that predictable ones are found quickly. Thus we have to modulate the learning speed according to the online reinforcement signal in a way that in rewarded situations the adaptation speed is reduced and in punished ones the speed in increased. At first glance it seems to be counterintuitive that we have to reduce learning speed in order to keep a behavior, but the self-organized search should be slowed down to find even better behaviors locally. Moreover, the controller is already able to produce the behavior at the time it is exhibited by the robot.

Let us define a reward signal $r(t) \in \mathbb{R}$ for each time t, which should be considered as a punishment for negative values and as a reward for positive values. To

R. Der and G. Martius: The Playful Machine, COSMOS 15, pp. 253–259.
springerlink.com

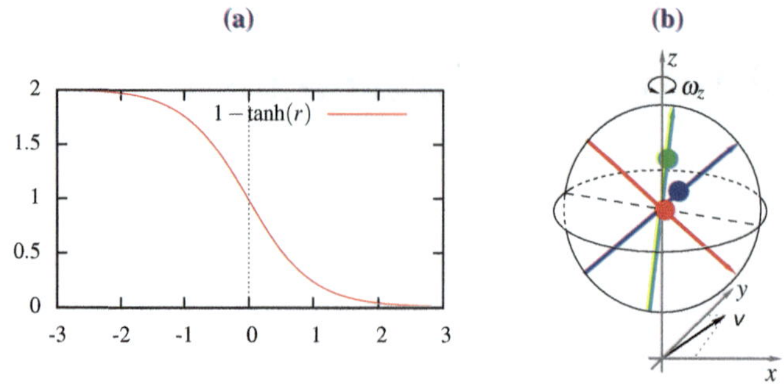

Fig. 14.1: Reward dependent factor and the SPHERICAL in world coordinates. **(a)** Factor for the modified error function, Eq. (14.1); **(b)** SPHERICAL with angular velocity (ω_z) around the z-axis and velocity vector v in x-y-plane. Note that x, y, z refer here to the axes of the world coordinate system.

incorporate the reward signal we define a new error function in the following way

$$E^{\mathrm{r}} = (1 - \tanh(r(t)))E \;, \tag{14.1}$$

where E is the usual TLE defined in Eq. (5.9). The factor $(1 - \tanh(r(t))$ approaches zero for large positive rewards and 2 for large negative rewards as depicted in Fig. 14.1(a). The multiplication of the error function is the same as multiplying the learning rate. Thus, we could similarly write $\varepsilon^r_{c,t} = (1 - \tanh(r(t)))\varepsilon_c$. The construction of the new error function Eq. (14.1) follows our considerations above. For a reward ($r > 0$) the factor $(1 - \tanh(r(t))$ is smaller than 1 and thus decreases the learning speed. For a punishment ($r < 0$) the learning speed is increased. We can expect that rewarded behaviors persists longer and punishment behaviors are left quicker. The hyperbolic tangent imposes constraints on effective values of $r(t)$, namely they have in the interval $(-3, 3)$ a differentiable effect on the learning dynamics. However, it is not required that the interval is being utilized by a particular reward function, nor does it do any harm to exceed the range.

Let us now apply the mechanism to shape the behaviors of the SPHERICAL. In the following sections we will demonstrate two different behaviors which can be effectively guided.

14.1 Reinforcing Speed

In the following experiment we will use the SPHERICAL, as described in Sect. 8.2. One of the simplest possible desired behaviors of this robot is to move fast. Let us construct the reward function for this goal. For small velocities the reward should be negative, thus causing a stronger change of behavior, whereas larger velocities should result in a positive reward. To achieve that, the reinforcement signal can be

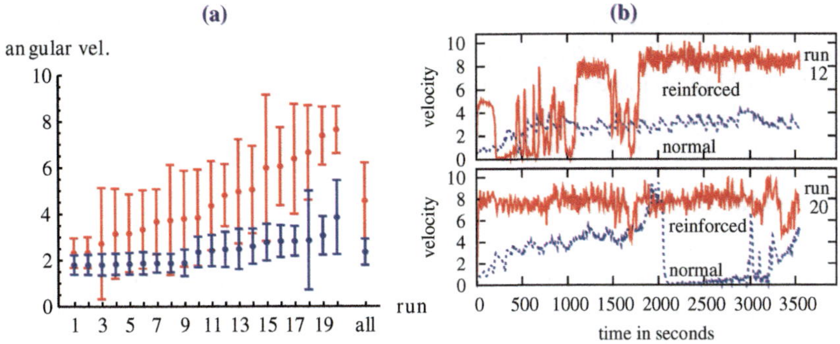

Fig. 14.2: Performance of the SPHERICAL rewarded for speed. **(a)** Mean and std. deviation of the velocity of the SPHERICAL for 20 runs each 60 min long with (**red**) and without (*blue*) speed reinforcement, sorted by velocity. The label '**all**' denotes the mean and std. deviation over the means of all runs, which is significantly ($p < 0.001$) higher for the reinforced runs. **(b)** Time course of the robot's velocity for run number 10 and 14, where *blue/dotted* shows the normal case and **red/solid** line shows the reinforced case.

expressed as

$$r(t) = \frac{1}{3}\|v_t\| - 1\,, \tag{14.2}$$

where v_t is the velocity vector of the robot, see Fig. 14.1(b). In order to compare the results with the unguided case the reward is shifted, such that it is zero for the average velocity of normal runs. The scaling is done to keep the reward within the effective range.[1]

We conducted 20 trials with the SPHERICAL with reinforcement and 20 trials without reinforcement, all with random initial conditions, each for 60 min in simulated real time on a flat surface without obstacles. The robot also experiences rolling friction, so that fast rolling really requires continuous motor activity. In Fig. 14.2 the mean velocity (measured at the center of the robot) for each simulation is plotted and the velocity trace of the robot for two reinforced and two normal runs are displayed as well. The simulations are sorted by performance and plotted pairwise for comparison. As desired, the mean velocities of the reinforced runs are larger than the ones of the normal runs. This is especially evident in the overall mean (mean of means marked with '**all**' in Fig. 14.2(a)), which is significantly different. The null hypothesis that the set of means of the reinforced runs and of the normals runs have an indistinguishable mean was rejected with $p < 0.001$ using the t-test. However, since straight and also fast rolling modes are easily predictable and active they are also exhibited without reinforcement for a long time. It is important to note that the fast rolling modes are also found again, after the robot was moving slower, see Fig. 14.2(b).

[1] Note that here the minimal reward is -1, so we do not fully use the effective range of punishment.

The guidance of the homeokinetic controller using a reward for fast motion has shown to increase the average speed of the robot significantly. Although there are also trials where no increased speed was found.

14.2 Reinforcing Spin

In a different setup we want the robot to follow curves and spin at the spot. We use the angular velocity ω_z around the z-axis of the world coordinates system, which is perpendicular to the ground plane, as depicted in Fig. 14.1(b). The reward function is now given by

$$r(t) = \frac{1}{3}\|\omega_z\| - 1\,. \tag{14.3}$$

Again the reward is scaled and shifted to be zero for normal runs and to be in an appropriate interval. Positive reward can be obtained by rolling in a curved fashion or by entering a pirouette mode. The latter can be compared to a pirouette done by figure-skaters—with some initial rotation the masses are moved towards the center, so that the robot spins fast in place. The robot also experiences rolling friction, so that fast pirouettes are not persistent.

Again, we conducted 20 trials with reinforcement and 20 trials without reinforcement, each for 60 min simulated real time on a flat surface without obstacles. In Fig. 14.3(a) the mean angular velocity ω_z for each simulation is plotted, again sorted by performance. The time evolution of the angular velocity for two reinforced and two normal runs are displayed in Fig. 14.3(b). In this scenario the differences be-

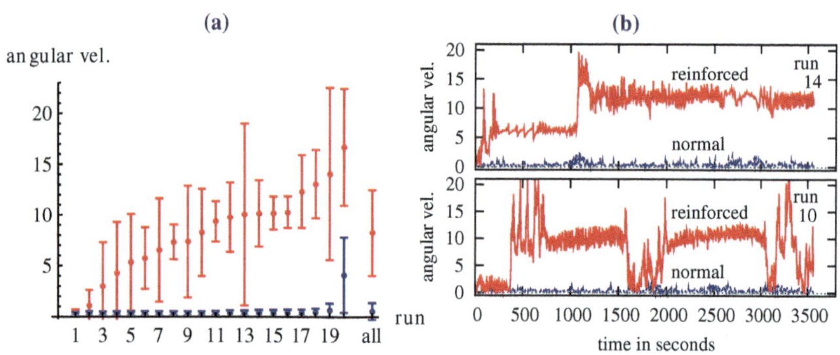

Fig. 14.3: Performance of the SPHERICAL rewarded for spin. **(a)** Mean and std. deviation of the angular velocity ω_z of the SPHERICAL for 20 runs each 60 min long with (**red**) and without (*blue*) spin reinforcement, sorted by angular velocity. The label '**all**' denotes the mean and std. deviation over the means of all runs. **(b)** Time course of the velocity for run number 12 and 20, where *blue/dotted* shows the normal case and **red/solid** line shows the reinforced case.

tween the normal runs and the reinforced runs are remarkable. Nearly all reinforced runs show a large mean angular velocity. The reason for this drastic difference is that these spinning modes are less predictable and therefore quickly abandoned in the non-reinforced setup. The traces show that the robot in a normal setup rarely performs spinning motion, whereas the reinforced robot performs, after some time of exploration, very fast spinning motions, which are persistent for several minutes. In this setup it can also be seen that the rewarded behaviors are found again after they were lost, see Fig. 14.3(b).

The mechanism to modulate the learning speed by a reward signal showed a strong effect on the behavior of the SPHERICAL. When controlled by the homeokinetic controller without guidance the robot rarely exhibits narrow curves or spinning behavior. In contrast the guided controller engaged the system into curved motion most of the time. One might wonder how it is possible that this technique is able to reach a behavior that is normally not exhibited. The reason is that when the robot is starting to follow a curve, then the learning rate of the controller goes down, although the forward model is still learning normally. In the unguided case the prediction error rises (because it is a new behavior) and thus the controller will quickly leave this behavior. This actually happens before a fast spinning is reached. In the rewarded case the forward model is able to capture the behavior before it is left (because of the slower drift), which in turn enables the control system to enter modes of more narrow curves.

14.3 Discussion

In this section we want to summarize and discuss the mechanisms of guidance. In Chap. 12 we proposed two methods to guide the otherwise freely self-organized behavior with teaching signals. We defined an appropriate error function that allows for directly specifying a desired motor pattern and verified its functioning with a simple robot experiment. The balance between self-organized behavior and target behavior can be adjusted with a single parameter. In order to be able to work with teaching signals in sensor space we used the internal forward model to transform them into motor teaching signals. In this framework, the SPHERICAL was taught to prefer the rotation around one particular axis solely by requesting a zero value of the sensor value measuring the axis orientation of that particular axis.

The teaching mechanisms formed the basis for further investigations that aim at higher level guiding mechanisms, which we considered in Chap. 13. We introduced cross-motor teachings to specify relations between motor channels, such as in phase or antiphase activity. This induces soft constraints and therewith reduces the effective dimensionality of the system. An example with the TWOWHEELED showed, that by requesting an in phase activity of left and right wheel, the behavior changes qualitatively to straight motion. Nevertheless, the explorative character of the controller is not destroyed. The resulting behaviors are less constrained than in the case of direct teaching. For example the TWOWHEELED can drive forward or backward at its own choice, whereas in the direct teaching setup the direction of driving is specified externally. It is also interesting that the robot remains sensitive to perturbations and changes its behavior from time to time so that the available environment is thoroughly scanned. Especially visible is the effectiveness of guidance with the high-dimensional ARMBAND. A cross-motor teaching with only one connection per joint leads to a fast and coordinated locomotion behavior. The direction of motion was rapidly and reliably changed by an altered connection setup. Interestingly the performance and speed of learning is almost independent of the dimensionality of the system, at least in the here considered case up to 40 DoF. The discrete cross-motor connections offers a good way for higher level control structures to direct the behavior of the robot.

In Chap. 14 we proposed a simple method to guide the self-organizing behavior using online reward signals. In essence the original time-loop error is multiplied by a strength factor, obtained from the reward signal. The approach was applied to the SPHERICAL with two goals, fast motion and curved rolling. In both cases the performance was significantly increased, although the reinforcement of fast motion showed only a small effect, because the homeokinetic controller achieves fast rolling modes already without guidance. When reinforcing curved rolling the results are distinctive. In this case we find fast spinning modes, which are not observed otherwise. Notably, the exploratory character of the paradigm still remains intact.

Let us shortly compare guided self-organization (GSO) using cross-motor teachings and rewards with other approaches to learning of autonomous robot behavior, namely evolutionary algorithms [113] and reinforcement learning [166].

We will consider several dimensions along which they are compared: *(i)* the achievement of the specified goal, *(ii)* the learning time, *(iii)* the flexibility of the solution, and *(iv)* the universality of the method. Evolutionary algorithms can for instance optimize the parameters of the controller (e. g. a neural network) and can possibly achieve the behaviors demonstrated here. There are many impressive results and systems of similar dynamical complexity have been successfully controlled, see for example [21, 29, 79, 100, 110] and references in [57]. In high-dimensional systems, however, identical subcomponents are typically used and long learning time are required (many generations with many individuals). Here we see the main strength of our system: As a matter of fact, the desired behaviors are found very fast even in high-dimensional and dynamically complex systems—we have very fast online-learning. Another difference is that the finally evolved controller is typically static, such that it only works in the conditions it was evolved in. Evolutionary algorithms are universal, which is a big plus, but the representation of the problem in the artificial genome and the definition of genetic operators requires great care, see e. g. [57, 156, 186]

When considering reinforcement learning (RL) we have to distinguish how the control is organized. Traditionally RL uses a set of discrete actions, which would be an intractably large set in the case of the 13-DoF ARMBAND. RL would not yield a good performance in this case. Since several years there are continuous space RL algorithms [47] with very powerful variants such as the natural actor-critics [130] with policy gradient [88] and recently with a path integral approach [170]. These allow for a continuous state and action space and are currently the most efficient RL algorithm for high-dimensional systems [1]. A great application is for instance the control of a seven DoF robotic arm to hit baseball and table tennis balls appropriately [129]. In this applications an initial supervised learning by human demonstration was used though. The learning times are still very long in comparison to the approaches proposed here. However, it is not a fair comparison, since the task is specified at a different level. RL deals with delayed rewards since it solves the credit assignment problem and in this way accounts for long-term effects. To the contrary the reward schema used here (Chap. 14) require online rewards.

In the case of the cross-motor teachings the specification of the desired behavior is indirect and limited in comparison to the possibilities of reward functions in RL and the fitness functions in evolutionary algorithms. So RL and evolutionary algorithms are universal and very flexible in the specification of the task whereas the GSO methods are very restricted in this respect. Also RL was proven to converge to the optimal solution under certain conditions [166] where with GSO the desired behaviors are only partially followed and no guarantee can be given. This is due to the underlying self-organization, which causes an ongoing exploration around the desired behaviors. Concerning the flexibility, the resulting controllers are highly adaptive and can change quickly to new situations, which is one of the major strengths of the self-organizing system.

To conclude, the guided self-organization methods offer a fast development of goal-oriented behaviors in high-dimensional continuous robotic systems from scratch, which cannot be achieved with other learning control systems so far. However, the implementation of goals is comparably limited. The reward-based guidance allows any reward signals, but no time delays are tolerated and it is not guaranteed that the reward is maximized. The cross-motor teaching method is suitable to select a subset of behaviors, but cannot be generalized to all behaviors. A combination of both methods is also possible, namely using cross-motor teachings to be very effective in high-dimensional systems and additionally using rewards to give a fine grain control over the behavior.

In the current setup the controller and the forward model will forget past behaviors, so that there is no long term effect of the reinforcement. Another problem for practical applications is that we achieve only a tendency to a certain behavior and cannot guarantee its successful execution. Theses problems can be solved with a framework that extracts behavioral primitives. This can be done for instance with a set of competing expert neural networks, see [102]. These expert networks represent controllers for a certain behavior and because of the closed loop setup and the generalization property of neural networks they provide often smooth transients into the specific attractor behaviors. With classical reinforcement learning these behavioral primitives can be used as discrete actions to solve more complex tasks [102].

Chapter 15
Algorithmic Implementation

Abstract: This chapter presents a unified algorithm implementing the homeokinetic learning rules including a number of extensions partly discussed already in earlier chapters of this book. We continue with some guidelines and tips on how to use the homeokinetic "brain." We discuss techniques and special methods to make the self-supervised learning of embodied systems more reliable from the practical point of view. This includes the regularization procedures for the singularities in the time-loop error and different norms of the error for the gradient descent. The internal complexity of the controller and the model is extended by the generalization to multilayer networks. Apart from that the computational complexity of the learning algorithm will be reduced essentially by easing non-trivial matrix inversions. This is important for truly autonomous hardware realizations.

The general homeokinetic learning rules, as given in Sect. 5.2, have been modified and generalized in the applications done so far in order to give them more flexibility and numerical stability. Particularly important are the specificities of different numbers of motors and sensors, which requires to keep an eye on defining pseudoinverses. The following sections will describe these modifications to eventually introduce the universal learning rule, which contains all of these extensions in a canonical way.

Later in this chapter we will provide some algorithmic details and some recipes on how to effectively use the homeokinetic controller with different kinds of robots. Eventually the learning rules for an extended architecture of the controller and model are derived.

15.1 The `sox` Algorithm

This section describes an extended version of the homeokinetic controller called `sox` algorithm (**s**elf-**o**rganizing e**x**tended) in the simulator. We will proceed in the following way. We start from the general homeokinetic learning rule and then consider the extensions one by one to obtain finally the new learning rules. Interestingly, all modifications will reduce to some minor changes in the auxiliary vectors, while the form of the learning rule stays the same.

R. Der and G. Martius: The Playful Machine, COSMOS 15, pp. 261–291.
springerlink.com

In order to avoid too much scrolling we will repeat the original learning rules as defined in Sect. 5.2 here. In the simulator these are implemented in the `sos` algorithm (**s**elf-**o**rganizing **s**imple or **so**–**s**imple ;-) with the standard implementation for the sensorimotor dynamics ψ given by

$$\psi(x) = M(x, K(x)) = Ag(Cx+h)+b.$$

The standard time-loop error (TLE) is defined as (5.14)

$$E = \xi^\top \frac{1}{LL^\top} \xi = v^\top v = \chi^\top \xi \tag{15.1}$$

with the auxiliary vectors $v \in \mathbb{R}^n$ and $\chi \in \mathbb{R}^n$ given by

$$v = \frac{1}{L}\xi, \ \chi = \frac{1}{L^\top} v \tag{15.2}$$

L being the Jacobian matrix of ψ. If necessary, the inverses are understood in the generalized sense, see Sect. 5.G and Sect. 15.4 (p. 277).

The update rule for the parameters of the controller $(\{C,h\})$ is given by the gradient descent on the TLE (5.16) whereas the parameters of the forward model $(\{A,S,b\})$ are obtained by gradient descending the prediction error $E^{\text{pred}} = \xi^\top \xi$.

The learning rules for the forward model are, in the compact matrix notation see Eqs. (4.5, 4.6),

$$\Delta A = \varepsilon_A \xi y^\top \tag{15.3}$$

$$\Delta b = \varepsilon_A \xi \tag{15.4}$$

where ε_A is the learning rate. The rules for the controller are given by Eqs. (5.25–5.27), i. e.

$$\Delta C = \varepsilon_c \mu v^\top - \epsilon y x^\top \tag{15.5}$$

$$\Delta h = -\epsilon y \tag{15.6}$$

$y = g(z)$ being the output of the controller and $g(z) = \tanh(Cx+h)$ is to be taken componentwise with $y \in \mathbb{R}^m$ and $z \in \mathbb{R}^m$. ε_c is the overall learning rate of the controller and ϵ the matrix of the channel dependent learning rates given by its matrix elements as

$$\epsilon_{ij} = 2\varepsilon_c \alpha \delta_{ij} \mu_i \zeta_i \tag{15.7}$$

with δ_{ij} the Kronecker delta (ϵ is a diagonal matrix). The vectors $\zeta \in \mathbb{R}^m$ and $\mu \in \mathbb{R}^m$ are defined as

$$\mu = G' A^\top \chi, \ \zeta = Cv \tag{15.8}$$

where G' is the diagonal matrix defined as $G'_{ij} = \delta_{ij} g'(z_i)$.

We will now consider the following new ingredients:

1. sensibility,
2. modifications of the forward model,

3. creativity,
4. harmonizing,
5. pseudoinverse and motor space, and
6. different error types.

Finally we will collect all modifications into the new update rules in Sect. 15.1.2, which look structurally very similar to the ones above.

15.1.1 Extensions and Corresponding Meta-Parameters

15.1.1.1 Choosing the Sensibility (Meta-Parameter **sense**)

The sensibility of the controller to perturbations can be adjusted by a meta-parameter called `sense` in the `sox` algorithm. The effect of this parameter consists in shifting the working regime of the system. This has been illustrated in the case of the pitchfork bifurcation and the Neimark-Sacker bifurcation earlier, see the discussion on the effective bifurcation point in Sect. 3.3.4. In those cases, the effect was clearly defined in terms of the effective bifurcation point. In more complicated systems the effect of the parameter is less clear but the idea of changing the sensibility of the robot may serve as a guideline when playing with the parameter.

The integration of the `sense` parameter was already considered explicitly in Eq. (5.24) so that we repeat it here only for completeness, see Eq. (15.7). The parameter weights the strength of the confinement term in Eq. (15.5). Default value is $\alpha = 1$. For larger values the sensibility is increased and for smaller values it is decreased. Note that for $\alpha \neq 1$ the controller outputs y sent to the motors of the robot are scaled roughly by a factor of $\frac{1}{\sqrt{\alpha}}$ as it can be inferred from the discussion of the fixed point flow in the 1-D loop, see Sect. 3.3.4. In applications the motor values may be rescaled if required.

15.1.1.2 Details of the Model (Meta-Parameters **causeaware** and **useS**)

For the implementation of the forward model we have two choices that can be selected in the `sox` algorithm with the flag `useS`. The most simple implementation corresponds to Eq. (4.2), i. e.

$$M(x,y) = Ay + b \tag{15.9}$$

and is selected with `useS=0`. The full version with the sensor branch, as introduced in Sect. 9.2.2, is given by

$$M(x,y) = Ay + Sx + b \tag{15.10}$$

and is enabled with `useS=1`.

In Chap. 9 different aspects and pitfalls of the simultaneous learning of forward model and controller have been discussed. Especially if the sensor branch (term Sx) is included there is an ambiguity in the causality chain since the forward model relates the new sensor values both back to the motor and the previous sensor values. We argue that it is reasonable for an autonomous agent to have an increased awareness of the causal relation between actions and sensations, see Sect. 9.2.3. This bias can be controlled with the meta-parameter ρ_A implemented into the learning rule for A as

$$\Delta A = \varepsilon_A \chi \hat{y}^\top \tag{15.11}$$

with

$$\hat{y} = y + \rho_A \eta \quad \text{and} \quad \eta = G'Cv\,, \tag{15.12}$$

the derivation being given in Eq. (9.14).

The new learning rule may be simplified and made more stable by exchanging χ by ξ in the Eq. (15.11). This can be motivated by writing Eq. (15.11) as

$$\Delta A = \varepsilon_A \Lambda \xi \hat{y}^\top$$

where $\Lambda = \left(LL^\top\right)^{-1}$ is a positive matrix (case of regular L). The effect of this matrix is simply to rescale the size of the gradient step along the eigenvectors of Λ so that the gradient is always pointing downhill independently of the size of the (always positive) factors. Thus, the matrix Λ can also be replaced with the unit matrix without inverting the direction of the gradient. In a wider context, the choice of Λ corresponds to choosing the Riemannian metric for the gradient descent. This is well known in the theory of the natural gradient, see for instance [4].

In our special case, the scaling factors introduced by Λ may vary over orders of magnitude due to the g'^{-1} factors. In the experiments done so far we have realized that this may give rise to instabilities in the model learning so that we propose to drop the Λ matrix from the learning rule altogether, leaving us with the new learning rule for the model matrix A as

$$\Delta A = \varepsilon_A \xi \hat{y}^\top\,. \tag{15.13}$$

If we set ρ_A to zero the original learning rule (15.3) is restored. Note that we keep the learning rules for the other parameters (S, b) unchanged. The meta-parameter ρ_A is implemented in the `sox` algorithm as `causeaware` with default value 0.01 if `useS=1` and 0 otherwise.

15.1.1.3 Choosing the Creativity (Meta-Parameter `creativity`)

Now we want to introduce a meta-parameter that may help in the creation of new modes and the amplification of latent but very subtle motion patterns. It is called `creativity`, which is to be understood in a metaphoric sense and should not be taken as an attempt to mimic true creativity of humans.

The experiment with the BARREL, considered in Sect. 8.5, is a good example to understand the meaning and implementation of this meta-parameter. There the BARREL was put into the upright position such that all sensor values (the angles of the axes with the ground) are equal to zero. Thus the activity in the sensorimotor loop breaks down and only small motor values (defining the position of the weights on the axes) will be created. Thus the model learning turns A rapidly towards zero as can be observed in the parameter panels in Video 8.11 and Video 8.13, so that any sensor noise (which is still present to a small amount) is amplified in the reconstruction process by as much as two orders of magnitude. The idea is then to use the reconstructed sensor values $\hat{x}_t = x_t + v_t$ as input into the controller. In the case of the BARREL this had a tremendous effect on the behavior, it created a new strategy of shaking the robot enabling it to find a mode to escape the impasse situation. Metaphorically we can say that the robot's "brain" increased its creativity for dealing with completely new situations.

This strategy is realized by replacing the input x_t into the controller by

$$\hat{x}_t = x_t + \beta v_t \tag{15.14}$$

where the meta-parameter β is introduced in order to control the effect. The reason for the weighting is that v_t may become very large if the response of the sensor to the motor signal is weak. The meta-parameter $\beta \in (0,1)$ is called `creativity` in the `sox` algorithm and can be used in order to tune the effect depending on the situation.

The new input to the controller has also a positive effect in more general situations. Without the `creativity` term, one often observes that the agent goes into some dynamical regime which seems to be an attractor in the combined dynamics of system and controller parameters. In the high dimensional systems it is not quite clear why the system does not leave these regions despite the destabilizing nature of the learning dynamics. For instance, our SNAKE, when lying on the ground, often comes to a complete rest although the homeokinetic learning actually destabilizes the "do nothing" behavior. The reason for that is not yet clear but the creativity term offers an effective way for getting out of the situation. In fact, using the `creativity` option proved more effective and resilient than just adding noise or increasing the learning rate, strategies that in the cases considered were a subtle deal needing some care.

The success of the procedure is also understood by considering the difference of the `creativity` option to the strategy of just adding noise to the sensor values. If using `creativity`, the input into the controller contains the shift v instead of a pure sensor noise. However, v is reconstructed from the prediction error ξ that is not a pure noise, especially if new behavioral modes are about to start, since then the forward model is not yet accurate and the prediction error contains systematic components of the motion. Feeding this into the controller, as it is done via Eq. (15.14), helps in many cases to amplify such latent modes.

15.1.1.4 Enhancing Dynamical Harmony (Meta-Parameter **harmony**)

Homeokinesis rests essentially on the simultaneity of the dynamics of the physical system (robot + environment) with that of the internal parameters. The tie between these two is given by the map ψ, which defines the internal representation of the sensorimotor dynamics. In general, there is a large complexity gap between ψ and the true world dynamics, which leads to vastly different characteristics of the prediction error ξ as a function of the behavior.

In fact there are two extreme attitudes in viewing the role of the error ξ. On the one hand one stipulates that the model captures all knowable dependencies on the behavior, i. e. the derivative

$$\frac{\partial}{\partial p}\xi_{t+1} \approx 0 \tag{15.15}$$

yielding the learning rule, used so far (5.16),

$$\Delta p_t = \varepsilon \chi^\top \frac{\partial L}{\partial p_t} v$$

where $p \in \mathbb{R}^1$ is any of the parameters of the controller $(\{C, h\})$. In this interpretation, any systematic dependence on p is already contained in ψ. The term ξ is considered to be more of an erratic nature so that its derivative will average out over time.

The other extreme is given by the case that the transition from x_t to x_{t+1} is captured only very qualitatively by ψ so that the derivative of ξ is essential. In that case, the difference between internal and external dynamics is vast, harmony between them is far away.

Then, instead of trying to improve ψ (what one still is up to, anyway) we can try to modify control in such a way that the controller learns better to produce actions that lead to predictable (under the present model) consequences and thus to a better agreement with the events going on in the world. This can be achieved by considering the new sensor values x_{t+1} as given from outside like in supervised learning. Then, instead of Eq. (15.15) we use

$$\frac{\partial}{\partial p}\xi_{t+1} = \frac{\partial}{\partial p}\left(x_{t+1} - \psi(x_t)\right) = -\frac{\partial}{\partial p}\psi(x_t) \ . \tag{15.16}$$

In this way, the controller is driven towards behaviors which are more in agreement with what is known internally so that the harmony between internal and external world is increased. Writing the error as $E = \chi^\top \xi$ (15.1) and using the product rule we immediately see that the inclusion of Eq. (15.16) yields

$$\frac{1}{\varepsilon_c}\Delta p = \chi^\top \frac{\partial L}{\partial p} v + \chi^\top \frac{\partial}{\partial p}\psi(x) \ . \tag{15.17}$$

Interestingly, in our specific setting of Sect. 5.2.2 (p. 85) we can integrate this into the update rules for the controller parameters by introducing the auxiliary vector

$$\hat{v} = v + \gamma_h x \tag{15.18}$$

as

$$\Delta C = \varepsilon_c \mu \hat{v}^\top - \epsilon y x^\top \tag{15.19}$$

$$\Delta h = \varepsilon_c \gamma_h \mu - \epsilon_i y_i \,. \tag{15.20}$$

The meta-parameters $\gamma_h \geq 0$ controls the strength of the additional term. For stability reasons it is necessary to choose $\gamma_h \ll 1$. In the learning rule for C the only difference to the original expressions Eq. (15.5) consists in replacing $v \to \hat{v}$. In the `sox` algorithm this meta-parameter is called `harmony`, which is 0 by default.

Stability Issues

One of the essential features of homeokinetic learning, namely the mixing of time scales for learning and behaving, requires some precautions in using this rule. The point is that the learning rates are kept at or at least close to the critical value where the time discrete gradient step swaps from descent to ascent leading to a runaway of the parameters, see Sect. 15.B for details. The new term $\hat{v}$ (15.18) combines two terms of vastly differing size: v is often very small (since it is the error term) while x is of the order of 1. As a consequence, we had to introduce an additional factor $\gamma_h \ll 1$.

In practical applications one must be a little careful when using this option, not only because of the different magnitudes of x and v but also due to the fact that the forward model can be far away from reality. In this case, obviously the behavior should better not be adapted to converge towards the model dynamics. Actually, it is difficult to give a formal description how and when to enable this extension. Often it helps to get more rich but still body inspired behaviors, but in extreme cases one obtains instead boring behaviors or a runaway of the parameters of the controller. The latter case is a real danger in this process since the harmony term may heavily interfere with the actual dynamics of the homeokinetic learning. So, when using the term one should monitor the parameter dynamics and turn γ_h down as soon as the runaway threatens to start. It is not very difficult to get some feeling for riding this horse and it is rewarding since by the parameter control one can influence the "character" of the developing system to some extent.

15.1.1.5 Choosing the Generalized Pseudoinverse (Flag **pseudo**)

Besides these directly visible modifications of the learning rules we can look at the matrix inversion of L featuring in χ and v. There is some freedom in the type of pseudoinverse which may generally be chosen as

$$L^+_{PQ} = Q \frac{1}{PLQ} P \,. \tag{15.21}$$

Later, a detailed presentation of the matter is given in Sect. 15.4 (p. 277).

The `sox` algorithm in the simulator offers several possibilities that can be used in particular if $m < n$ (more sensors than motors):

a) $P = A^\top, Q = C^\top$ is the normal (sensor space) case enabled by `pseudo=1`,
b) $P = A^\top, Q = A$ is the motor space version as given in Sect. 15.6 selected with `pseudo=2`, and
c) $P = C, Q = A$ is a further variant enabled with `pseudo=3`.

In contrast to `pseudo=0`, where the normal Moore-Penrose inverse of L is used, all variants (a-c) decrease the size of the inversion to an $m \times m$ matrix. Sect. 15.4.1 discusses the consequences of the different options in a simple example corresponding to one motor and many sensors. It turns out that in the case of more sensors than motors (as happens in many realistic situations, the motor-space approaches (b) and (c) are preferable. The flag `pseudo` can be switched during runtime.

15.1.1.6 Choosing the Error Type

In the definition of the TLE the Euclidean norm was used to determine the size of the reconstruction error v. The quantity v may change over orders of magnitude, since it is obtained by the inverse of L. Especially in high-dimensional systems, the robot may get stuck in some impasse or has ridden itself into some entanglement from which it can not so simply escape. In these cases, both sensor and motor values are constant so that the forward model after a very short time learns these constant correlations and the prediction error ξ goes to zero and so does the reconstruction error v.

In practice it often helps to shake the robot mechanically so that the sudden inflow of new sensor values leads to an increase in the error ξ. Another way is waiting (and possibly adding some sensor noise) until the deprivation effect, Sect. 9.1, together with creativity, see the Sect. 15.1.1.3, may lead to a drastic increase of the TLE by the inversion.

In these and many other cases, the drastic decrease of the errors may lead to prohibitively long periods of stagnation on the one hand, and to a destabilization of the learning dynamics (due to the overshooting effect of the gradient dynamics, see Sect. 15.B) on the other hand. We have tested a large number of adaption strategies for the learning rate ε for both model and controller. The most robust and effective one turned out to be a different norm by using instead of $E = v^\top v$ variants like $E^{\mathrm{sqrt}} = \sqrt{E}$ or $E^{\log} = \log E$. The `sox` algorithm contains a flag to switch between them with the flags `errortype=0,1,2` corresponding to the Euclidean (0), square root (1), or the logarithmic norm (2).

The implementation is rather straightforward since we can write

$$\frac{\partial E^{\mathrm{sqrt}}}{\partial p} = \frac{1}{2\sqrt{E}} \frac{\partial E}{\partial p} \quad \text{and} \quad \frac{\partial E^{\log}}{\partial p} = \frac{1}{E} \frac{\partial E}{\partial p}, \tag{15.22}$$

such that the different norms are realized by a variable learning rate factor.

Noteworthy are the following consequences of the different norms on the behavior. Both non-linear norms increase the learning speed in low error conditions and decrease it for high errors. The differences between the situations are reduced maximally by the logarithmic norm and less strongly by the square root norm. This implies that the difference in the life time of behaviors is less dependent on their predictability.

15.1.1.7 Numerical Tricks

Before putting the extensions together, let us briefly introduce some numerical tricks we are using in order to improve the stability of the algorithm.

Restricting the Update (Clipping)

If the learning rate is chosen too high in a particular situation or the Jacobian matrix comes close to a singular point, there is still an ultimate measure for avoiding a runaway of the parameters. The most simple solution is to restrict the size of the parameter update term componentwise. We use a constant clipping size of 0.1 for the model parameters and 0.05 or less for the controller parameters. As a consequence the parameters cannot change arbitrarily fast, but still fast enough to follow behavioral changes. For instance a change in value from 1 to -1 would require 20 or 40 steps, which is below 1 sec in our applications.

Dealing with Singular Denominators

In the algorithm, inverses of matrices are either done as Moore-Penrose pseudoinverses (Sect. 5.G) or in the sense of generalized pseudoinverses (Sect. 15.4). The latter case is interesting, because the inverted matrices are not regularized. However, the maximum dimension that can occur is the one given by the number of motors (or the minimum number of neurons in an intermediate layer of the network ψ in the multilayer case described in Sect. 15.3). Given the structure of the TLE landscape with singularities corresponding to the null space of L, the learning dynamics drives the matrices away from the peaks in the error landscape caused by small eigenvalues in the matrices, see for instance the discussion in Sect. 9.1.3 for more details. Hence, we may in general expect that the involved matrices are invertible. We have nevertheless introduced a strategy called `secureInverse` which checks the matrices for regularity and uses the regularized Moore-Penrose inverse if the test fails.

15.1.2 Putting it Together — The Universal Learning Rules

Collecting the results of the previous modifications leads to the new learning rules implemented in the simulator in the sox algorithm. The model parameters are updated according to

$$\Delta A = \varepsilon_A \xi \hat{y}^\top \tag{15.23}$$
$$\Delta S = \varepsilon_A \xi x^\top \tag{15.24}$$
$$\Delta b = \varepsilon_A \xi \ . \tag{15.25}$$

The learning rules for the controller are

$$\Delta C = \varepsilon_c \mu \hat{v}^\top - \epsilon y x^\top \tag{15.26}$$
$$\Delta h = \varepsilon_c \gamma_h \mu - \epsilon y \tag{15.27}$$

where the diagonal matrix of channel depending learning rates is (tanh case)

$$\epsilon_{ij} = 2\varepsilon_c \alpha \delta_{ij} \mu_i \zeta_i \ .$$

The definitions of the auxiliary vector remain unchanged as

$$v = L^+ \xi, \qquad \chi = L^{+\top} v, \tag{15.28}$$
$$\mu = G' A^\top \chi, \qquad \zeta = C v, \tag{15.29}$$

where $G'_{ij} = \delta_{ij} g'(z_i)$ and $g(z) = \tanh(z)$. The controller reads now

$$y = g(z), \qquad z = C(x + \beta v) + h, \tag{15.30}$$

and the modified shift and motor values are defined as

$$\hat{v} = v + \gamma_h x, \qquad \hat{y} = y + \rho_A \eta \tag{15.31}$$

with $\eta = G'Cv$. Additionally all synaptic weights (Eqs. (15.23–15.27)) are also subject to a small damping, i. e. for $p \in \{A, S, b, C, h\}$ we have

$$\Delta p = \cdots - \gamma_d p. \tag{15.32}$$

In comparison to the original learning rules of Sect. 5.2 only small modifications have been necessary to implement the extensions. These modifications are included into the learning rules in a systematic way such that they can be easily generalized to the case of multilayer networks for both controller and model, considered in Sect. 15.3.

All meta-parameters with their respective ranges are listed in Table 15.1. By choosing these parameters appropriately, one may tune the "character" of the playful machine to a certain extend. Some guidelines for choosing them are given in the following section.

Table 15.1: Meta-parameters of the sox algorithm.

symbol	meta-parameter	typical range	default	description
ε_c	`epsC`	0 – 1	0.1	Sect. 15.2.3
ε_A	`epsA`	0 – 1	0.1	Sect. 15.2.3
γ_h	`harmony`	0 – 0.1	0.01	Sect. 15.1.1.4
β	`creativity`	0 – 1	0.1	Sect. 15.1.1.3
α	`sense`	0.1 – 5	1	Sect. 15.1.1.1
ρ_A	`causeaware`	0 – 1	0	Sect. 15.1.1.2
γ_d	`damping`	$0 - 10^{-4}$	10^{-5}	Sect. 15.1.2
	`pseudo`	$\{0,1,2,3\}$	0	Sect. 15.1.1.5
	`errortype`	$\{0,1,2\}$	0	Sect. 15.1.1.6

15.2 Cookbook for Playful Machines

In this section we want to give some short guidelines and recipes for using the homeokinetic "brain" in robot control. A certain universality is one of the advantages of our approach. The ambition is that the controller can be connected to any robotic system, letting behaviors develop in a kind of free play. This has been tested in many practical cases so far and seen to work fine if some basic conditions are met. For instance the sensors and motors should "speak comparable languages" and the body has to provide enough feedback in its sensory response to the actions of the controller.

15.2.1 Type of Robots — Mechanical Setup

Let us first characterize again the types of robots suitable best for homeokinetic control. Good robots for behavioral self organization are compliantly engineered machines with strong embodiment effects. Compliance is understood in the way that actions are not perfectly executed due to the influences of external forces and intrabody couplings, see the ECCEROBOT example in Chap. 1. In this way the current mode of behavior influences significantly the response of the system to the motor commands. In our virtual robots, compliance is often a direct result of underactuated control, realized by either weak forces or by the presence of several to many passive degrees of freedom (like in the SLINGING SNAKE). In detail we have to consider the following.

Underactuatedness: The forces of a single actuator are not sufficient to move the body without the cooperation of the other degrees of freedom. This implies that whole body motions require a coordination of degrees of freedom. Once cooperation has started, as a side effect the control costs[1] may be significantly lower in the collective modes of the robot, which we hope to excite by homeokinetic control.

[1] The control cost is meant in terms of required internal complexity of the control structures.

Sensors and Motors: Sensors should be installed that measure the consequences of the actions in a more or less direct way. Typical examples are proprioceptive sensors like the joint angle sensors or the wheel velocity sensors. We found that exteroceptive sensors are also suitable, if the quantity they measure is modulated by the actions in some systematic way. For example infrared sensors allow a better adaption to the environment (Sects. 8.4 and 8.2.4). An interesting option is also to extend the body virtually by the infrared sensors, see Sect. 3.5. Acceleration sensors can be used nicely if a change in body pose is expected and desired (Sects. 8.1 and 8.5.2). Motor current sensors provide a very good feedback on the physical load depending on the interaction with the environment, see Sect. 8.1, but they require some preprocessing as discussed below.

In any concrete case sensors either are to be selected according to the motors or have to be adapted to the control of the latter. As a rule of thumb, motor signals and sensor readings should relate to the same physical property. Avoid for instance to measure the velocity of some mechanical part if it is actuated in a position control setting. However, sometimes quite obscure combinations yield surprisingly good results, e. g. current sensors and velocity/voltage control, see Sect. 8.1.

Preprocessing and Noise: Typically no preprocessing of the sensor values is required. However, it is advisable to have the sensor values roughly in the between -1 and 1. The scaling is not strictly required, but allows to use standard learning rates and ensures a similar integration of different sensors. Neither is a zero-mean of the values necessary, for instance our infrared sensors provide values in the range $[0,1]$. Nevertheless, there are cases where a preprocessing beyond simple scaling is required, for instance when current sensors are used. The current is always positive regardless of whether the motor is turning clockwise or counterclockwise, which cannot be handled by our pseudo-linear control structure. A simple remedy is to multiply the sensor values with the sign of the motor values, as done in Sect. 8.1.1.

The sensor noise is an important aspect of the homeokinetic approach. If there is absolutely no noise, then the system will possibly not start the sensitization process or later on might get stuck in metastable configurations for a very long time. Hence it is often helpful to add a small artificial sensor noise to the raw sensor signals. With our simulated sensors we typically add uniform white noise with the strength between 0.01 and 0.1. It is important to stress that adding (weak) sensor noise is totally different from choosing random actions.

15.2.2 Initialization Procedures — Motor Babbling

Once the appropriate sensors and motors are selected, the success still depends to some extent on the initialization of both the controller and the forward model. In the proprioceptive sensor case, we always initialize both C and A with the scaled unit matrices. This is only valid if the sensor-motor configuration is also in this one-to-one correspondence. With more general sensors the unit matrix may be totally wrong since the sensor and motor channels are perhaps permuted or the "dark side" of the configuration space is reached, see the discussion in Sect. 5.1.4 (p. 81).

We have implemented in the simulator a general initialization procedure in order to get a first understanding of the sensor motor relations by sending periodic signals y_t^{ext} of varying frequencies and phase relations to the motors. Optionally the robot is suspended in the air during this phase, so that it can move its different parts freely. The motor values y_t^{ext} and the incoming sensor values are then used to

1. train the controller network with vector x_t of sensor values as input and y_t^{ext} as target output and
2. learn the forward model with (x_t, y_t^{ext}) as inputs and x_{t+1} as target output.

After sufficiently many steps the initialization phase should be stopped, continuing with homeokinetic learning as usual. When working with hardware robots, this procedure is particularly helpful since sensors and motors can be connected arbitrarily to the "brain" and may even be of widely different nature.

15.2.3 Choosing the Meta-Parameters

At first glance it seems that there is a vast amount of meta-parameters to tune, but in practice only few meta-parameters have to be selected, the rest being optional or kept at their default values. The most important meta-parameters are the learning rates (ε_A,ε_c), the type of the internal model, and of the pseudoinverse. If the system has more sensors than motors use `pseudo=2` (or 3). If the world has a dynamics of its own influencing the prediction of the next sensor values, use the extended model with `useS=1` and set `causeaware=0.01` or higher.

Now to the learning rates: We typically keep both of them identical and start with 0.1. In order to select the meta-parameters appropriately it is helpful to observe the parameter dynamics online. The LpzRobots software package comes with tools to do that, see Sect. 16.4.5, which can be also used for studies without the simulation environment. The speed by which the parameter values change is a good indicator for the right learning rate and the error type. On the one hand, the learning rate is too low if the parameters do not visibly change. On the other hand, the learning rate is too high if the parameters are seen to jump heavily. In particular, segments of a steep linear increase in the parameter plots are an indication of heavy clipping (Sect. 15.1.1.7) happening if the learning rate is exceedingly large. In these cases an immediate shrinking of the learning rate by at least an order of magnitude is mandatory.

An indication to choose the square root or logarithmic error type is given if the robot shows long periods of inactivity followed by short highly active ones accompanied by a very strong parameter variations. The square root or logarithmic error norm reduces the difference in learning speed between active and inactive situations, such that the choice of the learning rate is easier, see also Sect. 15.1.1.6. We found the use of the latter error types especially helpful in high-dimensional systems.

15.2.4 Parameter Runaways — The Dark Side

The stability of the learning algorithm depends very much on the embodiment, the parameter selection, and above all on the physical situation the robot is in. As long as the robot can execute its motor commands relatively freely, there is no problem with the learning dynamics if the parameters are chosen not too wrong. If, however, the robot or parts of it are blocked for a long time or its motion is apparently not controlled by its own actions, the homeokinetic learning may drive the parameters into pathological configurations hindering the robot to return to a "normal" behavior even if the constraints are lifted, see for instance Fig. 8.15 in Sect. 8.3. The most problematic configuration arises if the Jacobian matrix develops eigenvalues with negative real parts. This corresponds to an inversion in sign of the sensor values (in the direction of the respective eigenvector) in each time step. The same is true for the motor commands so that the robot might be driven to self-destruction. This is why we call this region the "dark side" in funny terms, see also Sect. 5.1.4 (p. 81).

Actually, the TLE has a singularity at the zeros of the eigenvalues, so that—once at the right side—the learning dynamics should keep us way from this point and crossing the line should not be possible. However, through high frequency oscillations the system can get there via the complex plane without having to overcome the singularity. Another scenario is a potential collapse of the forward model, e. g. if actuated degrees are blocked, driving eigenvalues of A towards zero such that small fluctuations in the matrix elements can bring us to the negative side.

How to prevent it? Use the initialization procedure above to make sure you start at the right side. Reduce the learning rates and/or use a different error norm. Make sure that you use `causeaware>0` if the model matrix A is seen to collapse (especially if the extended model is used).

15.3 Increasing Internal Complexity — The `soml` Algorithm

The extreme simplicity of both controller and forward model is a key feature of the homeokinetic approach. The idea is that the complexity is not in the internal structures but in the physics of the body embedded into the world; and all we have to do is to find a sensible way of feeding energy into the system so that the most fundamental modes are getting excited. This helps us to find truly emerging modes, putting as few as possible into the behavior from outside. The flip side of the coin is the transient nature of the behaviors. A typical case is the humanoid in the pit, see Sect. 10.5. Often the emergence of climbing patterns is observed, which may repeat several times and even shape out into a sequence of elementary motions so that seemingly the robot develops a strategy for escaping from the pit. After few repetitions, however, the climbing mode fades away and the robot seems to become "interested" in new motions patterns. One interpretation is that longer sequences of elementary movements can not be kept "alive" by the oversimplified brain structure.

Having said that, increasing the internal complexity might help lifting the system to a higher complexity level. Besides increasing the structural complexity of the "brain," an effect is already expected by the additional parameters introduced in this way into the controller. The effect of additional parameters has been demonstrated already by the introduction of the bias (Sects. 6.3 and 7.3). The bias dynamics becomes a constitutive component of the behavior generation and the interplay between the bias and the physical dynamics has been shown to be the reason for the emerging cooperation under decentralized control (Sect. 6.4.2) and many other effects. We want to continue this path by increasing the number of neurons and layers in the controller network. It will be sufficient to introduce the method by simply adding one hidden layer into the controller network. The extension of the forward model will be briefly discussed at the end of this section. In the simulator the extended setting is implemented in the `soml` algorithm, where an arbitrary number of layers for the forward model and the controller can be used.

15.3.1 Controller Architecture

We have introduced in Sect. 3.6.3 (p. 52) the general structure of a feed-forward network. In the case of a single hidden layer this reads

$$y_i^{(k)} = g\left(\sum_j \left(C_{ij}^{(k)} y_j^{(k-1)}\right) + h_i^{(k)}\right), \quad \text{for } k = 1,2\,, \tag{15.33}$$

where $y^{(0)} = x$ is the input, $y^{(1)}$ the output vector of the hidden layer, and $y^{(2)}$ the output of the network. The latter is just the output of the controller $y = K(x)$. We may write this more compactly as

$$y = g\left(C^{(2)} g\left(C^{(1)} x + h^{(1)}\right) + h^{(2)}\right) \tag{15.34}$$

assuming for simplicity of presentation that all neuron have the same activation function $g(z)$, which is taken componentwise. We restrict ourselves here to the tanh function, thus $g_i(z) = \tanh(z_i)$.

15.3.2 Learning Rule

Keeping the simple forward model as given in Eq. 4.2 (p. 65) we obtain

$$\begin{aligned} \psi(x) &= Ay + Sx + b \\ &= Ag\left(C^{(2)} g\left(C^{(1)} x + h^{(1)}\right) + h^{(2)}\right) + Sx + b\,. \end{aligned} \tag{15.35}$$

The learning rules from gradient descending the TLE are given for general ψ by (5.16)

$$\Delta p = -\varepsilon_c \chi^\top \frac{\partial L}{\partial p} v \,. \tag{15.36}$$

The Jacobian matrix L of ψ is, based on Eq. (15.34),

$$L = AG'\left(z^{(2)}\right) C^{(2)} G'\left(z^{(1)}\right) C^{(1)} + S \tag{15.37}$$

where

$$z^{(l)} = C^{(l)} y^{(l-1)} + h^{(l)},$$

$G'_{ij} = \delta_{ij} g'_i$, and $y^{(0)} = x$.

In Sect. 15.D we derive, starting from the general rule Eq. (15.36), the following update rules for the matrices of the controller network:

$$\Delta C^{(l)} = \varepsilon_c \mu^{(l)} v^{(l-1)\top} - \epsilon^{(l)} y^{(l)} y^{(l-1)\top} \tag{15.38}$$

$$\Delta h^{(l)} = -\epsilon y^{(l)} \tag{15.39}$$

where

$$\mu^{(2)} = G'^{(2)} A^\top \chi, \quad \mu^{(1)} = G'^{(1)} C_2^\top \mu^{(2)} \tag{15.40}$$

and

$$v^{(1)} = G'^{(1)} C^{(1)} v^{(0)}, \quad v^{(0)} = v = L^+ \xi \tag{15.41}$$

so that $v^{(1)}$ is obtained from the reconstruction error v by forward propagation the latter to the next layer of neurons.

The matrix ϵ of the channel dependent learning rate is given by its matrix elements as

$$\epsilon_{ij}^{(l)} = \varepsilon_c \delta_{ij} \mu_i^{(l)} \zeta_i^{(l)} \,, \tag{15.42}$$

where $\zeta^{(l)}$ is obtained from $v^{(l-1)}$ as

$$\zeta^{(1)} = C^{(1)} v^{(0)}, \; \zeta^{(2)} = C^{(2)} v^{(1)} \,. \tag{15.43}$$

This can again be modified by introducing the `sense` meta-parameter (α) in order to control the working regime of the system.

The rules can be interpreted as if each layer has its inputs and outputs from and into the world. Once the vectors v and χ are known, all quantities are obtained by propagating them forward or backward, respectively, to the appropriate inputs or outputs of the synapses in each layer. Given these auxiliary quantities, the learning rules look identical for all layers. A pseudocode is given in the appendix Sect. 15.D.

Increasing the complexity of the forward model is straightforward. Consider any function $M(x,y)$ for the model, we can use the rule stated above putting

$$A = \frac{\partial}{\partial y} M(x,y) \tag{15.44}$$

$$S = \frac{\partial}{\partial x} M(x,y). \tag{15.45}$$

Note that both quantities may be state and time dependent, for instance in non-linear and recurrent networks. In order to maintain an appropriate mapping from motors to sensors (motor branch) in the forward model the methods introduced in Sect. 9.2 should be considered.

15.4 Generalized Pseudoinverse*

In the derivation of the learning rules, Sects. 5.2 and 15.1.2, we did not yet explicitly discuss how the matrix inversion is to be done in general. It turns out that we have different choices which have rather important implications on the parameter dynamics.

The learning rules given so far are still open to interpretation if the inverse of L does not exist, like in the case that the sensors outnumber the motors.[2] The usual way is to use the well known Moore-Penrose pseudoinverses, see Sect. 5.G for an introduction. However, we want to introduce a generalization of that concept by way of a certain "sandwiching" technique, defining the inverse of a square matrix L as

$$L^{+}_{PQ} = Q \frac{1}{PLQ} P \tag{15.46}$$

choosing the matrices P and Q appropriately so that the inversion of PLQ becomes possible in the classical sense. This is trivial if all matrices P, Q, L are of full rank so that automatically $L^{+}_{PQ} = L^{-1}$. In all other cases the matrices P and Q have to be chosen appropriately so that the inverse of PLQ exists.

The replacement of the (possibly nonexistent) inverses in Eq. (5.15) by the generalized pseudoinverses deserves still justification. In fact, we did not derive so far the learning rules of Sect. 5.2 if using the generalized pseudoinverses. This is done in the Appendix, see Sect. 15.A, where we show in relevant special cases that the generalized pseudoinverses can well be used in the learning rules even if the matrices P and Q are depending on the parameters of the system. This result is a little astonishing but very helpful since it allows to introduce various pseudoinverses in the learning rules without having to modify the structure of the rule itself. Instead, all one has to do is to redefine the vector v and χ (Eqs. (5.15, 15.2)) as

$$v = L^{+}_{PQ} \xi, \qquad \chi = \left(L^{+}_{PQ}\right)^{\top} v. \tag{15.47}$$

By way of example, in our standard case with $L = AG'C$, we choose $P = A^{\top}$ and $Q = C^{\top}$ so that

[2] Considered in the case with the simple forward model where $L = AG'C$ (5.12).

$$L^+_{A^\top C^\top} = C^\top \frac{1}{A^\top A G' C C^\top} A^\top = C^\top \frac{1}{CC^\top} \frac{1}{G'} \frac{1}{A^\top A} A^\top \tag{15.48}$$

(assuming the inverses of $CC^\top$ and $A^\top A$ exist;[3] as we may in our applications). Obviously this corresponds to a strategy of defining the pseudoinverse of L step by step like

$$v = L^+_{A^\top C^\top} \xi = C^\top \frac{1}{CC^\top} \frac{1}{G'} \eta = C^+ \zeta \tag{15.49}$$

$$\chi = \frac{1}{G'} A^{+\top} \frac{1}{CC^\top} \zeta \tag{15.50}$$

with the Moore-Penrose left and right pseudoinverses A^+ and C^+, respectively, and the auxiliary vectors $\eta = A^+ \xi$ and $\zeta = G'^{-1} \eta$. The vectors arise from the projection of the prediction error ξ back, in a first step, to the controller outputs (η) and, in a second step, to the membrane potential of the controller neurons (ζ).

This scheme can be carried over immediately to the case of more complicated networks, see Eq. (15.37). It simply amounts to taking the pseudoinverse of each layer in the network structure and do the inversion by taking the (minimal dimension) pseudoinverse. In this way, the complexity of the matrix inversions is always given by the number of motors and/or the minimal number of neurons in the two layers connected by a matrix $C^{(l)}$ in the case of multilayer networks, see eqneqn:SensMotLoop:ffn1 in Sect. 3.6.

However, the choice of the sandwiching process in Eq. (15.48) is only one of several that may seem appropriate. For instance we may also use the pseudoinverse as

$$L^+_{CA} = A \frac{1}{CLA} C = A \frac{1}{RG'R} C \tag{15.51}$$

where $R = CA$ is an $m \times m$ matrix of rank 1. Ignoring the non-linearities this matrix maps the motor values to the next (in time) motor values and it is close to the unit matrix in many applications. This point will be used for an approximate learning rule avoiding matrix inversions altogether, see Sect. 15.5.1 below. Another form that will be identified with the motor-space version of the controller in Sect. 15.6 is given by

$$L^+_{A^\top A} = A \frac{1}{A^\top L A} A^\top = A \frac{1}{A^\top A G' R} A^\top . \tag{15.52}$$

All of these variants are available in the sox algorithm, see Sect. 15.6. Let us now consider an analytical example where we will find out that the different pseudoinverses lead to qualitatively different dynamics.

[3] Note that the learning dynamics drives the matrices away from the singularities.

15.4.1 Examples

The freedom in defining the pseudoinverses and the influence on the learning dynamics can be demonstrated clearly in the extreme case of only a single motor ($m = 1$) but $n > 1$ sensors. Now, $C \in \mathbb{R}^{1\times n}$ is a row, $A \in \mathbb{R}^{n\times 1}$ is a column matrix, and the Jacobian matrix $L \in \mathbb{R}^{n\times n}$ of $\psi(x) = Ay = Ag(Cx+h)$ is of course singular. We will study two different ways of choosing the pseudoinverse in the general learning rules given by Eq. (5.22) or Eq. (15.5). We start with Eq. (15.48) and put, without restriction of generality, $\|A\| = 1$ and $\|C\| = c$ and find[4]

$$v = L^{+}\xi = \frac{\|\xi\|}{c^2 g'} \cos\left(A^{\top},\xi\right) C^{\top}$$

$$\mu = g' A^{\top}\chi = g' A^{\top} A \frac{1}{c^2 g'} C v = \frac{\|\xi\|}{c^2 g'^2} \cos\left(A^{\top},\xi\right)$$

so that both μ and $\zeta = Cv = c\|v\|$ are scalars. Absorbing common factors into the learning rate ε the learning rule is now (using $x = Ag$ by assuming $\xi \to 0$)

$$\frac{1}{\varepsilon}\Delta C = C - 2c^2 g^2 A^{\top} \tag{15.53}$$

so that the update is a combination of the row vectors C and $A^{\top}$. The dynamics is stationary if $C = uA^{\top}$ with a certain value of u determined by the fixed point of the state x.

When using the pseudoinverse given by Eq. (15.51) we get instead $v \propto A$, $\chi \propto C^{\top}$, and $Cv = c\|v\|\cos(C,A)$ so that

$$\frac{1}{\varepsilon}\Delta C = \left(1 - 2c\cos(C,A)\, g^2\right) A^{\top}\,. \tag{15.54}$$

Actually the same update rule is obtained if Eq. (15.52) is used for the pseudoinverse, even though the intermediate steps and the common factors are different. In both cases the update is solely in the direction of $A^{\top}$ so that it is completely defined by the latter. With a small decay term, the C dynamics has a stable equilibrium if $C \propto A^{\top}$, i. e. $c_i \propto a_i$ meaning that sensor channels strongly responding to the motor signals are also strongly contributing to the motor commands.

Discussion

At first glance both rules have the same stationary situation, however, when considering the stability of the fixed point we realize that Eq. (15.54) is stable whereas Eq. (15.53) is unstable in the stationary state. This can be seen from a linear stability analysis of Eq. (15.53). By way of example we consider Eq. (15.53) for $n = 2$ at the stationary point where $C = (c_1, c_2) = u(a_1, a_2)$ with $u = 2c^2 g^2$. Considering

[4] $\cos(a,b)$ denotes the cosine of the angle enclosed by the vectors a and b. It is defined as $\cos(a,b) = \frac{\langle a,b\rangle}{\|a\|\|b\|}$.

a perturbation $\delta C = (\delta c_1, \delta c_2)$ taken to be orthogonal to C so that $C^\top \delta C = 0$ we find that the norm $\|C + \delta C\|^2 = c^2 + O\left(\|\delta C\|^2\right)$ so that in leading order of $\|\delta C\|$ Eq. (15.53) becomes

$$\Delta \delta C = \varepsilon \delta C + O\left(\|\delta C\|^2\right)$$

meaning that the perturbation δC moving C away from the equilibrium point is getting amplified. Note, however, that the norm of the C matrix will not diverge since the length c of the row matrix C is squared in the confinement (second) term. This point is interesting since it shows how the system manages to become more and more sensitive without violating the confinement condition that keeps the neurons out of saturation. The "trick" homeokinetic learning detects is to drive the matrix elements of C to larger values in the direction orthogonal to A so that the postsynaptic potential $z = Cx = CAy$ (ignoring noise) remains finite.

In numerical simulations we confirmed this result. Assuming the following world $x = Ay + \varsigma$ with $A^\top = (0.5, 0.5)$. In the first case Eq. (15.48) (`pseudo=1` in `sox` algorithm) the C vector develops to $C = (d, -d)$ with $d \sim 10$, the exact value depends on the noise strength and the sign depends on the initial conditions. With the second or third version, Eqs. (15.51, 15.52) (`param=3,2`), we obtain indeed $C = (d, d)$ with $d \sim 1.2$ if both channels have the same noise strength. Otherwise the lower the noise the higher the factor and vice-versa. This is different from the case of independent channels, where the channel with more noise gets stronger coupled, see Sect. 5.3.2.

The third variant (15.52), also called *motor-space* version, seems to be more appropriate in the case with more sensors than motors. Note, however, that during learning the matrix A may change a lot due to the fact that the behavior changes and the way A reflects the correlations between motors and sensors is depending on the modes the system gets in. Thus, this version may give the system more robustness but is prone to run into dead locks if A is too far from reality. To avoid the latter see Sect. 15.1.1.2.

15.5 Simplified Algorithms

The broad applicability of the homeokinetic learning algorithms opens many perspectives for applications in high dimensional systems and, in particular, in real robots, giving them a high degree of autonomy. The inversion of the Jacobian matrix is the main obstacle in these goals, also in relating the approach to biological systems. We are now going to present modifications which either avoid inversion altogether or make it at least less time consuming. We start by the most radical simplification of the algorithm.

15.5.1 Avoiding Matrix Inversion

The inversion of the Jacobian matrix is the major discrepancy from the vision to develop a method of self-organization and/or self-regulation that is as close as possible to the biological world. Moreover, the inversion can make trouble when trying to implement the parameter dynamics in electronic hardware or on low performance embedded devices. Fortunately, the generalized pseudoinverse technique suggests an approximation that avoids nontrivial matrix inversions altogether. Consider the standard model where $L = AG'C$ and use the generalized pseudoinverse as (15.51)

$$L^{+} = A\frac{1}{CLA}C = A\frac{1}{RG'R}C$$

where $R = CA$ is an $m \times m$ matrix, m being the number of motors. In our dynamics

$$x_{t+1} = Ay_t + \xi_{t+1} = Ag\left(Cx_t + h\right) + \xi_{t+1} = Ag\left(Ry_{t-1} + C\xi_t + h\right) + \xi_{t+1}$$

we find, ignoring the bias and the nonlinearity, that $y_t \approx Ry_{t-1}$. In this way R connects the motor vectors at two subsequent times. With a realistic update rate, the y_t change only slowly so that R is approximately the unit matrix (times some factor due to the nonlinearity), which is also typically observed in practice. Absorbing factors into the learning rate, we may approximate

$$L^{+} = A\frac{1}{G'}C$$

so that the inversion of L reduces to the trivial problem of inverting the diagonal matrix G'. Then, in the learning rules the vector v and χ Eqs. (15.2, 15.28) may be defined as

$$v = A\frac{1}{G'}C\xi \quad \text{and} \quad \chi = C^{\top}\frac{1}{G'}A^{\top}v,$$

which is a great simplification reducing the time complexity of the learning rule to $O\left(m^2\right)$ as compared to $O\left(m^3\right)$ when really inverting L.

15.5.2 Matrix Inversion by Iteration

Since the inverted matrices are only involved in the vectors v and χ these can be obtained in a different way. We note at first that $v = L^{\top}\chi$ so that actually only χ is to be found. This can be realized iteratively by the update rule

$$\frac{1}{\rho}\Delta\chi = \left(\xi - LL^{\top}\chi\right) - \lambda\chi \tag{15.55}$$

where ρ controls the convergence and λ the degree of regularization. Eq. (15.55) has to be repeated in each time step t until convergence is (essentially) reached.

Upon convergence we have $\Delta\chi = 0$ so that

$$\chi(\lambda) = \frac{1}{LL^\top + \lambda\mathbb{I}}\xi \,. \tag{15.56}$$

If L is regular (no eigenvalues equal to zero), we may take the limit of $\lambda \to 0$ obtaining the desired χ. In general, keeping λ finite yields a regularized solution which is valid even if L is singular. Note also that

$$\lim_{\lambda\to 0} L^\top \chi(\lambda) = L^+\xi$$

where L^+ is the pseudoinverse of L.

Once these vectors are obtained, the update step in the parameter dynamics is realized without any matrix inversion. Otherwise, the latter is not such a big problem in practical applications as it might seem. Thanks to the generalized pseudoinverse techniques, the size of the matrices to be inverted has the dimensions of the number of motors. Thus in our applications with robots of up to 20 independently controlled motors, the normal inversion is absolutely not time critical on a conventional PC running the system in real time. However, replacing the inversion task with the iteration procedure reduces the implementation effort and makes it possible to run the system on an embedded controller board or even realize the system in hardware, for instance on an FPGA. Practical experiences also show that the relevant part of the χ vector is varying slowly so that the number of updates of χ can be kept small with the potential side effect of preventing the system from parameter runaway.

15.6 Motor-Space Approach

In Chap. 5 we introduced the time-loop error (TLE) with single step reconstruction. In the interaction representation, see Chap. 11, this is extended to arbitrarily many steps corresponding to the reconstruction of previous sensor values in a larger time horizon. Here we will discuss an extension to one and a half steps, meaning instead of reconstructing last sensor values we want to reconstruct previous motor values. Interestingly this motor-space approach can be implemented solely by choosing the right pseudoinverse in the normal one-step algorithm, such that this is a nice demonstration of the usefulness of the generalized pseudoinverse concept introduced in Sect. 15.4. It has been observed in many experiments that the motor-space approach is more suitable for controlling systems with more sensors than motors.

15.6.1 Reconstructing Previous Motor Values

In the general formulation of the sensorimotor dynamics we can write, using $\psi(x) = M(x,y)$,

$$x_{t+1} = \psi(x_t) + \xi_{t+1} = \phi(x_t, y_{t-1}, x_{t-1}) + \xi_{t+1} \tag{15.57}$$

where (ignoring the ξ_t term)

$$\phi(x_t, y_{t-1}, x_{t-1}) = M(x_t, K(M(x_{t-1}, y_{t-1}))) \ .$$

In the motor-space approach we want to reconstruct the motor values (y_{t-1}), which can be done in different ways, depending on the role x is given in this process. The most simple way is to keep x unchanged and try the reconstruction by finding the best estimate y^{recon} of y, i. e. minimize the distance $\left\|x_{t+1} - \phi\left(x_t, y_{t-1}^{\text{recon}}, x_{t-1}\right)\right\|$ between the predicted (output of ϕ) and the true sensor vector x_{t+1}. Following the arguments of Sect. 5.1.1 we obtain the reconstruction error in motor space as

$$q_{t-1} = J^{-1}\xi_{t+1}$$

where

$$J = \frac{\partial}{\partial y_{t-1}} \phi(x_t, y_{t-1}, x_{t-1})$$

is the Jacobian matrix of ϕ with respect to y.

Let us consider our standard setting, Eqs. (4.2, 4.3), where $\phi(x_t, y_{t-1}, x_{t-1}) = Ag(CAy_{t-1} + h) + b$ so that the Jacobian matrix is

$$J = AG'CA = AG'R \ . \tag{15.58}$$

The advantage of this formulation is that $R = CA$ is an $m \times m$ matrix which is favorable if the sensors outnumber the motors, so that $n > m$.

Assuming the $m \times m$ matrix $AG'R$ is regular, the TLE in motor space is

$$\begin{aligned} E = q^\top q &= \xi^\top \frac{1}{R^\top G' A^\top} \frac{1}{AG'R} \xi \\ &= \eta^\top \frac{1}{R^\top G'} \frac{1}{G'R} \eta = \zeta^\top \frac{1}{RR^\top} \zeta \end{aligned} \tag{15.59}$$

with the auxiliary vectors $\eta = A^+\xi$ (Moore-Penrose left inverse) and $\zeta = G'^{-1}\eta$.

15.6.2 Learning Rules

Gradient descending the error E yields the explicit learning rules, which stay in structure the same as in Eqs. (5.22, 5.23)

$$\Delta C = \varepsilon_c \mu v^\top - \epsilon y x^\top \tag{15.60}$$

$$\Delta h = -\epsilon y \tag{15.61}$$

but[5] with[6] $\mu = \left(RR^\top\right)^{-1}\zeta$, $q = R^\top \mu$, $R = CA$, and $v = Aq = A\left(CA\right)^{-1}\zeta$. The latter can be understood as the shift in the sensor values obtained from the motor output shift q propagated forward by the model matrix A. As before, the channel dependent learning rate is given in terms of μ and ζ by

$$\epsilon_{ij} = 2\varepsilon_c \delta_{ij} \mu_i \zeta_i \,. \tag{15.62}$$

Note that in the case that the matrix A is square and invertible, the only difference is in the definition of v since then $Aq = C^{-1}\zeta$.

This approach has been extensively tested in [102]. Interestingly, the learning rules have proven to be more favorable in many situations, in particular producing more smooth behaviors and stability against parameter runaway. One possible reason is that, forgetting about the nonlinearities, the matrix $R = CA$ describes the signal flow of the controller outputs over one time step. With any reasonable behavior, this is essentially a unit matrix (also observed in the majority of experiments) so that the inversions featuring in μ and v are well behaved. This may not be the case in the inversion of the C matrix alone, in particular in the $m < n$ case the inversion is different from using the Moore-Penrose pseudoinverse.

15.6.3 Using the Generalized Pseudoinverse

The motor space approach can be used as a further example for the usefulness of the generalized pseudoinverse introduced in Sect. 15.4. In fact, the Eqs. (15.60, 15.61) can be obtained as a special case of the general learning rule (in sensor space) by defining the inversion of $L = AG'C$ as

$$L^+_{A^\top A} = A \frac{1}{A^\top L A} A^\top \tag{15.63}$$

---

5 Derivation: Using $\chi = \left(JJ^\top\right)^{-1}\xi$, $q = J^{-1}\xi$, and $E = \chi^\top \xi$ we get

$$\begin{aligned}
-\frac{1}{2}\frac{\partial}{\partial C_{ij}}\chi^\top \xi &= \chi^\top \frac{\partial J}{\partial C_{ij}} q = \chi^\top A \frac{\partial G'C}{\partial C_{ij}} Aq \\
&= \left(G''A^\top \chi\right)_i (CAq)_i x_j + \left(G'A^\top \chi\right)_i (Aq)_j \\
&= -2\mu_i \zeta_i y_i x_j + \mu_i v_j
\end{aligned}$$

the latter line being obtained for the case that $g(z) = \tanh(z)$ and by using

$$\mu = G'A^\top \chi = \frac{1}{R^\top}\frac{1}{J}\xi = \frac{1}{RR^\top}\zeta \quad \text{and} \quad \zeta = G'^{-1}A^{-1}\xi = CAq \,.$$

6 An alternative definition is $q = J^{-1}\xi$, $\chi = J^{\top -1} q$, $\mu = G'A^\top \chi$ which makes the correspondence to the sensor space rules more transparent.

so that the shift v becomes

$$v = L^+_{A^\top A}\xi = A\frac{1}{A^\top LA}A^\top\xi = A\frac{1}{A^\top AG'CA}A^\top\xi \tag{15.64}$$
$$= A\frac{1}{CA}\zeta\,,$$

where $\zeta = G'^{-1}\eta$ and $\eta = A^+\xi$ with $A^+$is the Moore-Penrose pseudoinverse. Using v of Eq. (15.64) in $E = v^\top v$ yields immediately the motor space rule Eqs. (15.60, 15.61), the only difference being in the fact that all inversions are now directly obtained in the classical sense (since the denominator now consists of a product of $m \times m$ matrices) whereas above we had to invoke the interpretation of A^{-1} as Moore-Penrose inverse. For the implementation in the simulator see Sect. 15.1.1.5.

Appendix 15.A Learning with Generalized Pseudoinverse

In order to justify the use of the sandwiching strategy, introduced in Sect. 15.4, in the general learning rule, we have to consider

$$\frac{1}{2}\frac{\partial}{\partial p}v^\top v = v^\top\frac{\partial}{\partial p}Q\frac{1}{PLQ}P\xi\,.$$

Let us assume for the moment that Q and P are independent of C so that we are left with

$$v^\top Q\left(\frac{\partial}{\partial p}\frac{1}{PLQ}\right)P\xi = -v^\top Q\frac{1}{PLQ}P\left(\frac{\partial}{\partial p}L\right)Q\frac{1}{PLQ}P\xi$$
$$- -\chi^\top\left(\frac{\partial}{\partial p}L\right)v$$

which is the desired result, see Sect. 5.E for hints on the derivative. The general proof is a little less trivial, but, as a hint, in case that all matrices P, L, and Q are invertible, the expressions do not depend on P and Q at all so that any explicit dependency of P and Q on p must cancel out and the above result is recovered. Fortunately, the canceling takes place also for the most central cases considered in the algorithm. In particular, consider the generic case $L = AG'C$ (5.12) with the pseudoinverse chosen as $L^+_{A^\top C^\top}$ so that, see Eq. (15.49),

$$v = C^\top\frac{1}{CC^\top}\zeta = C^+\zeta\,. \tag{15.65}$$

As a fist step, we will now use this form of v and find the derivative of the TLE $E = v^\top v$ directly. Introducing the auxiliary vector

$$\mu = \frac{1}{CC^\top}\zeta \tag{15.66}$$

we find

$$\begin{aligned}\frac{1}{2}\frac{\partial}{\partial C}v^{\top}v &= v^{\top}\left(\frac{\partial}{\partial C}C^{\top}\frac{1}{CC^{\top}}\right)\zeta = \mu v^{\top} + v^{\top}C^{\top}\left(\frac{\partial}{\partial C}\frac{1}{CC^{\top}}\right)\zeta \\ &= \mu v^{\top} - v^{\top}C^{\top}\frac{1}{CC^{\top}}\left(\frac{\partial}{\partial C}CC^{\top}\right)\frac{1}{CC^{\top}}\zeta \\ &= \mu v^{\top} - \mu^{\top}\left(\frac{\partial}{\partial C}CC^{\top}\right)\mu \\ &= \mu v^{\top} - \mu v^{\top} - \mu v^{\top} \\ &= -\mu v^{\top}\end{aligned} \tag{15.67}$$

using

$$v^{\top}C^{\top}\frac{1}{CC^{\top}} = \left(\frac{1}{CC^{\top}}Cv\right)^{\top} = \left(\frac{1}{CC^{\top}}CC^{\top}\frac{1}{CC^{\top}}\zeta\right)^{\top} = \mu^{\top}$$

and $a^{\top}C^{\top}b = b^{\top}Ca$ for the derivatives of the $C^{\top}$ terms.

This equivalence can not be proven for all forms of the generalized pseudoinverse. We propose to deal with the problem in an approximate way by just ignoring the possible dependencies in the P and Q matrices. This is less troublesome as it might seem since the update for the parameters is given by the matrix element of the derivatives of L with respect to the vectors χ and v. Mild changes of these vectors may not reorient the gradient and this is all we need for the gradient descent. Again, this strategy is exact if the matrices P and Q are invertible.

Appendix 15.B Stability of the Gradient Descent

As mentioned already at several points, the high speed of the parameter dynamics is an important ingredient of homeokinetic learning. However, the parameter dynamics is a gradient process so that the learning rate must be controlled appropriately. In order to demonstrate this point let us consider the most simple case of a quadratic error function $E = ac^2$ with gradient dynamics

$$\Delta c = -\varepsilon ac \tag{15.68}$$

and solution

$$c_t = \kappa^t c_0 \tag{15.69}$$

with $\kappa = 1 - \varepsilon a$ meaning monotonous convergence for $0 < \kappa < 1$, convergence with alternating sign for $-1 < \kappa < 0$ and divergence if $\kappa > 1$.

In the TLE, the factor a can loosely be associated with the product of the vector components of μ and v, which may change over orders of magnitude on the time scale of the behavior. Counteracting this effect would mean to change ε all over the time. This is unreasonable in a realistic application. Remedies are the different error norms introduced in Sect. 15.1.1.6 and the following regularization of the nonlinearities.

15.B.1 Regularization of g′ Terms

The jumping of the error over orders of magnitude is caused essentially by the inversion of the matrices. Actually, since we are using inverses in the generalized sense, there is no problem from the strict mathematical point of view. However, the gradient descent is (and has to be) performed with a fixed step width Δt, such that a too steep gradient can lead to instabilities, as sketched in Eq. (15.69). In our system, the problem is that the quantities like μ and v may become very large if one of the matrices in the denominators is close to being singular. This may well happen with the G' matrices, which go to zero if the neurons are in the saturation region.

We will discuss a convenient regularization making the decay of the $g'(z)$ terms sufficiently soft in the tails. We propose in particular, in the tanh case, to use the second order approximation $\tanh'(z) = 1 - z^2 + O(z^4)$ so that

$$\frac{1}{g'(z)} = 1 + z^2 + O(z^4) \tag{15.70}$$

which is exact for infinitesimally small z but additionally guarantees a soft decay of the inverse $1/g'(z)$ in the tails. A somewhat more realistic approximation is obtained when approximating g' in fourth order

$$\frac{1}{g'(z)} = 1 + z^2 + \frac{1}{3}z^4 + O\left(z^6\right) . \tag{15.71}$$

One might argue that the use of such an approximation is inconsistent in view of the fact that, in order to get Eqs. (15.38, 15.39), the G' terms in $\partial L/\partial p$ in Eq. (15.36) have been treated exactly while we use the approximation in the vectors χ, ζ, and v. We take the attitude that the direction of the gradient is essentially given by the $\partial L/\partial p$ term, whereas the auxiliary vectors play more or less the role of weighting factors, which are mostly important for the speed of the parameter dynamics in the different channels. Thus, they may well be relevant for the composed dynamics (system + parameters) but in generic cases the approximations have been found to influence the behavior only marginally.

These regularizations can also be circumvented by using the regularized pseudoinverses, see Sect. 5.G, such that the `sox` algorithm does not regularized the g' terms.

Appendix 15.C Learning Rules for Arbitrary Neuron Types

So far we have given the explicit formula for the parameter dynamics assuming the activation function of the neurons is tanh. This is well suited for the case of the single layer controller and with motor activities ranging essentially from -1 to $+1$. However, this choice may be unfortunate in other applications and also for the hidden layer neurons in multilayer networks, see Sect. 15.3, as is well known from many applications of artificial neural networks. Therefore, we derive here the changes in the rules for different neuron types. The activation functions matter only in the derivation of the G' matrix with matrix elements $G'_{ij} = \delta_{ij} g'_i$ where $g = g(Cx + h)$. In all cases the Jacobian matrix L is of the form $L = \ldots G'C \ldots$ so that we have to consider

$$a^\top \left(\frac{\partial}{\partial C_{ij}} G'C \right) b = \left(G''a\right)_i (Cb)_i x_j + \left(G'a\right)_i b_j \tag{15.72}$$

(see also Sect. 5.F). In order to write this in terms of the two auxiliary vectors $\mu = G'a$ and $\zeta = Cb$, we introduce the matrix ϵ of the channel dependent learning rates by its matrix elements

$$\epsilon_{ij} = -\varepsilon_c \delta_{ij} \frac{g''_i}{g_i g'_i} \mu_i \zeta_i \tag{15.73}$$

writing $y = g(z)$, so that in matrix notation

$$\begin{aligned} a^\top \frac{\partial G'C}{\partial C} b &= -\epsilon y x^\top + \mu v^\top \\ a^\top \frac{\partial G'}{\partial h} b &= -\epsilon y \,. \end{aligned} \tag{15.74}$$

Eq. (15.73) generalizes the tanh case considered in Eqs. (5.24, 5.27). The α parameter can be introduced here as well.

In the case $g(z) = \tanh(z)$ the fraction in Eq. (15.73) is just -2 so that the original expressions are reobtained. In the case of a Fermi function

$$g(z) = \frac{1}{1 + e^{-z}} \tag{15.75}$$

we have $g'_i = (1 - g_i) g_i$ and

$$-\frac{g''(z)}{g'(z)\, g(z)} = 2 - \frac{1}{g(z)} = 1 - e^{-z} \tag{15.76}$$

which changes sign at $z = 0$ so that the term may be interpreted as a Hebbian or anti-Hebbian learning term depending on the state z of the neuron.

In the present version of the algorithms, the tanh and the Fermi function are explicitly available as neuron activation functions.

Appendix 15.D Derivation of the Multilayer Learning Rule

In this section we will derive the learning rules for the multilayer controller introduced in Sect. 15.3 (p. 274). The canonical learning rules (Sect. 5.2) are expressed in terms of the vectors v and χ

$$v = L^{+}\xi \text{ and } \chi = L^{+\top}v \tag{15.77}$$

where L^{+} is a Moore-Penrose pseudoinverse or a generalized pseudoinverse $L^{+} = Q\,(PLQ)^{-1}\,P$, see Sects. 5.G and 15.4. The only difference between the single layer and the multilayer case is the definition of L, see Eq. (15.37). We have to consider the general update for a parameter p (5.16) given by $\chi^{\top}\frac{\partial L}{\partial p}v$ where (two-layer case for simplicity)

$$L = AG_2'C^{(2)}G_1'C^{(1)}\,. \tag{15.78}$$

Let us introduce

$$\mu^{(2)} = G_2'A^{\top}\chi$$

obtained by backpropagating χ back to inside of the upper layer and

$$v^{(1)} = G_1'C^{(1)}v^{(0)}$$

obtained by forward propagating $v = v^{(0)}$ to the inputs of the upper layer. Moreover, we introduce

$$\zeta^{(2)} = C^{(2)}v^{(1)}\,.$$

Consider first the derivative [7] w. r. t. $C^{(2)}$

$$\begin{aligned}
\chi^{\top}\frac{\partial L}{\partial C_{ij}^{(2)}}v &= \chi^{\top}A\left(\frac{\partial}{\partial C_{ij}^{(2)}}G_2'C^{(2)}\right)G_1'C^{(1)}v \qquad (15.79)\\
&= \left(G_2'A^{\top}\chi\right)_i\left(G_1'C^{(1)}v\right)_j + \left(G_2''A^{\top}\chi\right)_i\left(C^{(2)}G_1'C^{(1)}v\right)_i y_j^{(1)}\\
&= \mu_i^{(2)}v_j^{(1)} + \left(G_2''G_2'^{-1}\mu^{(2)}\right)_i\zeta_i^{(2)}y_j^{(1)}\\
&= \mu_i^{(2)}v_j^{(1)} - \epsilon_i^{(2)}y_i^{(2)}y_j^{(1)}
\end{aligned}$$

where

$$\epsilon_i^{(2)} = -\varepsilon_C\frac{g_i^{(2)\prime\prime}}{g_i^{\prime(2)}g_i^{(2)}}\mu_i^{(2)}\zeta_i^{(2)}\,.$$

This is valid for any neuron activation function. In case that g is given by the tanh function, we have in particular

$$\epsilon_i^{(2)} = 2\varepsilon_C\mu_i^{(2)}\zeta_i^{(2)}$$

as it should be.

---

[7] Taking the derivatives follows the general scheme outlined by Eq. (15.72).

Analogously, introducing

$$\mu^{(1)} = G_1' C^{(2)\top} G_2' A^\top \chi$$

we obtain

$$\begin{aligned}
\chi^\top \frac{\partial L}{\partial C_{ij}^{(1)}} v &= \chi^\top A G_2' C^{(2)} \left(\frac{\partial}{\partial C_{ij}^{(1)}} G_1' C^{(1)} \right) v \\
&= \left(G_1' C^{(2)\top} G_2' A^\top \chi \right)_i v_j + \left(G_1'' C^{(2)\top} G_2' A^\top \chi \right)_i \left(C^{(1)} v \right)_i x_j \\
&= \mu_i^{(1)} v_j^{(0)} + \left(G_1'' G_1'^{-1} \mu^{(1)} \right)_i \zeta_i^{(1)} x_j \\
&= \mu_i^{(1)} v_j^{(0)} - \epsilon_i^{(1)} y_i^{(1)} y_j^{(0)}
\end{aligned}$$

where $x = y^{(0)}$, $v = v^{(0)}$ and $\mu_i^{(1)}$ is obtained by propagating χ back until inside the neurons of the lower layer. Moreover,

$$\epsilon_i^{(1)} = -\varepsilon_c \frac{g_i^{(1)\prime\prime}}{g_i^{\prime(1)} g_i^{(1)}} \mu_i^{(1)} \zeta_i^{(1)}$$

$$\zeta^{(1)} = C^{(1)} v^{(0)} .$$

We can create the auxiliary vectors also in an iterative manner by

$$v^{(1)} = G_1' C^{(1)} v^{(0)}$$

$$\mu^{(2)} = G_2' A^\top \chi, \ \mu^{(1)} = G_1' C^{(2)\top} \mu^{(2)} .$$

This are the common formula for the forward and back propagation of the error signals v and χ, respectively. Moreover,

$$\epsilon_i^{(l)} = \varepsilon_c \zeta_i^{(l)} \mu_i^{(l)} = \varepsilon_c \mu_i^{(l)} \left(C^{(l)} v^{(l-1)} \right)_i$$

is the general rule for the channel dependent learning rates in the layers.

15.D.1 Algorithmic Realization

For performance reasons the algorithmic realization uses a row-wise multiplication denoted by $\circ$, which is defined for any two vectors $a, b \in \mathbb{R}^k$ and $(n \times k)$ matrix M as

$$(a \circ b)_i = a_i b_i \quad \text{and} \quad (a \circ M)_{ij} = a_i M_{ij} .$$

A pseudocode for the learning reads as follows.
In each time step do:

1. Feed-forward sweep of the state x_t through the network $\psi(x_t)$ to get the activations of the neurons. Compute for $l = 1, 2$

$$y^{(l)} = g\left(z^{(l)}\right)$$

   with $y^{(0)} = x_t$ and

$$z^{(l)} = C^{(l)} y^{(l-1)} + h^{(l)}$$

   and find the auxiliary vectors

$$g'^{(l)} = g'\left(z^{(l)}\right) .$$

2. Determine L according to Eq. (15.78) and determine the generalized pseudoinverse $L^{+} = Q(PLQ)^{-1} P$, see Sect. 15.4 with conveniently defined matrices P and Q.
3. Determine the vectors $v = L^{+}\xi$ and $\chi = L^{+\top} v$.
4. Create the auxiliary vectors

$$v^{(1)} = g'^{(1)} \circ \left(C^{(1)} v^{(0)}\right), \quad v^{(0)} = v$$

   and

$$\mu^{(2)} = g'^{(2)} \circ \left(A^{\top}\chi\right), \quad \mu^{(1)} = g'^{(1)} \circ \left(C^{(2)\top} \mu^{(2)}\right)$$

   Moreover,

$$\epsilon^{(l)} = \varepsilon_c \mu^{(l)} \circ \left(C^{(l)} v^{(l-1)}\right)$$

5. Update

$$C^{(l)} = C^{(l)} + \varepsilon_c \mu^{(l)} v^{(l-1)\top} - \left(\epsilon^{(l)} \circ y^{(l)}\right) y^{(l-1)\top}$$

$$h^{(l)} = h^{(l)} \quad c^{(l)} \circ y^{(l)}$$

Chapter 16
The LPZROBOTS Simulator

Abstract: In this chapter we describe our robot simulator. We start with the overall structure of the software package containing the controller framework, the physics simulator and external tools. The controller framework makes it very easy to develop and test our algorithms, be it in simulations or with real robots. The physics simulator can handle rigid bodies with fixed geometric representation that are connected by actuated joints. Particular efforts have been undertaken to develop an elaborated treatment of physical object interactions including friction, elasticity, and slip. The chapter also briefly discusses the generation of virtual creatures, the user interface and the most important features of the LPZROBOTS simulation environment.

Realistic computer simulations are very important not only for experimental scientists but also for theoretic studies. They allow one to quickly check hypotheses and algorithms and verify generalizations and approximations that have been done in the course of analytical derivations. This is especially fitting for robotics, where the hardware is normally error-prone and requires rather intense maintenance. However, many argue that robot experiments must be performed with real robots only. This harsh opinion is rooted in the fact that software controllers tested in simulations often have not been able to reproduce the same results in reality. In our case we have a different setting. We do not develop static controllers, but instead aim at the self-organizing of behavior. If self-organization works with complex simulated systems then it will also work in real systems. The control algorithms adapt to the specific hardware during the behavior. In this way the behavior will be quantitatively different on a real robot but the qualitative properties will be the same. Fortunately, the gap between reality and simulation is also shrinking because we can nowadays perform physically realistic computer simulations, as you have seen in the numerous videos and experiments so far.

We developed a simulation software called LPZROBOTS that comes with a realistic physics simulation and is particularly suitable for our research. There are several distinct features as we will discover below. One of them is the support for custom materials and their proper interaction. Let us now take a closer look at the back-stage of our virtual world of playful machines. In the next section we focus on the overall structure of the simulator. Afterwards the framework for developing con-

R. Der and G. Martius: The Playful Machine, COSMOS 15, pp. 293–308.
springerlink.com

trollers is introduced, which is independent of the physics simulator that is described in Sect. 16.4. It follows a list of highlighted features (Sect. 16.5). A comprehensive documentation with technical details and the source code is available on the project website [107].

16.1 Structure

The simulation software is divided into the following parts:

Controller framework: Implementation of the controllers, neural networks, matrix library, introspection, and support functions (SelfOrg).

Physics simulator: Rigid body dynamics and graphical rendering (OdeRobots).

External tools: Visualization tools GuiLogger and MatrixViz.

The control algorithms are located in an independent package (SelfOrg) that can be used in other simulators or with hardware robots. For the development of our algorithms it is important to be able to observe the evolution of internal parameters online and to change some control parameters like learning rates during the runtime.[1] For that reason the framework allows for quick controller development and their flexible connection to robotic systems as well as the ability to connect to external tools. The next section will introduce this framework. Since the software is written in C++ we used the concepts of object-oriented programming. To understand the following text only a few terms need to be known, such as *class* that refers to an object type, *interfaces* which is an abstract object type to specify only the signature (available functions), and *subclass* or *inheritance* for the mechanism to define a more specific class based on an existing one.

16.2 Controller Framework

The SelfOrg framework is designed for connecting a controller to any system, be it a real robot, a simple academic program, or our full-fledged robot simulator. A `controller` is a class that has essentially a single function that receives a vector of sensor values and returns a vector of motor values. The most important part in the framework is the `wired controller`, consisting of a `controller` and a `wiring` paired with some utilities to log, plot and configure the system. The wiring allows for the preprocessing of sensor and motor values, making the connection to different systems very easy. For example a subset of sensor values can be selected or the derivative of a sensor value can be added. Likewise the motor values can be preprocessed, e. g. to protect the robot hardware from dangerous configurations. The

[1] The modification of parameters is only necessary during the test phase. In the robot experiments we report on the parameters are not changed manually, except states otherwise.

wired controller might be directly integrated into another program, e. g. into a real robot control program. Alternatively it might be used within the class agent together with the representation of a robot. A particular robot class can implemented the details needed for the simulation of the robot or it can implemented the communication to a real robot. Figure 16.1 depicts the structure and the information flow within an agent. All parts of the framework are specified using clear interfaces, such that controllers, wirings and robots can be interchanged easily. To make the development of the control algorithms convenient the SELFORG package contains a matrix library that is described in the following.

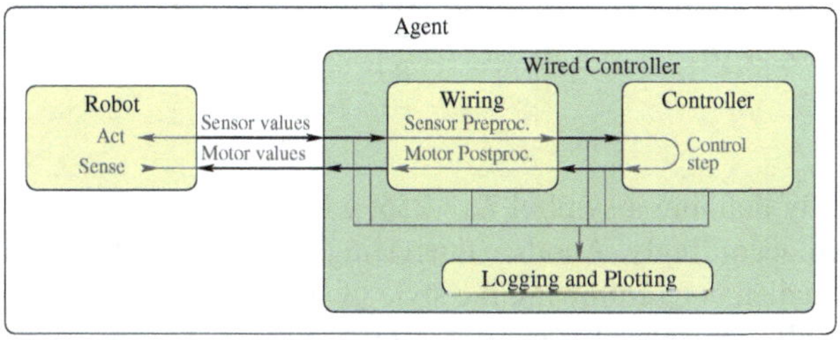

Fig. 16.1: Core architecture of an agent with wired controller and a robot in the sensorimotor loop. The arrows denote the information flow during one simulation step.

16.3 Matrix Library

The matrix library is particularly suitable for the development of our controllers but is nevertheless of general nature. The design is geared to simplicity. Mathematical formulas can be written almost in a one-to-one fashion in the program code. This requires a concise syntax and convenient operators.

The main features of the matrix library are automatic memory management, operator overloading, and pure operations. Vectors are treated as matrices with only one column or one row. For each operation there is a copy version and an in situ version. The latter can be used for optimization purposes, but require careful thinking. The copy operations work like their mathematical counterparts. Their operands are not changed and the result is a new matrix. All operations perform range checks and insure dimension compatibility, which can be globally switched off for performance reasons. Beside the usual arithmetic operations for matrices like $+,-,*$ we introduce multiplication with a scalar, the exponent and many more useful operations as listed in Table 16.1. Special attention should be given to the exponent operator, which, depending on the argument, is either the inversion (-1), the identity matrix (0), the matrix itself (1), the square (2), or the transposed (T). The latter is

Table 16.1: Matrix operations of the matrix library (excerpt).

function name	op.	meaning	description
`C(A)`	=	$C = A$	copy operation
`C.add(A,B)`	+	$C = A + B$	addition
`C.sub(A,B)`	-	$C = A - B$	subtraction
`C.mult(A,B)`	*	$C = A \cdot B$	multiplication
`C.mult(A,f)`	*	$C_{ij} = A_{ij} \cdot f$	scalar multiplication $f \in \mathbb{R}$
`C.exp(A,i)`	^	$C = A^i$	exponent $i \in (-1, 0, 1, 2, T)$
`C=A.pseudoInverse()`		$C = A^+$	pseudoinverse, see Sect. 5.G
`C=A.multrowwise(b)`	&	$C = A \circ b$	rowwise multiplication
`C=A.map(g)`		$C_{ij} = g(A_{ij})$	function application
`C=A.mapP(arg,g)`		$C_{ij} = g(\text{arg}, A_{ij})$	func. appl. with argument
`C=A.above(B)`		$C = \begin{pmatrix} A \\ B \end{pmatrix}$	vertical concatenation

implemented by defining a symbol $T = 255$, which is therefore just a number and can be treated accordingly. Another interesting operation is the function application operator called `map` following the style of functional programming. It applies a function to all elements of a matrix componentwise. For the exponent and the rowwise multiplication we reuse the operators ^ and & that are originally used for bit-wise operations. Somewhat counterintuitive is their operator precedence (order of evaluation), which is lower than for the other arithmetic operations. Since the precedence cannot be changed in C++ more parentheses must be used.

The following illustrative code example shows the generation of a random matrix

```
Matrix M(20,5); // 20 time 5 matrix initialized with 0
M = M.map(random_minusone_to_one)*0.01 // assign random values
```

where `double random_minusone_to_one(double)` returns a random number between -1 and 1. The program code for the equation $y = \tanh(Cx + h)$ is for instance

```
Matrix y = (C * x + h).map(tanh);
```

which looks very similar to the original mathematical formulation and is therefore quickly written and more easily understood.

The performance of the code is still high even though new matrices are created for intermediate results. We use high performance memory operation like `memcpy` and `memzero` to speed up initialization and copy procedures. Additionally, all range checks can be excluded at compile time after the code was successfully tested.

16.4 Physics Simulator

Let us now come to the actual simulator. In order to write a simulation, the user defines the constitutes of the virtual world, e. g. the environment, the obstacles, the agents and so on. Given that, the simulator enters the main loop and performs iteratively physical simulation steps using the "Open Dynamics Engine" (ODE) [154], which we will shortly discussed in the next section. A control step (Fig. 16.1) is performed every n-th iteration (specified by the parameter `controlinterval`). This allows one to select a certain update rate independent of the step size of the physical simulator. The update of the graphical display, which is done using the graphics library "Open Scene Graph" [116], is executed every k-th iteration, where k is calculated to achieve a proper frame rate of 25 fps. In order to obtain a smooth and continuous simulation the internal time of the simulation is synchronized with the real time. Of course, different factors are supported to speed up or slow down. There are also a variety of additional functions, which can be overloaded to have sufficient control over the simulation process, e. g. to define custom keystrokes and to trigger specific events. Let us now take a look behind the curtains to see how the simulated physics works.

16.4.1 Rigid Body Dynamics — Open Dynamics Engine

The physics simulation is based on the rigid body dynamics of the software library "Open Dynamics Engine," short ODE. Rigid body dynamics means that the simulated world consists of objects with a fixed geometric shape (rigid), e. g. spheres, boxes, and so forth, and a mass tensor that matches the inertial properties of a body with that shape. The objects can be connected by different joints, e. g. hinge joints, sliders, ball joints and so forth, see Fig. 16.2. The joints constrain the relative motion of the two connected objects. All object follow the Newtonian laws of physics, namely gravitation, inertial forces, gyroscopic forces and so on. The differential equations are integrated iteratively using a semi-implicit Euler method. In contrast to an explicit Euler method this does not lead to an accumulation of the errors in way that very unphysical effect occur. Nevertheless, it is a first order method and thus not very accurate. In order to improve the accuracy a small simulation time steps can be used, which makes the system slower. Our own experiences show that the simulations run very stable and accurate. For instance a snake robot (Sect. 9.1.4) with 15 joints actuated by motors in space (no gravity) will stay at the same point in space for a long time (hours of simulation time) despite the heavy movement of the joints (induced by motors, see below). Eventually, however it will start to rotate or move away. These errors are fortunately not critical in our experiments.

One of the most important parts in the rigid body simulation is the detection and treatment of collisions in order to simulate the interaction of bodies. The ODE offers routines to check for collisions and proposes a number of so-called contact points. The treatment of collisions is done via a special contact joint, that is places at the contact point and exists only for one time step. The contact joints mimic surface interactions such as friction, elasticity, bouncing and slip. In the following we will have a closer look at the developed strategy for efficient collisions detection and the realization of material and surface properties.

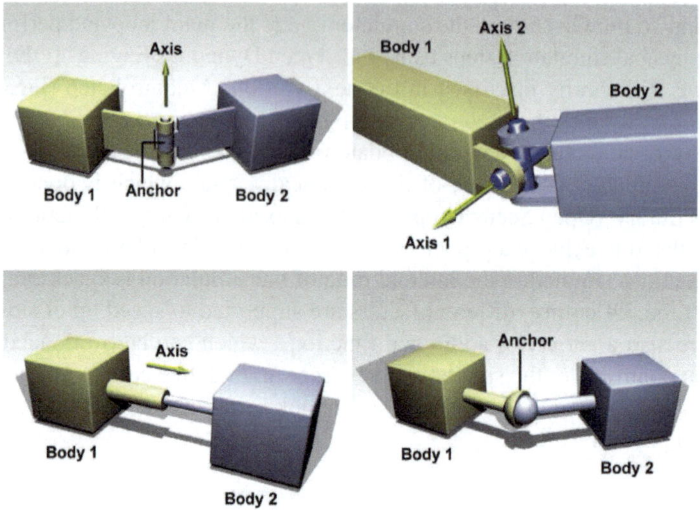

Fig. 16.2: Joints to connect rigid bodies. Hinge joint (upper left), universal joint (upper right), slider joint (lower left), and ball joint (lower right). Images taken from [154].

16.4.2 Efficient Collision Detection

To make collision detection practically computable also in larger systems ODE uses so called *collision spaces*, that group a number of preferably close geometric bodies together. Thus, robots usually have their own collision space. Collision detection is first performed on the level of collision spaces using their bounding box, i. e. checking for the intersection of the smallest cubes containing all bodies of a collision space. Only in the case of an intersection of the bounding boxes the geometric bodies of the two collision spaces must be pairwise tested. Additional collision tests within each space have to be performed as well.

Since not all collision spaces are supposed to treat internal collisions, there is a list of spaces to be checked. Further and more importantly not all geometric bodies are supposed to collide with each other. For instance, bodies connected with joints should typically not collide since they intersect by construction. To exclude any pair of bodies we introduced a hash set, which is checked for each potential collision.

16.4.3 Material and Surface Properties

In order to model complex scenarios the collision treatment of the ODE needs to be augmented. Normally collisions are treated in a global callback function where the two colliding geometric bodies are given. In order to distinguish between dif-

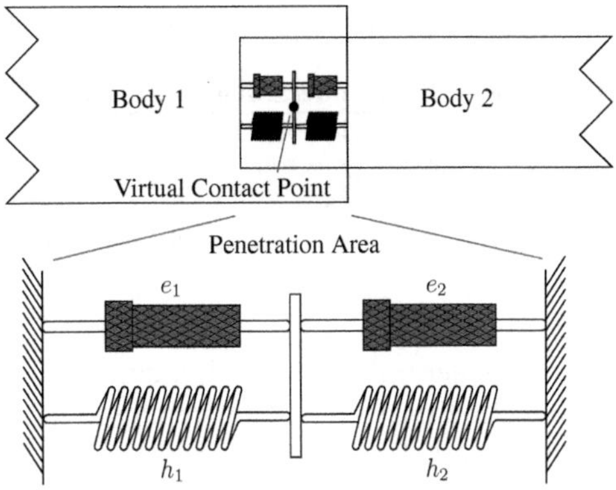

Fig. 16.3: Substance-interaction model with two serial spring-damper elements. For each body we have the elasticity *e* (damping) and the hardness *h* (spring constant) forming a spring-damper element.

ferent physical interactions, each geometric body object carries a `substance`[2] object. The interaction parameters are thus to be obtained through the combination of the two `substances`. We consider four different parameters k_p, k_d, μ, and *slip* to describe the interaction. Here k_p and k_d denote the spring constant and damping constant respectively, μ is the Coulomb friction constant, and *slip* is the force dependent slip (FDS) parameter. FDS is used to model non-coulomb friction that occurs for example when the wheels of a car starts to slide sideways. More formally, the two contacting surfaces slide past each other with a velocity proportional to the force that is being applied tangentially to the surface. This differs from normal (Coulomb) frictional effects since it does not cause a constant acceleration, but rather leads to a steady velocity.

In order to obtain these parameters from two substances we need a suitable description of the physical properties and a way to combine them. Our design for the substance parameters are roughness (r), slip (s), hardness (h), and elasticity (e). The Coulomb friction parameter μ is obtained by a multiplication of the roughness from both substances. This results in a high friction for two rough materials but in low friction, if one of the materials is very smooth (e. g. ice). The *slip* parameter is the sum of both slip parameters. The spring and damping constants are calculated using the schema of two spring-damper elements serially connected as illustrated in Fig. 16.3.

The spring constant of each collision side is given by the hardness h_1 and h_2. The spring constant k_p of the entire system is given by

$$\frac{1}{k_p} = \frac{1}{h_1} + \frac{1}{h_2}. \qquad (16.1)$$

[2] The name *substance* was chosen due to the fact that the possibly better fitting term, *material*, is already used by the graphics renderer to describe visual surface properties.

Table 16.2: Substance and interaction parameters.

Parameter	Range	Interaction Parameter	
roughness (r)	$[0,\infty)$	$\mu = r_1 \cdot r_2$	
slip (s)	$[0,\infty)$	$slip = s_1 + s_2$	
hardness (h)	$(0,\infty)$	$k_p = \frac{h_1 \cdot h_2}{h_1+h_2}$	(16.1)
elasticity (e)	$[0,1]$	$k_d = \frac{h_2(1-e_1)+h_1(1-e_2)}{h_1+h_2}$	(16.2)

The damping constant k_d is derived from the elasticities e_i of the substances, but is more difficult to compute. Considering the damping in the form of energy loss we can write the energy or work done by each spring as: $W_i = F \cdot x_i$, where x_i is the length by which the spring is compressed from the relaxation position (no collision). The new length of the spring is not known but we can calculate it from the force using $F = h_i x_i$, such that we obtain $W_i = \frac{F^2}{h_i}$. The energy loss through damping is $W_i^D = W_i(1-e_i)$. The work of the serially coupled system is given by the sum of the works. The final damping is now the relative loss of work:

$$k_d = \frac{W_1^D + W_2^D}{W_1 + W_2} = \frac{F^2(1-e_1)/h_1 + F^2(1-e_2)/h_2}{F^2/h_1 + F^2/h_2} = \frac{h_2(1-e_1) + h_1(1-e_2)}{h_1 + h_2} \tag{16.2}$$

Table 16.2 summarizes the parameters and their dependencies.

Now one needs to convert k_p and k_d to the parameters used by the ODE. There are two parameters that come directly from the regularization of the constraint equation and from the joint alignment procedure. According to the manual [154] the two parameters can be obtained as follows:

$$\mathrm{ERP} = \frac{\Delta t k_p}{\Delta t k_p + k_d} \qquad \mathrm{CFM} = \frac{1}{\Delta t k_p + k_d}$$

The above described approach works very well to mimic the interaction of very different materials. The simulator comes with a set of standard substances, see Ta-

Table 16.3: Standard substances implemented in LpzRobots. Each substance has only one free parameter. The interval of accepted values is given in the table. The values are defined ad hoc.

Material	roughness	hardness	elasticity	slip
Metal	$\in [0.1,1]$	200	0.80	0.01
Plastic	$\in [0.5,1]$	40	0.50	0.01
Rubber	3	$\in [5,50]$	0.95	0.00
Foam	2	$\in [1,30]$	0.00	0.10

Video 16.1: Illustration of the interaction of different materials. All spheres have the same size and mass. The substances are foam (yellow), rubber (dark brown), plastic (white), and metal (silver), see Table 16.3. Note the different penetrations (second picture) and bouncing heights. The video can be watched at `http://playfulmachines.com`.

ble 16.3, nevertheless custom substances are also possible. To give an impression the Video 16.1 shows the bouncing of balls made of the four standard substances.

In order to support also special cases the `substance` class has an optional callback function, which can overwrite the default collision treatment. For instance when a material has different properties in different directions, like the skin of a snake. Another example are the infrared distance sensors that are implemented with ray-objects and their sensor value is obtained from collision routines, however no collision treatment is required. Another example are contact or tactile force sensors.

16.4.4 Motors and Sensors

The joints connecting the rigid bodies can be actuated by different motors. The motors are connected to the joints and apply torques there. There are three types of motors, angular motors, PID servos, velocity controlled servos which are described in the following:

Angular motor: It controls the desired angular velocity of the joint and has a maximal torque it can apply. This velocity is realized in a single simulation time step, given the maximal torque is not exceeded. Internally this motor is implemented as a constraint in the equation of the dynamics, such that it virtually looks ahead in time to take into account all forces that are applied to that joint in the current simulation time step. In this way they are very stable and do not lead to jitter, especially in multi-joint systems. By the way, an electric motor with a strong gearbox behaves in a similar way.

PID servo: A classical PID (proportional-integral-derivative) controlled servo motor. The motor values provides the nominal position ($\in [-1,1]$) of the joint and the applied force/torque is calculated with the following equation:

$$f_t = P\left(e_t + D\dot{e}_t + I\sum_{\tau=0}^{t} e_\tau\right) \tag{16.3}$$

where e is the position error (nominal position minus current position), $\dot{e}$ is its temporal derivative and P,D,I are the gain factors. We use deliberately the factor P to scale the entire force, such that the total power of the motor can be adjusted. The remaining factors D and I are now relative and easier to adjust. The factor D scales the derivative, such that large values avoid overshooting, but also make the servo slower. For this reason it is also called the damping parameter. The integral factor I is to cancel offsets due to e. g. constant external forces. All parameters have to be tuned empirically. In a multi-joint system it is very difficult to tune them in order to obtain a smooth behavior. For example the damping term cannot be increased too much because of the discrete nature of the simulation.

Velocity controlled servo: These servos combine the stability of the angular motors with the position control of the PID servos. Hence the motor values specify the nominal position of the joint from which a velocity is computed as

$$v_t = Ve_t \tag{16.4}$$

where V is the maximal velocity of the servo. In this way the angular velocity of the joint is smaller, the smaller the position error, which in turn avoids overshooting. Since the system is iterated discretely V is not allowed to be too large, i. e. $V_{\max} = |e_t|/\Delta t$. The maximal force the motor can apply in the current time step is given by

$$f_t = F\tanh(1/C + |e_t|) \tag{16.5}$$

where F is the maximal torque of motor in general and $C > 0$ is a compliance term. For small C the motor has its full power independent of the position error. This leads to a stiff system without compliance such that only little feedback from the physical system will be visible in the sensor values. In this configuration the velocity controlled servo is a good model of a traditional electrical servo motor. On the other hand if C is large, the motor has very little power at the target point, such that it behaves very compliant. Forces that act on the joint will be visible in small deflections from the target point, much like an elastic system of muscles and tendons. The Dynamixel servo motors of Robotis [146] can be configured to have a maximal torque and provides also a position sensor, such that Eq. (16.5) can be implemented in hardware.

The velocity controlled servos are very useful to build compliant machines, which allow the controller to develop a "feeling" of the body. Another actuator is a `speaker` that can emit sound signals at different frequencies.

Let us now consider the available sensors. The following list starts with the proprioceptive sensors (measuring the internal state) and continues with exteroceptive sensors that sense the external world.

Joint position: Each joint can supply the current joint position and angular velocity. These sensors provide the most important feedback to the controller.

Orientation: An orientation sensor measures the orientation of a part of the body. There are different representations available: The full orientation matrix (3×3), the vector of one internal axis, or the projection of the internal axes onto the z-coordinate of the global coordinate system (z axis looks upwards).

Velocity: The velocity sensor measures the translational and rotational velocity of a part of the body.

Infrared: Active infrared sensors are popular in robotics to measure the distance to objects. They emit a light beam in the infrared spectrum and measure the amount of reflectance, which is proportional to the distance. In the simulation these sensors are implemented with ray objects that virtually collide with other objects, such that the distance can be computed. The sensor has a maximal range where the sensor value is 0. For close objects the sensor value is 1 and the characteristics in between can be selected to be linear or non-linear. In order to get a visual feedback of the measurement we render the sensor with a colored line.

Microphone: The microphone can detect sound signals emitted by a speaker. Different quantities can be measured: Frequency, amplitude, and direction of the sound source. The implementation is rudimentary and does not take into account reflections or damping of sound by other objects.

Camera: The simulator also supports a virtual camera. A picture is captured each time a control step is performed independent of the frame rate of the graphical window. The resolution and the field of view can be configured. Additionally a chain of preprocessors can be defined, such as color filters, motion detection and so on.

16.4.5 User Interaction

The user interaction with the simulator has different aspects. Operations concerning the graphical representation of the scene, e. g. camera position, display style or video recording are accessible through the graphical window. The camera can be manipulated with the mouse in combination with the mouse buttons in different modes, which are shortly introduced in Sect. 16.5. Also external forces can be applied to the robots using the mouse together with the <Ctrl> key. A transparent head-up display shows the simulation time and the simulation speed with respect to real time. All available keystrokes can be displayed on demand at a help screen. The second way of interacting with the simulator is via a console on the terminal window. It provided an interactive console where parameters can be set and for instance controllers can be saved. The console features a history, auto-completion and many more characteristics of a UNIX shell. The interface was intentionally uncoupled from the graphics in order to be usable in real robot experiments or non-graphical applications. Both interfaces are depicted in Figure 16.4. Finally, the user can display internal parameters online, such as sensor values and synaptic weights with different custom tools like our GuiLogger and MatrixViz as displayed in Figure 16.5 and 16.6.

(a)

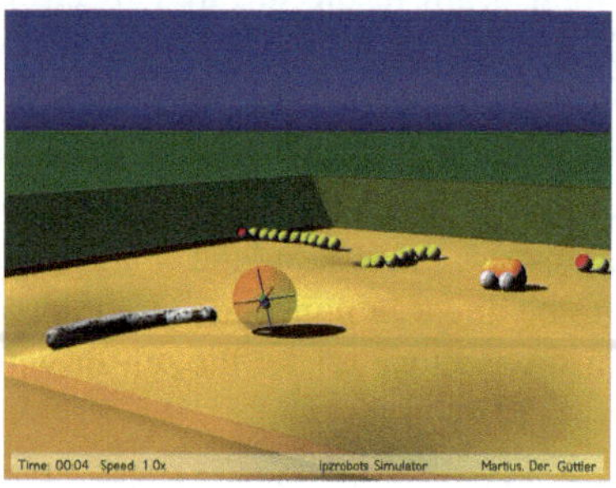

(b)

```
Welcome to LpzRobots
-----------------------------------------------------
...
[Simulation Environment]
 simstepsize=          0.010000  stepsize of the ...
 gravity=             -9.810000  strength of gravity
...
[SineController]
 period=              50         period of oscillation
 phaseshift=           1         phaseshift of channels
...
######
Type: ? for help or press TAB
>
? help  list  load  ls quit  set  show  store  view
> ls
Agents ---------------(for store and load)
ID: Name
 1: Barrel (Controller and Robot)
Objects --------------(for set and show)
ID: Name
  1: Simulation Environment
  2: SineController
  3: Barrel
> period=10
 period = 10.0000
```

Fig. 16.4: User interface of the LpzRobots simulator. **(a)** Graphical simulation window; **(b)** Terminal with console interface.

16.4.6 Creating the Virtual World

This section will give an overview of how to build up virtual worlds in LpzRobots.

While designing the simulator we had to combine physical, geometrical and graphical representations of the objects in the virtual world into one structure. This structure is called `primitive` and can have all of these properties. There are also cases where no physical body or geometric representations are required or wanted. For example, static objects in the world, like walls or the floor, do not need a mass and impulse because they are considered to be unmovable. Likewise a weight in-

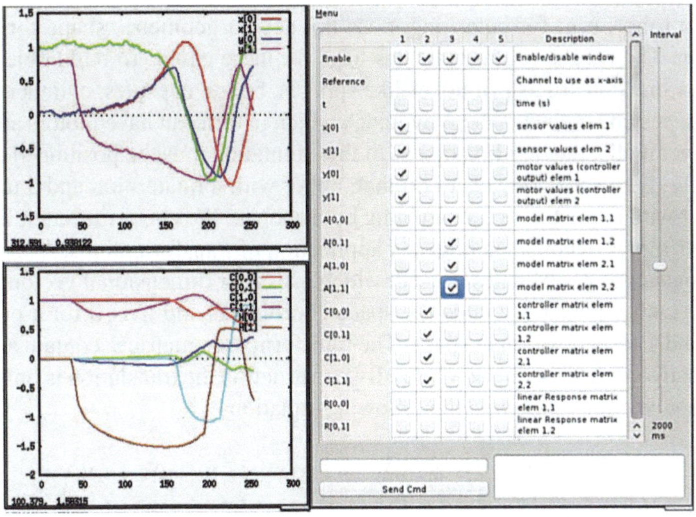

Fig. 16.5: GUILOGGER with two controlled Gnuplot windows. In the main window (**right**) a set of channels is selected. Their temporal evolution is shown in the subwindows (**left**), here sensor values and motor values, and synaptic weight of the controller.

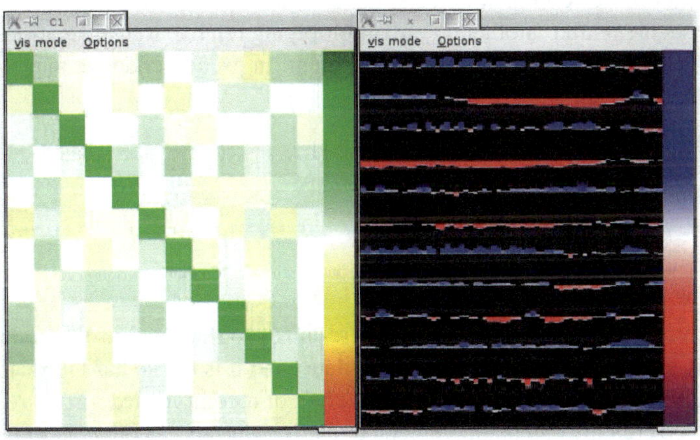

Fig. 16.6: Visualization tool MATRIXVIZ. The (**left**) window displays the entries of a 12×12 controller matrix ($C1$) and the (**right**) window shows the time evolution of the sensor value vector x.

side of a robot, e. g. for balancing, does not need a geometric shape for collision detection. The `primitive` class has flags for these cases. To build objects in the simulator one constructs primitives like spheres, boxes, capsules, cylinders or combined shapes and connects them by joints, which in turn can have motors attached to them. Eventually, the building blocks of the simulation must be positioned correctly. Generally there are two ways to do that, either with a quaternion and a translation vector or with a 4×4 matrix containing both at once. We chose the latter, because it is much simpler in concatenation and application of transformations. A special case of homogeneous coordinates is used, which uses four dimensional vectors containing (x, y, z, w) where x, y, z code the space coordinates and w is 0 for a orientation vector and 1 for a position in space. The transformation matrices contain a 3×3 rotation matrix and a translation vector. By construction the translation is only applied to position vectors as seen on the following equation

$$\begin{pmatrix} x & y & z & w \end{pmatrix} \begin{pmatrix} r_{11} & r_{12} & r_{13} & 0 \\ r_{21} & r_{22} & r_{23} & 0 \\ r_{31} & r_{32} & r_{33} & 0 \\ t_x & t_y & t_z & 1 \end{pmatrix} = \begin{pmatrix} r_{11}x + r_{21}y + r_{31}z + t_x w \\ r_{12}x + r_{22}y + r_{32}z + t_y w \\ r_{13}x + r_{23}y + r_{33}z + t_z w \\ w \end{pmatrix}^{\top} . \tag{16.6}$$

The matrix representation enables to concatenate transformations via simple matrix multiplication. For complex objects like a multi-segment arm one can recursively add one segment after another by only multiplying relative transformations with the transformation matrix of the previous segment. In pseudo code we may write for a linear chain of segments:

$m \leftarrow$globalPose
createSegmentAt(m)
for all $l \in$ localTransformations **do**
  $m \leftarrow l * m$
  createSegmentAt(m)
end for

The robots and other complex objects are thus programmed in the code directly. This may sound cumbersome at the first glance but it is not so hard if local transformations are used. In this way the construction of a complicated virtual robot like a dog-shaped creature can be implemented relatively easily. Another aspect is that the programmed robot is much more flexible than one created with a graphical editor. For example the legs of a robot can be generates in a for-loop and a parameter can determine the number of legs. Many of the robots in our simulator have a set of parameters to configure them to the needs of the experiment.

To set up complicated wall formations it is of course very useful to have a graphical editor. We make use of the 2D-CAD program `xfig` [153], with which the walls can be drawn. The program also supports a depth value for each line, which is used for the height of the walls. Figure 16.7(b) shows an environment created in this way.

16.5 Highlights

Apart from the above mentioned features the simulator has many interesting properties, some of which will be listed in the following.

Flexibility: The robots and controllers can be flexibly connected, such that a controller can be tested on many robots. The simulations are programmed in C++ such that customizations are easily implemented, for instance keystrokes, changes in the environment or certain scheduled events. New components, like new robots, controller, servos or the like can also be added by the user in its own simulation.

Simulation speed: The speed of the simulation can be controlled by a single factor, determining the ratio between simulation time and real time. The simulation is synchronized with the clock so that a homogeneous time flow is achieved independent of the currently available CPU resources, provided that they are sufficient. This synchronization can also be turned off in order to run at maximal speed. Depending on the complexity of the simulation the speedup is typically above five and up a few hundred fold on a nowadays PC.

Camera control: The user can operate the camera in different modes and control its movements with the mouse. The modes include a static camera that can be swayed and tilted up and down and moved in all directions. Another mode is the *following* mode that automatically moves the position of the camera relative to a selected robot. The *TV* mode points the view of the camera towards a selected robot. All camera movements are additionally smoothed to avoid unnatural jumps for details see [59].

Video recording: To capture the simulation in a movie, the user can at any time start and stop the recording of a video. This might slow down the simulation speed, however the video is kept at the right rate.

Multi-threading: Since computers nowadays have several processing cores it is important to parallelize the execution of code. The simulator can optionally run the graphics, physics, and controllers in separate threads. To be able to reproduce results, the handling of random numbers was modified to cope with the parallel processing.

Playgrounds: In order to create an appropriate environment for robots there are several obstacles and rectangular and round arenas available. Additionally, a customized wall configuration can be created with the common 2D-CAD program `xfig` [153] and imported into the simulator, see Figure 16.7.

Terrains: For more challenging environments, a terrain using a height-map bitmap can be created, see Figure 16.4.

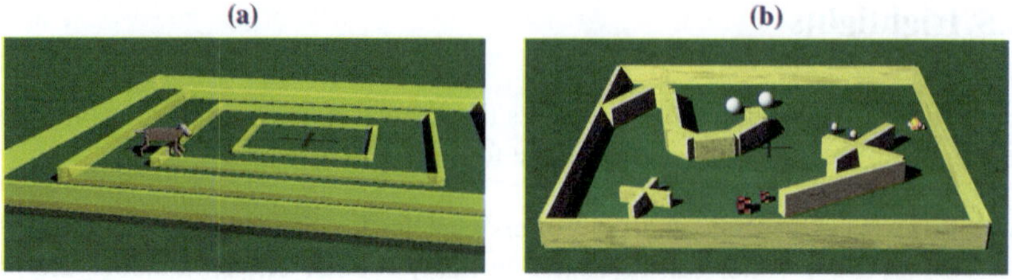

Fig. 16.7: Different Environments with obstacles. **(a)** A set of rectangular arenas stacked into each other (simple for-loop); **(b)** A wall configuration drawn with `xfig`.

16.6 Summary

This chapter provided a brief overview of the robot simulator LpzRobots, featuring much more than could be highlighted here. Complex virtual robots and environments can be set up in a short amount of time and efficient simulations are possible. We developed a material interaction model that allows the user to specify intuitive material properties and therewith simulate more realistic simulations.

16.6.1 Credits

The design and the major part of the implementation was conducted by one of the authors (G. M.). Further credits and thanks go to Frank Güttler, Frank Hesse, Marcel Kretschmann, Dominique Schneider, Jörn Hoffmann, Jörg Weide, and Antonia Siegert.

Chapter 17
Discussion and Perspectives

This book started with a question. Now, more than 300 pages and, if you were eager, nearly 30 experiments later, let us try to draw a balance. Have we solved the problem of getting robots into controlled activities without telling them what to do, without giving a task, a goal, or any other external pressure for development? The answer to this question is a clear yes. We have formulated a general principle—homeokinesis—that was seen to provide, in a natural and unbiased way, the desired drive to activity while still "keeping things under control."

The second part of our original question—what may the emerging behaviors look like—is of a more difficult nature. If there is nothing prescribed from outside, can we expect any systematics at all in the behaviors? And in what way, if any at all, are behaviors related to the physical dynamics of the body in its environment?

Results

In order to find answers we started with the investigation of simple one- and two-dimensional systems. The results are very clear there. In the most simple cases, we find that homeokinetic learning drives the agent into a specific working regime[1] where it is actively taking decisions while still being very sensitive to the influence of the environment. If the controller is enlarged in its functionality, we observed the emergence of a searching behavior that can be described as a systematic self-exploration of the behavior space. This has many interesting consequences such as the emerging cooperation in high-dimensional systems with decentralized homeokinetic control (chain of wheeled robots and ARMBAND). When connected to a complex real or virtual robot, homeokinetic learning was observed in many experiments to recognize and excite the natural modes of the system in a rather straightforward way—both the BARREL and the SPHERICAL were found to self-explore the variety of their rolling modes in a seemingly systematic way, the SLINGING SNAKE was surprisingly driven by its two-motor head into a slinging whole-body motion, and the SPHERICAL adapted its rolling patterns to the geometry of the environment just by its rolling activities. In a similar way, embodiment-related motion patterns

[1] At the so-called **effective** bifurcation point.

R. Der and G. Martius: The Playful Machine, COSMOS 15, pp. 309–312.
springerlink.com

were also excited in several hardware machines like the SEMNI and the ROCKING STAMPER with different and partly strange sensor-motor combinations. In those examples homeokinetic learning revealed another of its astonishing properties—the emergence of a tight sensorimotor coordination that can be traced back to the sensitization effect.

Another feature is the element of surprise, hinting at a certain creativity, evident for instance in the case of the BARREL by the emergence of the precession mode, a new mode that is latent in the physical system, waiting so to say for getting recognized and being excited by the controller. This is a particularly nice example of what we call body-inspired control. In a sense, we observe how the "brain" learns to "feel" the physics of the body and how to excite its most fundamental modes by getting into a kind of functional resonance with the body.

Importantly, the emerging self-exploration is not just a scanning of the full parameter space but is characterized by a systematic, self-induced restriction to the subspace of embodiment related behaviors. Moreover, this emerging dimension reduction can be traced back to the symmetries and to the physical properties of the body.

Can we expect similar results in the high-dimensional systems—the actual target group of our high-flying ambitions? The picture is less clear in this case but it reveals already several of its facets. In principle there is no difference between low-dimensional and high-dimensional systems, but the characteristics are more obscured because of the dramatically increased complexity. Nevertheless, the emergence of specific, low-complexity motion patterns is quite obvious. The emerging dimension reduction has many far reaching consequences. Among them is the observed scalability of the homeokinesis approach—although we are dealing with a state-parameter space of several hundred dimensions without any prestructuring, all examples run on a normal PC in real time.

The examples chosen for investigation and proposed as experiments to the reader comprise various robots ranging from dog-like, to snake-like up to humanoid robots in different situations. The aim of the experiments is to understand how the controller learns to "feel" the specific physical properties of the body in its environment. These very high-dimensional systems do not lend themselves easily to analytical considerations so that we have chosen physical conditions that might help to better bring out the characteristics of the process. Examples of that kind of scaffolding are suspending the HUMANOID like a bungee jumper, the RHOENRAD, or the high bar. Interestingly, in all situations the robots develop whole-body motion patterns that seemingly are related to the specific environmental situation: the DOG starts playing with a barrier eventually jumping or climbing over it; the SNAKE develops coiling and jumping modes; we observe emerging climbing behaviors of a HUMANOID seemingly trying to get out of a pit; and witness wrestling like scenarios if a HUMANOID is encountering a companion. Eventually, in our robotic zoo all kinds of robots are brought together so that homeokinesis can prove its robustness against heavy interactions with other robots or dynamical objects.

While being many and varied and often unexpected, the emerging low-dimensional modes are not at all random but are controlled in a definite way in our tightly coupled sensorimotor loop so that they are reproducible under similar conditions. Identified and stored away with a context tag, these modes may be used as behavioral primitives in more complex behavioral architectures. Being long-lived but transient in nature, these modes may emerge and decay in a loose succession so that the homeokinetic system realizes a self-exploration of its dimension-reduced behavior space much in the same way as was made explicit in the low-dimensional systems.

How Can This be Useful?

Another of our ambitions is to be useful in real world robotics where task orientation prevails and self-organization still seems to be at a rather weak position. We have already given several possibilities how homeokinetic learning can become immediately useful, for instance in the self-rescue scenarios and by its inherent creativity to find new solutions in strange and unexpected situations. Another point is an impact on the control of highly compliant machines with their inherent whole-body involvement. In such systems, the emerging dimension reduction of homeokinetic learning might be of direct help for developing novel control strategies. A related point of immediate interest is the use of homeokinetic learning as an exploration strategy for reinforcement learning. The established reduction of the search space to the physically relevant subregions may help to escape the curse of dimensionality that jeopardizes any conventional search strategy.

On a more general level, we support a new direction of research, guided self-organization, treating three different scenarios in the book. The idea is to gently orient the self-organization process towards desired directions, much in the sense of a liberal child-raising. Guided self-organization is still in its infancy but, in the considered scenarios, we could show how the integration of external learning signals into the self-organization process can be done without perturbing the subtle features of the latter. The examples treated were reaching from direct prescriptions for the motor signals to the integration of unspecific reinforcement signals to directly hinting of how to break a certain symmetry, considering both low- and high-dimensional systems. The results are encouraging and suggest the use of homeokinesis with guidance as a kind of generator for basic sensorimotor patterns that may form building blocks in behavior-based, task-oriented robotics or in reinforcement learning.

Most of the applications done so far are restricted to simulated robots, but the most ambitious goal is the application to real robots, considered as the ultimate "acid test" for the method. The real robots applications (SEMNI, real BARREL, and others) done so far have all been restricted to the case of two motors so that the full complexity of the method could not been tested yet. But the perspectives are clear. Like the story told in the SEMNI case, the important advantage of our approach is its from scratch philosophy. Take the homeokinetic brain, connect it to the sensors and motors of any real robot (see cookbook for the details) and look what happens. While interesting results may not be expected if the robot is very rigid, just carrying out the prescribed motor patterns in a faithful way, it will generate most vivid and often surprising behaviors if the robot is underactuated and/or compliant to external influences and the intrarobot interactions. While a nightmare to classical methods of behavior generation, the uncertainties and contingencies prevailing in these systems are well in agreement with the basic working principle of homeokinesis. In fact, homeokinesis is feeding on the differences between the target and the true response of the body to its intended actions, in this way getting the information about the physical reactions of the body.

Speculative Ideas

Robots as helpmates of humans have to fulfill specific tasks. As it turned out over the decades of active robotics research, the way towards this goal is long, tedious, and full of obstacles. This book proposes a detour: let us forget for a while the concentration on special properties or skills like reaching, grasping, walking, jumping, or swimming. Instead, give the robots more autonomy, letting them unfold their behaviors in a playful exploration of their potential motion patterns outside of that narrow focus. Goal orientation will be introduced only in a second step by methods we also have addressed in this book.

Speculatively, the detour may even be the faster way to the goal, as is often the case. Taking the detour also may provide a fresh view on robotics, extending the expectation horizon from pure goal orientation to the more playful behaviors seen in young children playing, acrobatics, or modern dancing to give a few examples. Instead of making the robot a slave to our intentions, the new freedom for individual development given to the robot, letting it find out by itself about the best symbiosis of body, brain and environment, can also create a new, more friendly relationship between robots and humans.

Eventually, we want to speculate about the relation to biology. Homeokinesis has been demonstrated to provide an internal drive for the unfolding of behavior without any external pressure. Interestingly, a homeokinetic system is realized very simply. In the cases considered and tested in this book, all you need are two neural networks of very simple structure plus some standard backpropagation and backprojection procedures to provide the signals for a (kind of) Hebbian learning process. There is nothing in these mechanisms that should not be realizable in natural systems, in higher animals at least. Our question: could nature have discovered such a structure in evolution? If so, this might help to better understand the internal mechanisms underlying the drives for playing of animals and young children. This will be difficult to decide in biology but the robotic experiments can help to find the right questions and possibly also give some prototype patterns that could be matched against biological findings.

List of Figures

List of Videos

List of Experiments

References

[1] Aberdeen, D.: POMDPs and policy gradients. In: Proc. of the Machine Learning Summer School (MLSS), Canberra, Australia (2006)

[2] Abraham, R.H., Shaw, C.D.: Dynamics, The Geometry of Behaviour. Addison-Wesley (1992)

[3] Adami, C.: Introduction to artificial life. Springer, Heidelberg (1998)

[4] Amari, S.: Natural gradients work efficiently in learning. Neural Computation 10 (1998)

[5] Arnold, L.: Random Dynamical Systems, Corr. 2nd printing. Springer, Heidelberg (1998)

[6] Ashby, W.R.: The homeostat. Electronic Engineering 20, 380–388 (1948)

[7] Ashby, W.R.: The electronic brain. Radio Electronics (1949)

[8] Ashby, W.R.: Design for a Brain. Chapman and Hall, London (1954)

[9] Ay, N., Bernigau, H., Der, R., Prokopenko, M.: Information driven self-organization: The dynamical systems approach to autonomous robot behavior. Theory Biosci. (to appear, 2011)

[10] Ay, N., Bertschinger, N., Der, R., Güttler, F., Olbrich, E.: Predictive information and explorative behavior of autonomous robots. The European Physical Journal B - Condensed Matter and Complex Systems 63(3), 329–339 (2008)

[11] Bak, P., Tang, C., Wiesenfeld, K.: Self-organized criticality: an explanation of $1/f$ noise. Physical Review Letters 59, 381–384 (1987)

[12] Bayindir, L., Şahin, E.: A review of studies in swarm robotics. Turkish Journal of Electrical Engineering (2007)

[13] Beer, R.D.: A dynamical systems perspective on autonomous agents. Tech. Rep., Artificial Intelligence (1992)

[14] Beggs, J., Plenz, D.: Neuronal avalanches in neocortical circuits. Journal of Neuroscience 23(11), 167–177 (2003)

[15] Beni, G.: From Swarm Intelligence to Swarm Robotics. In: Şahin, E., Spears, W. (eds.) Swarm Robotics: State-of-the-art Survey. LNCS, Springer, Heidelberg (2000)

[16] Bereiter, C.: Towards a solution to the learning paradox. Review of Educational Research 55(2), 201–226 (1985)

[17] Bertschinger, N., Olbrich, E., Ay, N., Jost, J.: Information and closure in systems theory. In: Artmann, S., Dittrich, P. (eds.) Explorations in the Complexity of Possible Life. Proceedings of the 7th German Workshop of Artificial Life, pp. 9–21. IOS Press BV, Amsterdam (2006)

[18] Bertschinger, N., Olbrich, E., Ay, N., Jost, J.: Autonomy: An information theoretic perspective. Biosystems 91(2), 331–345 (2008)

[19] Bialek, W., Nemenman, I., Tishby, N.: Predictability, complexity and learning. Neural Computation 13, 2409 (2001)

[20] Bonabeau, E., Dorigo, M., Theraulaz, G.: Swarm intelligence: from natural to artificial systems. Oxford University Press, Inc., New York (1999)

[21] Bongard, J.C., Zykov, V., Lipson, H.: Resilient machines through continuous self-modeling. Science 314, 1118–1121 (2006); doi:10.1126/science.1133687

[22] Braitenberg, V.: Vehicles: Experiments in Synthetic Psychology. MIT Press (1984)

[23] Braun, C., Heinz, U., Schweizer, R., Wiech, K., Birbaumer, N., Topka, H.: Dynamic organization of the somatosensory cortex induced by motor activity. Brain 124(11), 2259–2267 (2001); doi:10.1093/brain/124.11.2259

[24] Brembs, B.: Towards a scientific concept of free will as a biological trait: spontaneous actions and decision-making in invertebrates. Proc. R. Soc. B 278, 930–939 (2011)

[25] Brooks, R.A.: A robust layered control system for a mobile robot. In: Tech. rep., Massachusetts Institute of Technology, Cambridge, MA, USA (1985)

[26] Brooks, R.A.: Intelligence without representation. Artificial Intelligence Journal 47, 139–159 (1991)

[27] Camazine, S., Franks, N.R., Sneyd, J., Bonabeau, E., Deneubourg, J.L., Theraula, G.: Self-Organization in Biological Systems. Princeton University Press, Princeton (2001)

[28] Cannon, W.B.: The Wisdom of the Body. Norton, New York (1939)

[29] Chemova, S., Veloso, M.: An evolutionary approach to gait learning for four-legged robots. In: Proc. IEEE IROS 2004, vol. 3, pp. 2562–2567 (2004)

[30] Chikazumi, S., Charap, S.H.: Physics of Magnetism. Krieger Pub. Co. (1978)

[31] Choi, J., Wehrspohn, R.B., Gösele, U.: Mechanism of guided self organization producing quasi-monodomain porous alumina. Electrochimica Acta 50(13), 2591–2595 (2005), doi:10.1016/j.electacta.2004.11.004

[32] Collins, S.H., Ruina, A.: A bipedal walking robot with efficient and human-like gait. In: IEEE Conf. on Robotics and Automation, ICRA 2005, pp. 1983–1988. IEEE Press (2006)

[33] Cybenko, G.: Approximations by superpositions of sigmoidal functions. Mathematics of Control, Signals, and Systems 2, 303–314 (1989)

[34] Dautenhahn, K., Nehaniv, C.L.: Imitation in animals and artifacts. MIT Press, Cambridge (2002)

[35] Davidson, P.R., Wolpert, D.M.: Widespread access to predictive models in the motor system: a short review. Journal of Neural Engineering 2(3), S313–S319 (2005)

[36] Dayan, P., Abbott, L.F.: Theoretical Neuroscience: Computational and Mathematical Modeling of Neural Systems. The MIT Press (2005)

[37] Deneubourg, J.L., Goss, S.: Collective patterns and decision making. Ethology, Ecology and Evolution 1(4), 295–311 (1989)

[38] Deneubourg, J.L., Theraulaz, G., Beckers, R.: Swarm made architectures. In: Bourgine, P., Varela, E. (eds.) Proceedings of the 1st European Conference on Artificial Life, pp. 123–133. MIT Press (1992)

[39] Der, R.: Self-organized acquisition of situated behaviors. Theory Biosci. 120, 179–187 (2001)

[40] Der, R., Güttler, F., Ay, N.: Predictive information and emergent cooperativity in a chain of mobile robots. In: Bullock, S., Noble, J., Watson, R., Bedau, M.A. (eds.) Proc. Artificial Life XI, pp. 166–172. MIT Press (2008)

[41] Der, R., Hesse, F., Martius, G.: Rocking stamper and jumping snake from a dynamical system approach to artificial life. Adaptive Behavior 14(2), 105–115 (2006); doi:10.1177/105971230601400202

[42] Der, R., Liebscher, R.: True autonomy from self-organized adaptivity. In: Proc. Workshop Biologically Inspired Robotics, Bristol (2002)

[43] Der, R., Martius, G.: Actions: Emerging Self-exploration in a Dynamical Systems Approach to Early Robot Development. In: Nolfi, S., Baldassarre, G., Calabretta, R., Hallam, J.C.T., Marocco, D., Meyer, J.-A., Miglino, O., Parisi, D. (eds.) SAB 2006. LNCS (LNAI), vol. 4095, pp. 406–421. Springer, Heidelberg (2006)

[44] Der, R., Martius, G., Hesse, F.: Let it roll – emerging sensorimotor coordination in a spherical robot. In: Rocha, L.M., Yaeger, L.S., Bedau, M.A., Floreano, D., Goldstone, R.L., Vespignani, A. (eds.) Proc. Artificial Life X, pp. 192–198. Intl. Society for Artificial Life, MIT Press (2006)

[45] Deutscher, J., Blake, A., Reid, I.: Articulated body motion capture by annealed particle filtering. In: IEEE Conf. on Computer Vision and Pattern Recognition, pp. 126–133 (2000)

[46] Dongyong, Y., Jingping, J., Yuzo, Y.: Distal supervised learning control and its application to CSTR systems. In: SICE 2000, Proc. of the 39th SICE Annual Conference, pp. 209–214 (2000); doi:10.1109/SICE.2000.889681

[47] Doya, K.: Reinforcement learning in continuous time and space. Neural Computation 12(1), 219–245 (2000); doi:10.1162/089976600300015961

[48] Elman, J.L.: Finding structure in time. Cognitive Science 14, 179–211 (1990)

[49] Ernst, U.A., Pawelzik, K.R., Sahar-Pikielny, C., Tsodyks, M.V.: Intracortical origin of visual maps. Nature Neurosci. 4, 431–436 (2001)

[50] Feldman, D.E.: Synapses, scaling and homeostasis in vivo. Nature Neuroscience 5, 712–714 (2002)

[51] Fenichel, N.: Geometric singular perturbation theory for ordinary differential equations. Journal Differential Equations 31, 53–98 (1979)

[52] Floreano, D., Nolfi, S.: God save the red queen! competition in co-evolutionary robotics. In: Genetic Programming 1997: Proceedings of the Second Annual Conference, vol. 5, pp. 398–406. Morgan Kaufmann (1997)

[53] Fodor, J.A.: Fixation of belief and concept acquisition. In: Piatelli-Palmerini, M. (ed.) Language and learning: The debate between Jean Piaget and Noam Chomsky, pp. 142–149. Harvard University Press, Cambridge (1980)

[54] Frégnac, Y.: Homeostasis or synaptic plasticity. Nature 391, 845–846 (1998)

[55] Gardiner, C.W.: Handbool of Stochastic Methods, 2nd edn. Springer, Heidelberg (1990)

[56] von Glasersfeld, E.: Scheme theory as a key to the learning paradox. In: Tryphon, A., Vonèche, J. (eds.) Working with Piaget: Essays in Honour of Bärbel Inhelder, pp. 139–146. Psychology Press, London (2001)

[57] Gong, D., Yan, J., Zuo, G.: A review of gait optimization based on evolutionary computation. Applied Computational Intelligence and Soft Computing (2010); doi:10.1155/2010/413179

[58] Gordon, J.W., Smith, J.O.: A sine generation algorithm for vlsi applications. In: Proc. of Int. Computer Music Conf., pp. 165–168 (1985)

[59] Güttler, F.: Realitätsnahe simulationsumgebung einer selbstorganisierenden roboterwelt. Master's thesis, University Leipzig (2007), `http://lips.informatik.uni-leipzig.de/pub/2007-8`

[60] Haken, H.: Synergetics: An Introduction. Nonequilibrium Phase Transition and Self-Organization in Physics, Chemistry, and Biology, 3rd edn. Springer, Heidelberg (1983); revised and enlarged edition

[61] Haken, H.: Advanced Synergetics. Springer, Berlin (1987)

[62] Haken, H.: Synergetics: Introduction and Advanced Topics. Springer, Heidelberg (2004)

[63] Haken, H.: Information and Self-Organization. A Macroscopic Approach to Complex Systems. Springer Series in Synergetics. Springer, Heidelberg (2006)

[64] Haschke, R.: Bifurcations in discrete-time neural networks – controlling complex network behaviour with inputs. Ph.D. thesis, Bielefeld University (2003)

[65] Haschke, R., Steil, J.J.: Input space bifurcation manifolds of recurrent neural networks. Neurocomputing 64C, 25–38 (2005)

[66] Hebb, D.O.: The Organization of Behavior: A Neuropsychological Theory. new edn. Wiley, New York (1949)

[67] Herrmann, M., Holicki, M., Der, R.: On Ashby's homeostat: A formal model of adaptive regulation. In: Schaal, S. (ed.) From Animals to Animats, pp. 324–333. MIT Press (2004)

[68] Hesse, F.: Self-organizing control for autonomous robots. Ph.D. thesis, University of Göttingen, Institute for Nonlinear Dynamics (2009)

[69] Hesse, F., Martius, G., Der, R., Herrmann, J.M.: A sensor-based learning algorithm for the self-organization of robot behavior. Algorithms 2(1), 398–409 (2009)

[70] Hild, M.: Neurodynamische Module zur Bewegungssteuerung autonomer mobiler Roboter. Ph.D. thesis, Humboldt-Universität zu Berlin, Mathematisch-Naturwissenschaftliche Fakultät II (2007)

[71] Hild, M.: Labor für Neurorobotik (2011), `http://www.neurorobotik.de/index_de.php`

[72] Hild, M., Kubisch, M.: Self-Exploration of Autonomous Robots Using Attractor-Based Behavior Control and ABC-Learning. In: Proceedings of the 11th Scandinavian Conference on Artificial Intelligence, SCAI 2011 (2011)

[73] Hornby, G.S., Fujita, M., Takamura, S., Yamamoto, T., Hanagata, O.: Autonomous evolution of gaits with the sony quadruped robot. In: Proceedings of 1999 Genetic and Evolutionary Computation Conference GECCO, pp. 1297–1304. Morgan Kaufmann (1999)

[74] Hülse, M., Pasemann, F.: Dynamical neural schmitt trigger for robot control. In: Dorronsoro, J.R. (ed.) ICANN 2002. LNCS, vol. 2415, pp. 142–783. Springer, Heidelberg (2002); doi:10.1007/3-540-46084-5_127

[75] Hülse, M., Wischmann, S., Manoonpong, P., von Twickel, A., Pasemann, F.: Dynamical systems in the sensorimotor loop: On the interrelation between internal and external mechanisms of evolved robot behavior. In: Lungarella et al. [96], pp. 186–195

[76] Iida, F., Dravid, R., Paul, C.: Design and control of a pendulum driven hopping robot. In: IEEE/RSJ International Conference on Intelligent Robots and Systems 2002, vol. 3, pp. 2141–2146 (2002); doi:10.1109/IRDS.2002.1041584

[77] Iida, F., Pfeifer, R.: Self-stabilization and behavioral diversity of embodied adaptive locomotion. In: Iida, F., Pfeifer, R., Steels, L., Kuniyoshi, Y. (eds.) Embodied Artificial Intelligence. LNCS, vol. 3139, pp. 619–629. Springer, Heidelberg (2004), doi:10.1007/978-3-540-27833-7_9

[78] Ijspeert, A., Nakanishi, J., Schaal, S.: Movement imitation with nonlinear dynamical systems in humanoid robots. In: Robotics and Automation, IEEE International Conference on Proceedings of ICRA 2002, vol. 2, pp. 1398–1403 (2002); doi:10.1109/ROBOT.2002.1014739

[79] Ijspeert, A.J., Hallam, J., Willshaw, D.: Evolving Swimming Controllers for a Simulated Lamprey with Inspiration from Neurobiology. Adaptive Behavior 7(2), 151–172 (1999), doi:10.1177/105971239900700202

[80] Ikegami, T., Suzuki, K.: From a homeostatic to a homeodynamic self. Biosystems 91, 388–400 (2008)

[81] Jaeger, H., Christaller, T.: Dual dynamics: Designing behavior systems for autonomous robots. Artificial Life and Robotics 2, 76–79 (1997)

[82] Jordan, M.I., Rumelhart, D.E.: Forward models: Supervised learning with a distal teacher. Cognitive Science 16(3), 307–354 (1992)

[83] Jung, T., Polani, D., Stone, P.: Empowerment for continuous agent-environment systems. Adaptive Behaviour (2010) (accepted)

[84] Kanamaru, T.: Van der pol oscillator. Scholarpedia 2(1), 2202 (2007)

[85] Kaplan, F., Oudeyer, P.-y.: Maximizing learning progress: An internal reward system for development. In: Iida, F., Pfeifer, R., Steels, L., Kuniyoshi, Y. (eds.) Embodied Artificial Intelligence. LNCS (LNAI), vol. 3139, pp. 629–629. Springer, Heidelberg (2004)

[86] Klaassen, B., Zahedi, K., Pasemann, F.: A modular approach to construction and control of walking robots. In: Robotik 2004, VDI-Berichte, vol. 1841, pp. 633–640 (2004)

[87] Klyubin, A.S., Polani, D., Nehaniv, C.L.: Empowerment: a universal agent-centric measure of control. In: Congress on Evolutionary Computation, pp. 128–135 (2005)

[88] Kober, J., Peters, J.: Policy search for motor primitives in robotics. In: Koller, D., Schuurmans, D., Bengio, Y., Bottou, L. (eds.) Twenty-Second Annual Conference on Neural Information Processing Systems, Curran, Red Hook, NY, USA, pp. 849–856 (2009)

[89] Kohonen, T.: Self-organization and associative memory. Springer-Verlag New York, Inc., New York (1989)

[90] Kuniyoshi, Y., Sangawa, S.: Early motor development from partially ordered neural-body dynamics: experiments with a cortico-spinal-musculo-skeletal model. Biological Cybernetics 95(6), 589–605 (2006); doi:10.1007/s00422-006-0127-z

[91] Kuniyoshi, Y., Yorozu, Y., Ohmura, Y., Terada, K., Otani, T., Nagakubo, A., Yamamoto, T.: From humanoid embodiment to theory of mind. In: Iida, F., Pfeifer, R., Steels, L., Kuniyoshi, Y. (eds.) Embodied Artificial Intelligence. LNCS (LNAI), vol. 3139, pp. 202–218. Springer, Heidelberg (2004)

[92] Kuznetsov, Y.A.: Elements of Applied Bifurcation Theory. Springer, Heidelberg (2004)

[93] Langton, C.G.: Artificial Life: An Overview. MIT Press, Cambridge (1995)

[94] Levina, A., Herrmann, J.M., Geisel, T.: Dynamical synapses causing self-organized criticality in neural networks. Nature Physics 3, 857–860 (2007); doi:10.1038/nphys758

[95] Liu, R.T., Liaw, S.S., Maini, P.K.: Two-stage turing model for generating pigment patterns on the leopard and the jaguar. Physical Review E 74(1), 011914 (2006), doi:10.1103/PhysRevE.74.011914

[96] Lungarella, M., Iida, F., Bongard, J.C., Pfeifer, R. (eds.): 50 Years of Aritficial Intelligence. LNCS (LNAI), vol. 4850. Springer, Heidelberg (2007)

[97] Lungarella, M., Metta, G., Pfeifer, R., Sandini, G.: Developmental robotics: a survey. Con nect. Sci. 15(4), 151–190 (2003)

[98] Lungarella, M., Pegors, T., Bulwinkle, D., Sporns, O.: Methods for quantifying the informational structure of sensory and motor data. Neuroinformatics 3(3), 243–262 (2005)

[99] Manoonpong, P.: Neural Preprocessing and Control of Reactive Walking Machines Towards Versatile Artificial Perception-Action Systems. In: Cognitive Technologies, Springer, Heidelberg (2007)

[100] de Margerie, E., Mouret, J.B., Doncieux, S., Meyer, J.A.: Artificial evolution of the morphology and kinematics in a flapping-wing mini UAV. Bioinspiration and Biomimetics 2, 65–82 (2007)

[101] Markelić, I., Zahedi, K.: An evolved neural network for fast quadrupedal locomotion. In: Xie, M., Dubowsky, S. (eds.) Advances in Climbing and Walking Robots, Proceedings of 10th International Conference (CLAWAR 2007), pp. 65–72. World Scientific Publishing Company (2007)

[102] Martius, G.: Goal-oriented control of self-organizing behavior in autonomous robots. Ph.D. thesis, Georg-August-Universität Göttingen (2010), `http://resolver.sub.uni-goettingen.de/purl/?webdoc-2459`

[103] Martius, G., Der, R.: Simulation software for The Playful Machine (2011), `http://playfulmachines.com/book/software`

[104] Martius, G., Herrmann, J.M.: Taming the beast: Guided self-organization of behavior in autonomous robots. In: Doncieux, S., Girard, B., Guillot, A., Hallam, J., Meyer, J.-A., Mouret, J.-B. (eds.) From Animals to Animats 11. LNCS, vol. 6226, pp. 50–61. Springer, Heidelberg (2010), doi:10.1007/978-3-642-15193-4_5

[105] Martius, G., Herrmann, J.M.: Tipping the scales: Guidance and intrinsically motivated behavior. In: Advances in Artificial Life, ECAL 2011, pp. 506–513. MIT Press (2011)

[106] Martius, G., Herrmann, J.M., Der, R.: Guided self-organisation for autonomous robot development. In: Almeida, F., Almeida e Costa, F., Rocha, L., Costa, E., Harvey, I., Coutinho, A. (eds.) Advances in Artificial Life 9th European Conference, ECAL 2007. LNCS, vol. 4648, pp. 766–775. Springer, Heidelberg (2007)

[107] Martius, G., Hesse, F., Güttler, F., Der, R.: LpzRobots: A free and powerful robot simulator (2010), `http://robot.informatik.uni-leipzig.de/software`

[108] Maslow, A.H.: Motivation and personality. Harper, New York (1954)

[109] Maturana, H., Varela, F.: Autopoiesis and cognition: the realization of the living. Boston studies in the philosophy of science. D. Reidel Pub. Co. (1980)

[110] Mazzapioda, M. G., Nolfi, S.: Synchronization and Gait Adaptation in Evolving Hexapod Robots. In: Nolfi, S., Baldassarre, G., Calabretta, R., Hallam, J.C.T., Marocco, D., Meyer, J.-A., Miglino, O., Parisi, D. (eds.) SAB 2006. LNCS (LNAI), vol. 4095, pp. 113–125. Springer, Heidelberg (2006)

[111] McGeer, T.: Passive dynamic walking. Int. Journal of Robotics Research 9(2), 62–82 (1990); doi:10.1177/027836499000900206

[112] Newman, M.E.J.: Power laws, Pareto distributions and Zipf's law. Contemporary Physics 46, 323 (2005); doi:10.1080/00107510500052444

[113] Nolfi, S., Floreano, D.: Evolutionary Robotics. The Biology, Intelligence, and Technology of Self-organizing Machines. MIT Press, Cambridge (2001)

[114] Nolfi, S., Parisi, D., Elman, J.L.: Learning and evolution in neural networks. Adaptive Behavior 3(1), 5–28 (1994); doi:10.1177/105971239400300102

[115] Nouyan, S., Campo, A., Dorigo, M.: Path formation in a robot swarm. Swarm Intelligence 2, 1–23 (2008); doi:10.1007/s11721-007-0009-6

[116] Open Scene Graph: Open source high performance 3D graphics toolkit (2008), `http://www.openscenegraph.org`

[117] Oudeyer, P.Y., Kaplan, F., Hafner, V.: Intrinsic motivation systems for autonomous mental development. IEEE Transactions on Evolutionary Computation 11(2), 265–286 (2007)

[118] di Paolo, E.: Homeostatic adaptation to inversion of the visual field and other sensorimotor disruptions. In: Meyer, J.A., Berthoz, A., Floreano, D., Roitblat, H., Wilson, S.W. (eds.) From Animals to Animats, SAB 2000, pp. 440–449. MIT Press (2000)

[119] di Paolo, E.: Organismically-inspired robotics: Homeostatic adaptation and natural teleology beyond the closed sensorimotor loop. In: Murase, K., Asakura, T. (eds.) Dynamical Systems Approach to Embodiment and Sociality, pp. 19–42. Advanced Knowledge International, Adelaide (2003)

[120] Pasemann, F.: Discrete dynamics of two neuron networks. Open Systems & Information Dynamics 2, 49–66 (1993)

[121] Pasemann, F.: Dynamics of a single model neuron. Intl. Journal of Bifurcation and Chaos 2, 271–278 (1993)

[122] Pasemann, F.: A simple chaotic neuron. Phys. D 104, 205–211 (1997)

[123] Pasemann, F.: Complex dynamics and the structure of small neural networks. Network: Computation in Neural Systems, 195–216 (2002)

[124] Pasemann, F., Hild, M., Zahedi, K.: SO(2)-networks as neural oscillators. In: Mira, J., Álvarez, J.R. (eds.) IWANN 2003. LNCS, vol. 2686, pp. 144–151. Springer, Heidelberg (2003)

[125] Paul, C.: Morphology and computation. In: Proc. Int. Conf. on Simulation of Adaptive Behavior, pp. 33–38. MIT Press (2004)

[126] PC Watch, http://pc.watch.impress.co.jp (2003)

[127] Pearl, J.: Causality. Cambridge University Press (2000)

[128] Pearson, K., Gordon, J.: Spinal reflexes. In: Kandel, E., Schwartz, J.H., Jessell, T.M. (eds.) Principles of Neural Science, 4th edn., pp. 713–736. McGraw-Hill, New York (2000)

[129] Peters, J., Schaal, S.: Natural Actor-Critic. Neurocomputing 71(7-9), 1180–1190 (2008)

[130] Peters, J., Vijayakumar, S., Schaal, S.: Natural Actor-Critic. In: Gama, J., Camacho, R., Brazdil, P.B., Jorge, A.M., Torgo, L. (eds.) ECML 2005. LNCS (LNAI), vol. 3720, pp. 280–291. Springer, Heidelberg (2005)

[131] Pfeifer, R., Bongard, J.C.: How the Body Shapes the Way We Think: A New View of Intelligence. MIT Press, Cambridge (2006)

[132] Pfeifer, R., Gómez, G.: Morphological Computation – Connecting Brain, Body, and Environment. In: Sendhoff, B., Körner, E., Sporns, O., Ritter, H., Doya, K. (eds.) Creating Brain-Like Intelligence. LNCS, vol. 5436, pp. 66–83. Springer, Heidelberg (2009), doi:10.1007/978-3-642-00616-6_5

[133] Pfeifer, R., Iida, F., Gómez, G.: Morphological computation for adaptive behavior and cognition. International Congress Series, vol. 1291, pp. 22–29 (2006); doi:10.1016/j.ics.2005.12.080

[134] Pfeifer, R., Lungarella, M., Iida, F.: Self-organization, embodiment, and biologically inspired robotics. Science 318, 1088–1093 (2007)

[135] Pfeifer, R., Lungarella, M., Sporns, O., Kuniyoshi, Y.: On the information theoretic implications of embodiment – principles and methods. In: Lungarella, et al. [96], pp. 76–86

[136] Pfeifer, R., Scheier, C.: Understanding intelligence. MIT Press, Boston (1999)

[137] Pisella, L., Mattingley, J.B.: The contribution of spatial remapping impairments to unilateral visual neglect. Neuroscience and Biobehavioral Reviews 28, 181–200 (2004)

[138] Pitti, A., Lungarella, M., Kuniyoshi, Y.: Exploration of natural dynamics through resonance and chaos. In: Arai, T., Pfeifer, R., Balch, T.R., Yokoi, H. (eds.) IAS, pp. 558–565. IOS Press (2006)

[139] Polani, D.: Foundations and formalizations of self-organization. In: Prokopenko, M. (ed.) Advances in Applied Self-organizing Systems, pp. 19–37. Springer, Heidelberg (2008)

[140] Polani, D., Sporns, O., Lungarella, M.: How information and embodiment shape intelligent information processing. In: Lungarella, et al.[96], pp. 99–111

[141] Popp, J.: Spherical robots (2004), http://www.sphericalrobots.com

[142] Prokopenko, M.: Design vs self-organization. In: Prokopenko, M. (ed.) Advances in Applied Self-organizing Systems, pp. 3–17. Springer, Heidelberg (2008)

[143] Prokopenko, M.: Guided self-organization. HFSP Journal 3(5), 287–289 (2009)

[144] Reynolds, C.W.: Flocks, herds, and schools: A distributed behavioral model. In: Proc. Computer Graphics (SIGGRAPH 1987), vol. 21(4), pp. 25–34 (1987)

[145] Rittscher, J., Blake, A., Hoogs, A., Stein, G.: Mathematical modelling of animate and intentional motion. Phil. Trans. R. Soc. Lond. B, 475–490 (2003)

[146] Robotis Ltd.: Robotis web page (2009), http://www.robotis.com

[147] Rodriguez, A.: Guided self-organizing particle systems for basic problem solving. Ph.D. thesis, University of Maryland (College Park, Md., USA) (2007)

[148] Salge, C., Polani, D.: Information-driven organization of visual receptive fields. Advances in Complex Systems 12(3), 311–326 (2009)

[149] Schaal, S., Ijspeert, A., Billard, A.: Computational approaches to motor learning by imitation, vol. 1431, pp. 199–218. Oxford university press (2004)

[150] Schmidhuber, J.: A possibility for implementing curiosity and boredom in model-building neural controllers. In: Proceedings of the First International Conference on Simulation of Adaptive Behavior, pp. 222–227. MIT Press, Cambridge (1990)

[151] Schmidhuber, J.: Driven by compression progress: A simple principle explains essential aspects of subjective beauty, novelty, surprise, interestingness, attention, curiosity, creativity, art, science, music, jokes. Anticipatory Behavior in Adaptive Learning Systems, 48–76 (2009)

[152] Schöner, G., Dose, M., Engels, C.: Dynamics of behavior: Theory and applications for autonomous robot architectures. Robotics and Autonomous Systems 16, 213–245 (1995)

[153] Smith, B.V.: Xfig drawing program for the X windows system (2008), http://www.xfig.org

[154] Smith, R.: Open Dynamics Engine – open source, high performance library for simulating rigid body dynamics (2008), http://ode.org

[155] Sporns, O., Lungarella, M.: Evolving coordinated behavior by maximizing information structure. In: Rocha, L.M., Yaeger, L.S., Bedau, M.A., Floreano, D., Goldstone, R.L., Vespignani, A. (eds.) Proc. Artificial Life X, pp. 323–329. Intl. Society for Artificial Life. MIT Press, Bradford Books (2006)

[156] Stanley, K.O., D'Ambrosio, D.B., Gauci, J.: A hypercube-based encoding for evolving large-scale neural networks. Artif. Life 15, 185–212 (2009); doi:10.1162/artl.2009.15.2.15202

[157] Steels, L.: The autotelic principle. In: Iida, F., Pfeifer, R., Steels, L., Kuniyoshi, Y. (eds.) Embodied Artificial Intelligence. LNCS (LNAI), vol. 3139, pp. 231–242. Springer, Heidelberg (2004)

[158] Steffe, L.P.: The learning paradox: A plausible counterexample. In: Steffe, L.P. (ed.) Epistomological Foundations of Mathematical Experience, pp. 26–44. Springer, New York (1991)

[159] Steingrube, S., Timme, M., Woergoetter, F., Manoonpong, P.: Self-organized adaptation of a simple neural circuit enables complex robot behaviour. Nature Physics 6, 224–230 (2010)

[160] Steinhage, A., Schöner, G.: Self-calibration based on invariant view recognition: Dynamic approach to navigation. Robotics and Autonomous Systems 20, 133–156 (1997)

[161] Stewart, B.H., Thompson, J.M.: Nonlinear Dynamics and Chaos. John Wiley and Sons (2002)

[162] Still, S., Precup, D.: An information-theoretic approach to curiosity-driven reinforcement learning. Theory in Biosciences (to appear, 2011)

[163] Stitt, S., Zheng, Y.F.: Distal learning applied to biped robots. In: Proc. of the IEEE Intl. Conf. on Robotics and Automation, pp. 137–142. IEEE Computer Society (1994)

[164] Storck, J., Hochreiter, S., Schmidhuber, J.: Reinforcement driven information acquisition in non-deterministic environments. In: Proceedings of the International Conference on Artificial Neural Networks, pp. 159–164 (1995)

[165] Strogatz, S.: Nonlinear Dynamics and Chaos. Addison-Wesley, New York (1994)

[166] Sutton, R.S., Barton, A.G.: Reinforcement learning: Past, present and future. In: SEAL, pp. 195–197 (1998)

[167] Suykens, J.A.K., Vandewalle, J.P.L., Moor, B.L.R.D.: Artificial neural networks for modelling and control of non-linear systems. Springer, New York (1996)

[168] Svennebring, J., Koenig, S.: Building terrain-covering ant robots: A feasibility study. Auton. Robots 16(3), 313–332 (2004); doi:10.1023/B:AURO.0000025793.46961.f6

[169] The Robot Studio sarl, (2011), http://www.therobotstudio.com

[170] Theodorou, E., Buchli, J., Schaal, S.: A generalized path integral control approach to reinforcement learning. Journal of Machine Learning Research 11, 3137–3181 (2010)

[171] Tononi, G., Cirelli, C.: Sleep function and synaptic homeostasis. Sleep medicine reviews 10, 49–62 (2006)

[172] Trianni, V., Dorigo, M.: Self-organisation and communication in groups of simulated and physical robots. Biological Cybernetics 95, 213–231 (2006); doi:10.1007/s00422-006-0080-x

[173] Turing, A.M.: The chemical basis of morphogenesis. Phil. Trans. Roy. Soc. 237, 37 (1952)

[174] Turrigiano, G.G.: Homeostatic plasticity in neuronal networks: The more things change, the more they stay the same. Trends in Neuroscience 22, 221–228 (1999)

[175] Turrigiano, G.G., Nelson, S.: Hebb and homeostasis in neuronal plasticity. Current Opinion in Neurobiology 10, 358–364 (2000)

[176] Turrigiano, G.G., Nelson, S.B.: Homeostatic plasticity in the developing nervous system. Nature Reviews Neuroscience 5, 97–107 (2004)

[177] Twickel, A.V., Pasemann, F.: Reflex-oscillations in evolved single leg neurocontrollers for walking machines. Natural Computing 6, 311–337 (2007); doi:10.1007/s11047-006-9011-y

[178] Ventre-Dominey, J., Vallee, B.: Vestibular integration in human cerebral cortex contributes to spatial remapping. Neuropsychologia 45, 435–439 (2007)

[179] Weng, J., McClelland, J., Pentland, A., Sporns, O., Stockman, I., Sur, M., Thelen, E.: Autonomous mental development by robots and animals. Science 291, 599–600 (2001)

[180] Werbos, P.J.: Beyond regression: new tools for prediction and analysis in the behavioral science.Ph.D. thesis. Harvard University, Cambridge, MA (1974)

[181] Wikimedia Commons (2010), `http://commons.wikimwedia.org`

[182] Wikipedia: Self-organization – wikipedia, the free encyclopedia (May 28, 2009)

[183] Wikipedia: Delta rule — wikipedia, the free encyclopedia (February 11, 2010)

[184] Wiley, K.: Gnat clouds, flocking algorithm (1999), `http://keithwiley.com/artificialLife/gnatCloud.shtml`

[185] Wischmann, S.: Neural dynamics of social behavior: An evolutionary and mechanistic perspective on communication, cooperation, and competition among situated agents.Ph.D. thesis. University of Bonn, Germany (2008)

[186] Withall, M., Hinde, C., Stone, R.: An improved representation for evolving programs. Genetic Programming and Evolvable Machines 10, 37–70 (2009); doi:10.1007/s10710-008-9069-7

[187] Wolpert, D.M., Kawato, M.: Multiple paired forward and inverse models for motor control. Neural Networks 11, 1317–1329 (1998)

[188] Wolpert, D.M., Miall, R.C., Kawato, M.: Internal models in the cerebellum. Trends in Cognitive Sciences 2, 338–347 (1998)

[189] Yaeger, L.: Computational genetics, physiology, metabolism, neural systems, learning, vision and behaviour or polyworld: Life in a new context. In: Artificial Life III. SFI Studies in the Sciences of Complexity, vol. XVII, pp. 263–298. AddisonWesley (1994)

[190] Yang, L., Epstein, I.R.: Oscillatory turing patterns in reaction-diffusion systems with two coupled layers. Physical Review Letters 90(17) (2003)

[191] Zahedi, K., Ay, N., Der, R.: Higher coordination with less control – A result of information maximization in the sensorimotor loop. Adaptive Behavior 18(3-4), 338–355 (2010); doi:10.1177/1059712310375314

[192] Zhabotinsky, A.M.: Belousov-zhabotinsky reaction. Scholarpedia 2(9), 1435 (2007)

[193] Ziemke, T.: On the role of emotion in biological and robotic autonomy. Biosystems 91, 401–408 (2007)

Index

Cognitive Systems Monographs

Edited by R. Dillmann, Y. Nakamura, S. Schaal and D. Vernon

Vol. 1: Arena, P.; Patanè, L. (Eds.)
Spatial Temporal Patterns for
Action-Oriented Perception
in Roving Robots
425 p. 2009 [978-3-540-88463-7]

Vol. 2: Ivancevic, T.T.; Jovanovic, B.;
Djukic, S.; Djukic, M.; Markovic, S.
Complex Sports Biodynamics
326 p. 2009 [978-3-540-89970-9]

Vol. 3: Magnani, L.
Abductive Cognition
534 p. 2009 [978-3-642-03630-9]

Vol. 4: Azad, P.
Visual Perception for Manipulation
and Imitation in Humanoid Robots
270 p. 2009 [978-3-642-04228-7]

Vol. 5: de Aguiar, E.
Animation and Performance Capture
Using Digitized Models
168 p. 2010 [978-3-642-10315-5]

Vol. 6: Ritter, H.; Sagerer, G.;
Dillmann, R.; Buss, M.:
Human Centered Robot Systems
216 p. 2009 [978-3-642-10402-2]

Vol. 7: Levi, P.; Kernbach, S. (Eds.):
Symbiotic Multi-Robot Organisms
467 p. 2010 [978-3-642-11691-9]

Vol. 8: Christensen, H.I.;
Kruijff, G.-J.M.; Wyatt, J.L. (Eds.):
Cognitive Systems
491 p. 2010 [978-3-642-11693-3]

Vol. 9: Hamann, H.:
Space-Time Continuous Models of Swarm
Robotic Systems
147 p. 2010 [978-3-642-13376-3]

Vol. 10: Allerkamp, D.:
Tactile Perception of Textiles in a
Virtual-Reality System
120 p. 2010 [978-3-642-13973-4]

Vol. 11: Vernon, D.; von Hofsten, C.; Fadiga, L.:
A Roadmap for Cognitive Development
in Humanoid Robots
227 p. 2010 [978-3-642-16903-8]

Vol. 12: Ivancevic, T.T.; Jovanovic, B.;
Jovanovic, S.; Djukic, M.; Djukic, N.;
Lukman, A.:
Paradigm Shift for Future Tennis
375 p. 2011 [978-3-642-17094-2]

Vol. 13: Bardone, E.:
Seeking Chances
168 p. 2011 [978-3-642-19632-4]

Vol. 14: Jakimovski, B.:
Biologically Inspired Approaches for
Locomotion, Anomaly Detection and
Reconfiguration for Walking Robots
201 p. 2011 [978-3-642-22504-8]

Vol. 15: Der, R.; Martius, G.:
The Playful Machine
336 p. 2011 [978-3-642-20252-0]

GPSR Compliance

The European Union's (EU) General Product Safety Regulation (GPSR) is a set of rules that requires consumer products to be safe and our obligations to ensure this.

If you have any concerns about our products, you can contact us on ProductSafety@springernature.com

In case Publisher is established outside the EU, the EU authorized representative is:

Springer Nature Customer Service Center GmbH
Europaplatz 3
69115 Heidelberg, Germany

Batch number: 10682211

Printed by Printforce, the Netherlands